William Maccall, Charles Jean Marie Letourneau

Biology

William Maccall, Charles Jean Marie Letourneau

Biology

ISBN/EAN: 9783337214852

Printed in Europe, USA, Canada, Australia, Japan

Cover: Foto ©berggeist007 / pixelio.de

More available books at **www.hansebooks.com**

BIOLOGY.

BIOLOGY.

BY

DR. CHARLES LETOURNEAU.

TRANSLATED BY

WILLIAM MACCALL.

"Pro Veritate."

WITH EIGHTY-THREE ILLUSTRATIONS.

LONDON: CHAPMAN AND HALL, 193, PICCADILLY.
PHILADELPHIA: J. B. LIPPINCOTT AND CO.
1878.

LONDON:
R. CLAY, SONS, AND TAYLOR, PRINTERS,
BREAD STREET HILL,
QUEEN VICTORIA STREET.

PREFACE.

THE word *Biology*, which seems to have been employed for the first time by Treviranus, is far from bearing in the scientific vocabulary a completely settled import. It may not be unprofitable to determine the sense of the word. Etymologically it signifies literally "science of life," and embraces everything relating, intimately or remotely, to the study of organised beings; that is to say, a whole group of sciences, among which is comprehended Anthropology, for instance. It is in this encyclopædical sense that Auguste Comte took the word "Biology," though as far as we ourselves are concerned we intend to give it a sense much more restricted. Under the designation "Biology," we merely place the exposition and the coordination of all the great facts and great laws of life, or nearly what is usually understood by "General Physiology," when this denomination is applied to the two organic kingdoms. In this volume we have simply attempted to state concisely what life is, and how organised beings are nourished, grow, are reproduced, move, feel and think.

Even while limiting ourselves to this comparatively restricted domain, we have had to consider, to group, to condense and to classify, an enormous mass of facts derived from all the natural sciences. Among these facts, numerous as the stars of the heaven and the sands of the sea, we have been compelled to make a choice, and to select as much as possible what was most important, most significative, most luminous.

We hope that the learned men who devote themselves to special subjects may find in our modest production some new combinations, perchance some of those general views which are sometimes lacking to certain men in other respects very distinguished, but who abide too closely in this or that district of knowledge, as happens often in this age when the division of scientific labour is carried to excess. Nevertheless, we do not write for scientific men. We wish especially to be read by the mass of enlightened people, whom our very incomplete system of public instruction has left almost unacquainted with Biology. In effect, our best establishments for secondary instruction limit their ambition to imparting sufficiently complete ideas of physics, and very incomplete ideas concerning chemistry; but they stop too timidly on the threshold of Biology, the mysteries of which are accessible only to a small number of special men. This is a defect exceedingly deplorable, exceedingly prejudicial to general progress. It is on account of this defect that so many false and even pernicious ideas continue to find acceptance and empire in public opinion; hence it is, in a great measure, that true philosophy, or rather that philosophy which is alone solid and sound, that which flows directly and legitimately from observation and experiment, has such difficulty in diffusing itself.

The object of our little book is to remedy this serious educational deficiency in those who are otherwise enlightened. It is therefore a work of vulgarisation. Certain scientific men, too strictly confined within their own circle, and whose horizon is bounded by the walls of their laboratories, pronounce, with disdain, though unjustly, this term *vulgarisation*. To find the truth is surely a noble labour; but what is the value of the discovered truth, if care is not taken to propagate it, to introduce it into the patrimony of general knowledge?

On the other hand, it must be granted that the work of popularising has been brought somewhat into disrepute by a crowd of pseudo-scientific publications, the authors of which, trusting too little to the intelligence of the reader, either administer only an infinitesimal dose of science, or think themselves obliged to dilute the main idea with a deluge of light or pleasing words, sacrificing thus at once to the most amiable and most dangerous of our national peculiarities. Science only deserves its name upon condition of preserving a somewhat austere nobleness. For our part we have taken care not to rob science of that which constitutes its strength, and for this we trust the reader will give us credit. In our opinion there is not a person of moderate intelligence who will not be able, at the cost of a slight mental effort, to read and comprehend this book; and we think also that by such a perusal of it, sufficiently clear and complete ideas of Biology will have been imparted.

This is not a polemical work, but rather an exposition

of facts. Nevertheless, amongst these facts are some which are indisputable ; also, when we have met with them, we have not hesitated to formulate the conclusions or inductions which resulted from them. We have always done this temperately and with brevity, and without having any other motive than the love of truth.

We trust that this volume may be read, and that profitably, and that it may awaken, in a large number of its readers, love of and respect for science, namely, that which alone, in these sad times, is at once a refuge and a hope.

CH. LETOURNEAU.

BIOLOGY.

BOOK I.

OF ORGANISED MATTER IN GENERAL.

CHAPTER I.

CONSTITUTION OF MATTER.—UNITY OF SUBSTANCE IN THE ORGANIC WORLD AND IN THE INORGANIC WORLD.

THE sciences of observation demand at the outset from him who wishes to cultivate them an act of faith. Though it is perfectly incontestable that the exterior world manifests itself to us solely by exciting in our mind an incessant series of phenomena of consciousness, of phenomena called *subjective*, we are nevertheless compelled, unless we wish to plunge into the doubt applauded by Pyrrho and Berkeley, to believe our senses as honest and sincere witnesses when they signalise to us the existence, apart from our own being, of a vast material universe, the elements whereof, without pause in movement, awaken in us, by acting on our organism, impressions, sensations, and consequently ideas and desires.

The exterior world exists independently of our conscious life; it was when as yet we were not; and it will be when we are no more. Without stopping, as a few years ago M. Littré did, to discuss the point whether the certitude of the existence of the external world is of first or second quality, leaving aside every metaphysical refinement, we must first firmly believe in the real existence of the external world; because all our senses cease not to cry to us in every tone that the *objective*, the *Non-Me* of the psychologists, is not a chimera, because the contrary opinion would strike with nullity all observation, all experience, all reasoning, all knowledge.

The reality of the exterior world once admitted, and man never having been led to doubt that reality, except through a species of intellectual depravation, people naturally inquired what could be the internal constitution of the substance of the universe. They suspected that behind the appearance infinitely mobile and varied of the exterior phenomena there might exist a general and related force. Our object in this work not being to pass in review the opinions or the reveries of the different philosophical schools, we make haste to expound the most probable theories and systems, those which observation has confirmed, and which by slow degrees have conquered in science their right of citizenship.

Leucippus seems to have been the first to have had the intuition of the most rational theory on the constitution of the universal substance. In his opinion this substance is a discontinuous mass of granules, solid, infinitely small, separated by void spaces. It is "the void mingled with solid," according to an expression of Bacon.

Democritus admitted that these primordial granules were full, impenetrable, moreover insecable, and, for this last reason, he called them *atoms*.[1] But the conception of atoms, full and dis-

[1] For what is it that Democritus says?—"That there are substances in infinite number, which are called *atoms*, because they cannot be divided, which are, however, different, which have no quality whatever, are impassible, which are dispersed here and there in the infinite void, which approach each other, gather themselves together, enter into conjunction; that from these assemblages one result appears as water, another as fire, another as tree, another

persed through the limitless void of the universe, did not furnish a sufficiently precise account of the constitution of bodies. Epicurus appeared, whose doctrine was so magnificently sung by the great poet Lucretius. He immensely improved the atomic theory of Leucippus and Democritus by vivifying atoms, and by supposing them endowed with spontaneous movement. From the mobility of atoms resulted their various aggregations and the dissemblances of bodies. According to Epicurus, atoms of necessity mingled together, intertwined, literally caught and clung to each other. A philosopher who had the talent to preach and to propagate in France the atomic theory without seeming to offend the orthodoxy of his epoch, which was still very suspicious, Gassendi, restored to honour the atomic doctrine of the ancients. He admits, according to the expression of Epicurus, that "that which is moves in that which is not," that is, that atoms are not in contact, but that they are separated by void spaces.

Thus then, according to this theory, the world is composed of an innumerable quantity of atoms, mobile, infinitely small, distant from each other. These atoms are in a perpetual state of movement, rushing toward each other, repelling each other, for they have their sympathies and antipathies. It is from the diversity of their affinities that result their exceedingly diversified modes of grouping and the variety of the external world. It is by their vibrations, their oscillations that they reveal themselves to man by impressing his organs of sense. They have as essential qualities inalterability, eternity. When they gather together, new bodies are formed; when they disaggregate, bodies previously existing dissolve and seem to vanish. They are unhewn stones which have passed, pass, and are

as man; that everything consists of atoms, which he also calls *ideas*, and that nothing else exists, forasmuch as generation cannot arise from that which is not while likewise what exists cannot cease to be, because atoms are so firm that they cannot change nor alter nor suffer."—PLUTARCH, *Miscellaneous Works: Against the Epicurean Colotes;* Amyot's translation, Clavier's edition, vol. xx., Paris, 1803.

destined evermore to pass from one edifice to another. Their totality constitutes the general substance of the universe, and, in reality, this general substance undergoes no other changes than modifications in the distribution of its constituent elements. All the phenomena, all the varied aspects, all the revolutions of the universe can be referred essentially to simple atomic displacements.

This grand theory, so admirably simple and seductive, would be nothing but a brilliant speculation, if facts, numerous and rigorously observed, did not now serve it as basis and demonstration.

We rapidly enumerate the most important of these facts, which belong for the most part to the domain of Physics and Chemistry.

Wenzel, Richter, Proust proved first of all that in chemical compositions and decompositions, bodies combine according to proportions rigorously defined. Dalton formulated the law of multiple proportions, and deduced therefrom naturally that matter is constituted by atoms extended, having a constant weight, and that those atoms are of various species.

When atoms of the same species come into juxtaposition, we have what we call *simple* bodies, such as hydrogen, oxygen, azote. On the contrary, the bodies called *compound* result from the juxtaposition of atoms of diversified nature, whence come acids, salts, oxides, and also all the unstable and complex compounds which constitute organic substances.

This is not all : to the law of Dalton the law of Avogadro and of Ampère is adjoined. This last law establishes that all gases, temperature and pressure being equal, have the same elastic force. But as this force is probably due to the shock of atoms or groups of atoms, molecules, on the sides of the vessels which imprison the gases, we must admit that in the conditions aforesaid all gases contain, under the same volume, the same number of molecules or of atoms.

Finally, Dulong and Petit have been able to show, experimentally, that the atoms of simple bodies all possess the same specific heat.

All these great laws, slowly evolved by observation and experiment, have transformed into a solid scientific theory the brilliant but vague intuition of the thinkers of ancient Greece. With ground so firm to rest on, chemistry has been able to particularise more, to study in some sort the individual character of atoms; in scientific language, it has arrived at the notion of *atomicity.*

Atoms have as general characteristics extension, impenetrability, indestructibility, and eternal activity. But these general characteristics exclude not a number of specific differences. The progress of chemistry will no doubt show us what amount of truth there is in the hypotheses of Dumas and of Lockyer, according to which the simple bodies of chemistry as it now exists are merely indecomposed bodies. According to this assumption our metals and our metalloids are simple modifications of a single substance, probably hydrogen, the atoms thereof forming different molecular groupings. In the present state of science, these ideas, as yet purely hypothetical, can be passed by; and relying for the present on the great laws of Dalton, Ampère, Dulong, and Petit, we have the right to consider the simple bodies of contemporary chemistry as representing groups of atoms identical among themselves in each simple body, but specifically different from one simple body to another. Now each of these atomic species has its individual energy, its own affinities. In the group of the other atomic species it has friends, it has indifferents, it has enemies. It willingly unites itself to the first, neglects the second, refuses, on the contrary, to combine with the last. Moreover, this faculty of attracting and of being attracted attains in each atomic species a different degree of energy. Whence we may conclude that there are in the different atomic species differences of mass and of form. In aggregating themselves thus, according to their affinities, atoms arrange themselves into small systems, having in each body a special structure. These atomic systems are called *molecules.*

The atoms of alkaline metals, such as potassium and sodium,

cannot fix each more than one atom of chlorine or of bromine; they are *monoatomic*, as, for instance, hydrogen. Calcium, barium, strontium, in order that their attractive power may be saturated, need to fix two atoms of chlorine; they are *diatomic*, as, for instance, oxygen. Phosphorus, which in the perchlorure of phosphorus succeeds in fixing five atoms of chlorine, is *pentatomic*.

It is these inequalities in the mode and the power of combination, in the capacity of saturation, which we call the *atomicity* of each atomical species, designating specially by that expression the maximum capacity of saturation. However, hereby is by no means implied that a pentatomic species, for instance azote, cannot combine with less than five atoms. Azote, which fixes five atoms in the chlorohydrate of ammonia (AzH^4Cl), is not more than *triatomic* in ammoniac gas (AzH^3), and is only *diatomic* in the bioxide of azote. For the sake of greater clearness, the denomination *atomicity* is reserved to designate the capacity of absolute saturation. The capacities of inferior saturations are called *quantivalences*. Thus then azote is pentatomic, but it is trivalent in gas ammoniac, and so on.

This notion of atomicity has thrown a great light on the ultimate texture of bodies, and also on the march hither and thither of atoms in various combinations. In effect, free or combined, every atom tends to saturate itself by the annexion of other atoms. If, for instance, a tetratomic atom has combined with two atoms only, it ceases not to tend to saturate its attractive force; it strives to fix two atoms more. But these two atoms once found, no other simple body can combine with our tetratomic atom, unless by displacing one or two of its atoms and becoming their substitute. If, for instance, we take from a carburet of saturated hydrogen an atom of hydrogen, the molecule thus mutilated can unite itself to an atom of chlorine. But the chlorine is monoatomic; this, however, does not hinder it from fixing the complex molecule of the carburet, impoverished to the extent of an atom of hydrogen. The reason is that certain atomic groups, certain molecules, can play in combinations the part of a single atom. They are what we call *com-*

pound radicals. This notion of compound radicals has a predominant importance in the chemistry of *organic* substances, so called because nearly altogether they constitute the substance of living bodies. It simplifies extremely their apparent complexity. It is thus that, according to Mulder, the formula of albumine is $10(C^{40}H^{31}N^{5}O^{12}) + S^{2}Ph$. If we limit ourselves to totalising the atoms, this formula gives $C^{400}H^{310}N^{50}O^{120} + S^{2}Ph$, a molecule of frightful complication. But if we admit a compound radical, proteine $(C^{40}H^{31}N^{5}O^{12})$, comporting itself as a simple atom, the molecular structure of albumine is enormously simplified: it approaches that which we are accustomed to meet in the chemistry called *mineral.* It is probably also from this notion of compound radicals that we must seek the explanation of what has been called *isomeria.* If bodies having the same elementary composition, such as the tartaric and paratartaric, malic acid, citric acid, sugars, gums, have nevertheless distinctive properties, we must probably attribute the dissimilarities to differences of molecular structure, to the existence, in the very heart of these isomeric bodies, of dissimilar compound radicals.

There is another notion not less important than that of compound radicals for the easy comprehension of the formulas of the chemistry called *organic,* the notion, namely, of autosaturation. In effect, the atomicity of a simple body does not always expend itself on atoms of a different species; it may manifest itself between atoms of the same species. The atoms of carbon, for instance, can saturate themselves. An atom of carbon, which is tetratomic, $-\overset{|}{\underset{|}{\mathrm{C}}}-$, may, by expending merely a quarter of its atomicity, unite itself with another atom of carbon, which, in its turn, will neutralise in this combination a quarter of its attractive energy; there will result thus therefrom a molecule hexavalent, that is to say, capable of still enchaining six atoms:

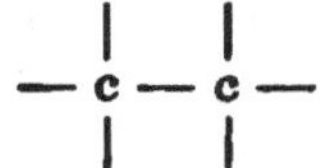

Let a third atom of carbon then unite itself to this molecule, we have an octovalent molecule :

```
    |     |     |
 —  C  —  C  —  C  —
    |     |     |
```

Finally, the adjunction of a fourth atom of carbon gives a decavalent compound :

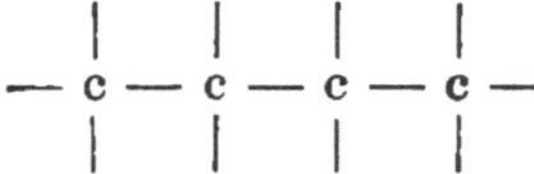

This notion of autosaturation has enabled us to systematise a quantity of facts of organic chemistry, to create rationally new compounds, to classify and to seriate groups. We owe to it the theory of alcohols, and that of hydrocarburets.[1]

The preceding pages contain the principal notions of general chemistry, which as we proceed we propose to apply. We must, however, before ending this chapter say a few words on what have been called *catalyses.* Certain bodies brought into contact with other bodies determine by their presence alone, and without taking any other part in the reactions, either combinations or metamorphoses or unfoldings. It seems as if in these cases the body, intervening, by its presence alone, brings into play an attractive force sufficient to disturb the atomicity of the body which it influences, without, however, being able to enter into combination with that body. For instance, platina determines, by its presence alone, the combination of oxygen and hydrogen, the formation of water; it transforms also alcohol into acetic acid by determining its oxidation.

These are catalyses of combination.

The albuminoidal substances introduced into the stomach impregnate themselves there with gastric juice, expand, and in

[1] Consult for further details, Wurtz, *Philosophie Chimique*, *Chimie nouvelle*, &c. Naquet, article, *Atomique* (*Théorie*) in the *Encyclopédie Générale.*

consequence the organic substance of the gastric juice achieves in these alimentary substances an isomeric modification, which renders them liquid, absorbable, in short, transforms them into albuminose. In the same way, under the influence of sulphuric acid diluted, cane-sugar, cellulose, gums, and fecules are metamorphosed first of all into dextrine, and then into glycose, or grape-sugar.

These are isomeric catalyses.

The hippuric acid of the urines of herbivorous animals unfolds itself, under the influence of the mucous elements modified by the air, into hippuric acid and sugar of gelatine or glycocoll.

That is an unfolded catalysis.

In sum, the universe must be regarded as a whole composed of atoms dissimilar, and variously grouped according to their affinities. These active atoms are the foundation, the substance, the cause of all things: to use the expression of Tyndall, they are giants travestied.

The various aspects of bodies result from the various modes of aggregation of the constituent elements.

"All the changes accomplished on the surface of the globe are due to combinations which are made or to combinations which are unmade." [1]

All chemical phenomena are consequently the expression of atomic combinations, and can be included in four general types:

1. Simple change of molecular structure, or isomeria.

2. Unfolding of compound molecules.

3. Adjunction, addition of atoms, or of molecules not yet saturated, or, inversely, subtraction of atoms.

4. Substitution of certain atoms, certain molecules for others in a compound body.

These general characteristics manifestly exclude all ultimate, all radical difference between living organised bodies and

[1] Dumas, *Traité de Chimie*, t. viii.

inorganic bodies. Is there sufficient reason, however, for distinguishing between an inorganic world and an organic world? What are the dissimilar qualities of these two grand groups? This is what we propose to examine in the following chapters.

[*Note.*—When in this chapter atoms are spoken of as *full*, it is in the sense of a *plenum* excluding a *vacuum.*—Translator.]

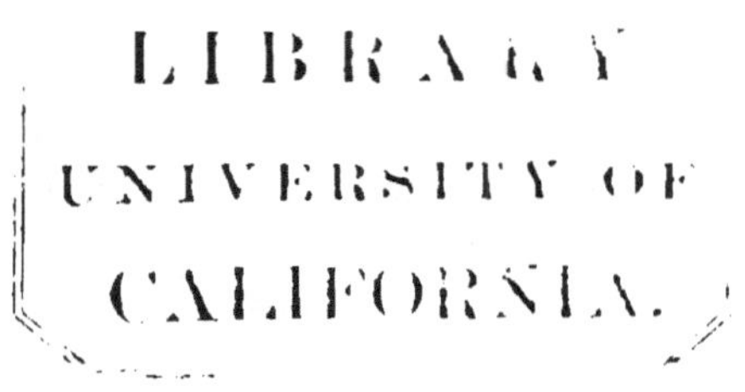

CHAPTER II.

ANORGANIC SUBSTANCES AND ORGANIC SUBSTANCES.

If, as results from the preceding exposition, the universe is a whole eternally unstable in form, eternally immutable in substance, it follows as a matter of course that living or organised bodies cannot be constituted of aught essentially special. An integrant part of the medium which environs them, they come forth from it only to return to it, and there is not an atom of their substance which does not participate of the eternity of universal matter, the basis of everything which exists. There is not one of these atoms which has not played an infinite number of parts in an infinity of organic and anorganic combinations, and which is not destined to play an infinite number more. Also, in analysing elementarily the most complex of animals, man, we, in normal conditions, find in him only fourteen simple bodies of mineral chemistry, the list of which is herewith given :—

Oxygen,	Phosphorus,	Calcium,
Hydrogen,	Fluor,	Magnesium,
Azote,	Chlorine,	Silicium,
Carbon,	Sodium,	Iron.
Sulphur,	Potassium,	

Moreover it must be stated that the mass of the human body is especially constituted of four of these simple bodies, namely, azote, carbon, hydrogen, oxygen.

If the results are accepted of elementary chemical analysis by decomposition, organised beings do not differ in substance from unorganised beings. But the analogy in substance does not exclude very important differences in form; for we know that the properties of bodies are intimately related to their composition, to the mode of aggregation of the substances which constitute them.

Let us remark, in passing, that of the four simple bodies occupying the first place in the composition of the body of man and of the animals, there are two whose affinities of combination are neither strong nor numerous, and which have a certain degree of chemical inertia. Carbon, which is completely inert in ordinary temperatures, unites itself only to a small number of substances, and often by only a feeble bond. Nevertheless it largely blends with the constitution of plants, and occupies a very important place in the constitution of animals. Azote, more indifferent still than carbon, is found in large quantities in the vegetal kingdom, in quantities still larger in the animal kingdom. It is this very inertia shared, though in a less degree by a third element, hydrogen, which renders these bodies suitable to figure in the chemical constitution of living beings.[1] In these beings, in effect, matter is in a state of extreme mobility; it is subject to a perpetual movement of combination and decombination; without repose, without truce, its elements go and come, have reciprocities of action, aggregate themselves, disaggregate themselves; there is a real whirl of atoms, in the very midst of which fixed compounds, with chemical elements solidly cemented together, can only figure in a secondary fashion. Here are needed compounds unstable, of a great molecular mobility, capable of forming, disaggregating, metamorphosing, themselves, of renewing the woof of the living tissues.

At the very outset the ternary compounds non-azotised, the aggregates of hydrogen, of oxygen, and of carbon, that is to say, the fixed oils, the fats, the gums, the starches, the resins, the sugars, and so on, the constituent principles of plants and

[1] See H. Spencer, *The Principles of Biology*, vol. i.

animals possess a great inertia and a notable instability; often they are susceptible of isomeria (sugars, dextrine, and the like).

As regards the more complex compounds, those in which carbon, hydrogen, azote, sulphur, phosphorus ally themselves to form the substances called albuminoidal, molecular instability is carried in them to the maximum; the unfoldings, the isomeric modifications are effected with extreme facility. Further on we shall, in reference to nutrition and digestion, signalise the important isomeric modifications which transform the insoluble albuminoidal aliments into soluble substances. Let us call attention also, by the way, to the still more curious and typical metamorphoses which the various kinds of virus and miasma produce in the albuminoidal substances of living bodies. It deserves remark, besides, that these last substances, when once modified isomerically, possess the murderous property of transmitting by simple contact to sound organic substances the molecular alteration they have themselves undergone.

But, finally, we may observe that these actions of contact are not peculiar to organic substances. They have, like the isomeric phenomena, their analogues in the chemistry called *anorganical.* In effect there is no radical difference, no abruptly settled frontier between organic chemistry and inorganic chemistry. The two kinds of chemistry study the same elementary bodies which are subject to the same laws. Organic substances proceed from anorganic substances, and return to them incessantly, to come forth from them anew. For the most part we merely find in organic substances greater complexity and instability. Also we see modern chemistry striving more and more to pluck from living bodies the monopoly of the fabrication of substances called *organic.*

Moreover, if we place in a graduated series the mineral and organic compounds, we discover between the two classes transitory groups, forming a point of union: these are the carburets of hydrogen, the alcohols, the ethers, ternary acids, fat bodies, the synthesis of which the chemist is now able to accomplish. Neither is there anything inalienable or special in the composition of organic products. We can succeed in substituting magnesia for lime in the shells of eggs. We can, in fats, replace hydrogen by chlorine, without modifying essentially the properties of the compound. Chemical synthesis has also tried to reproduce the simplest of the azotic organic substances. There has been a direct reproduction of urea, of taurine, of glycocoll in the laboratory; and if the true albuminoidal bodies have hitherto defied the efforts of synthetic chemistry, we may almost with certainty predict that they will not defy them always. It will then be possible to appropriate direct from the mineral world fibrine, albumine, caseine, and so on, that is to say, the special, the most needful aliments of man. This grand discovery must inaugurate for civilised communities a new era. It must be for man a real enfranchisement, by diminishing in a prodigious measure the sum of muscular labour, to which for more or less duration and progression he is now doomed.

These preliminaries settled, we are able to enumerate the various groups of simple or compound substances, which by their union constitute the bodies of organised beings. M. Ch. Robin has given an excellent classification of these substances or *immediate principles.*[1] According to him, the immediate principles are the ultimate solid bodies, liquid or gaseous, to which we can reduce the liquid or solid organised substance, the humours and the elements. But in order that these ultimate materials may merit the name of *immediate principles,* M. Robin thinks they must be obtained without chemical decomposition, by simple coagulations and successive crystallisations.

These bodies, which, by their innermost blending, their reciprocal dissolution, constitute the semi-solid organised substance, can be grouped into three classes :

1. The first class comprehends the crystallisable or volatile bodies without decomposition, having a mineral origin and coming forth from the organism as they had entered it (water, certain salts, and so on).

2. The immediate principles of the second class are also

[1] Charles Robin, *Leçons sur les Humeurs,* Paris, 1867.

crystallisable or volatile without decomposition, but they are found in the organism itself, and come forth from it direct as excrementitial bodies. They are acids, for instance, the acids tartaric, lactic, uric, citric; vegetal and animal alkaloids: creatine, creatinine, urea, caffeïne, and so on; fat or resinous bodies; sugars of the liver, of grape, of milk, of cane, and the like.

3. In the third class of immediate principles we find bodies not crystallisable or coagulable. They are formed in the organism itself; then, decomposed there, give birth to the immediate principles of the second class. The organic substances, properly so called, the substances of the third class, constitute the most important part of the body of organised beings (globuline, musculine, fibrine, albumine, caseïne, cellulose, starch, dextrine, gum, and some colouring matters, such as hæmatine, biliverdine.)

It is from the intense union, molecule by molecule, of substances appertaining to these three groups that organised substance results, formed of multiple elements, but constituted in great part of bodies complex, inert, unstable, easily decomposed, either through the play of chemical affinities, or the action of undulations calorific, luminous, electric.

So far we have occupied ourselves with the materials of organised bodies only from the point of view of their chemical composition; but it is quite as needful to take into consideration their physical state and physical properties.

An English chemist, justly celebrated, Graham, fell on the happy idea of grouping all bodies, according to their characteristic physical state, into two grand classes, that of *crystalloids* and that of the *colloids.*[1]

The crystalloids comprehend all the bodies which ordinarily form solutions sapid and free from viscosity. These bodies have furthermore the property of traversing by diffusion porous partitions.

The colloids have a consistency more or less gelatinous (gum, starch, tannin, gelatine, albumine). They diffuse themselves

[1] *Phil. Transactions*, 1861, p. 183; Moigno, *Physique Moléculaire.*

feebly and slowly, as the following table indicates, which gives the time of equal diffusion for some bodies taken in the two classes :—

Chlorohydric acid	1
Chlorure of sodium	2, 33
Cane-sugar	7
Sulphate of magnesia	7
Albumine	49
Caramel	98

Herefrom it is seen that chlorohydric acid traverses the porous membranes forty-nine times faster than albumine, and ninety-eight times faster than caramel. It is no doubt owing to this feeble diffusibility that the colloids are savourless when they are pure. Besides, these colloids do not comprehend merely the complex organic substances called *albuminoids;* certain bodies indubitably mineral, such as silica, hydrated peroxide of iron, can assume the colloidal condition. Both the one and the other, moreover, enter into the composition of organised bodies. A particular soluble form of the hydrated peroxide of iron, which normally is an element of the blood, gives, when we dissolve it in water in the proportion of 1 to 100, a red liquid, condensing into coagulum, into a sort of rutilant clot, under the influence of traces of acids, of alkalies, of alkaline carbonates, and of neutral salts.

Certain colloids, such as gelatine, gum arabic, are soluble in water; certain others, such as gum tragacanth, are insoluble therein. In any case they have as a general characteristic the power of absorbing a great quantity of water, of augmenting enormously in volume, and of then losing this water very rapidly by evaporation. It seems as if in this case they are merely subject to a sort of capillary imbibition. Yet it must be admitted that they incorporate more intensely with themselves a certain quantity of water as an integrant portion. Hence it appears that there is for colloids a *water of gelatinisation,* as there is for crystals a water of crystallisation.

There is not, however, any absolute incompatibility between colloids and crystalloids. If the colloids are for the most part complex organic compositions, we have seen above that very simple mineral compounds can assume the colloidal state; and on the other hand Reichert discovered in 1849 that albuminoidal substances can take the crystalloidal form. We shall be able to cite examples of this last case when speaking of plants. In almost all seeds, in effect, we find a white powder, finely grained, and presenting sometimes crystallised facets, square edged. The diameter of these particles is from $0^{m},00125$ to $0^{m},0375$. They are called *particles of aleurone.* They are composed of fibrine, of albumine, of legumine, of gliadine, of gum, of sugar, and so on. They are aliments in reserve.

These albuminoidal crystalloids are birefringent; they are all insoluble and unassailable in water and alcohol.[1]

We have signalised above the strong diffusibility of the crystalloids; it is so great that they can penetrate the colloids, blend with them as intensely as with water, while on the contrary the colloids can scarcely diffuse themselves into effective union with each other.

From this enormous difference of diffusibility between colloids and crystalloids it results that, if we separate by a porous membrane water and a colloid holding in solution a crystalloid, this last disengages itself from the colloid and traverses the membrane to dissolve in the water. It is thus that we can very easily with a membrane dialyser extract from a colloidal substance arsenious acid, digitaline, and so on. This process is made use of in certain toxicological researches, and also industrially to purify gums, albumine, caramel, and the like.

The reader has no doubt already the presentiment of the weight and worth of some preceding statements for the comprehension of biological facts. In effect every organised being is a compound of colloidal bodies holding in solution crystalloidal bodies. But this organised body is in a state of perpetual reno-

[1] Duchartre, *Botanique*, p. 69; Sachs, *Traité de Botanique*, p. 72.

vation. Unceasingly it plays, face to face with the exterior medium, the part of dialyser, either directly or by the aid of special apparatus. It forms nutritive soluble substances, and rejects waste substances, likewise soluble, at least in the liquids of the organism.

When, for instance, the residuum of the waste of the living tissues is composed of crystalloid bodies, these bodies can easily and rapidly traverse the colloidal substance of the tissues to be expelled from the organism; but their expulsion leaves in those same tissues a void which other soluble substances can come to fill by permeability; and in this fashion the losses undergone by the living machine are repaired without difficulty.

Finally, the colloidal state is the form the most suitable for the manifestation of the instability, the molecular mobility of the complex bodies which constitute organised beings. Under this form they are really in the dynamical state; they yield without difficulty to the shock, to the action of incident bodies. They can unmake and remake themselves, become the scene of a perpetual exchange of molecules and of atoms, in a word, of a vital progression and regression.

CHAPTER III.

CHEMICAL COMPOSITION OF ANIMALS AND PLANTS.

In the two living kingdoms, organised substance is, as we have already seen, constituted by three groups of bodies intimately blended, and which Chevreul was the first to call *immediate principles.* It is now needful to compare with each other the chemical species which enter into the composition of the plant and into that of the animal. We shall glance very rapidly at the immediate principles of the first category. In effect, water, which constitutes in weight the largest part of organised beings, mineral salts, atmospheric gases, are manifestly unable to furnish to us sufficiently distinctive characteristics. But that the results of the comparison may be the more striking we shall indicate first of all in bold outline what is the chemical composition of plants, and what is that of animals.

1. *Chemical Composition of Plants.*

Organised vegetal tissues, when submitted to desiccation, present a friable residuum, the weight of which is very variable. In the average of terrestrial plants this residuum is from a fifth to a third of the total weight; but it rises to eight-ninths if we take ripe seeds, and can descend to a tenth or a twentieth in aquatic plants and certain mushrooms. This residuum, desiccated, offers always to chemical analysis, carbon, hydrogen, oxygen, azote and sulphur, potassium, calcium, magnesium, iron, phosphorus. Often, moreover, we find therein sodium, lithium, manganese, silicium, chlore. Finally, in the marine plants we discover iodine and brome.

Such are the ultimate results of analysis; but, of course, during life, these bodies are not, for the most part, in a state of liberty; they are combined in various manners. The metals are usually in the state of salts, of sulphates, of phosphates, of carbonates, of oxalates, and so on. There is also a certain quantity of oxygen, of azote, of hydrogen and of carbonic acid dissolved in the liquids or impregnating the vegetal anatomical elements. But the true organic compounds are ternary or quaternary compounds. The ternary compounds are formed of carbon, of hydrogen, and of oxygen. They constitute the strongest part of the vegetal texture. Let us mention first of all cellulose, which forms almost alone the primary part of the vegetal cells, and then many substances which are isomeric to it, such as inuline and xylogen. The first of these isomers of cellulose, inuline, is found in decomposed roots, in colchicum bulbs, in dahlia tubers, and so on. As to xylogen, it is the substance which gives rigidity to ligneous tissues. Furthermore, in putting ourselves at the point of view of chemical composition we have to see the relation of cellulose to the starches, the sugars and the gums. To the type of sugars, the sugar of grape, or glycose, has long been given the formula $C^{12}H^{12}O^{12},2HO$. Starch is composed of $C^{12}H^9O^9,HO$. In reality, these ternary bodies have already in a large measure the characteristics of complexity and instability peculiar to organic substances, and their definitive formula is still a subject of discord among chemists. According to M. Wurtz, for instance, the formula of cellulose would be $C^6H^{10}O^6$, that of gum arabic $C^{12}H^{22}O^{11}$, that of starch $C\,H^{10}O^5$, and this formula would not vary by the isomeric transformation of starch into dextrine. Saccharine and amylaceous matters bear as chemical characteristic the inclusion of hydrogen and oxygen in such proportions that the oxygen could suffice exactly to saturate the hydrogen and to transform it into water. The general formula of these groups would therefore be $C^m(H^2O)^n$.[1]

[1] Wurtz, *Chimie Nouvelle.*

To complete the enumeration of the ternary vegetal compounds, we have to mention the fat vegetal bodies, the non-azotised oils, which are also compounds in complex molecules of carbon, of oxygen, and of hydrogen.

After the group of ternary organic substances comes a tribe of azotised compounds, wrongly, and in virtue of questionable chemical theories, called *quaternary* bodies. The molecules of these last bodies are, it is true, formed for the most part by atoms of carbon, of oxygen, of hydrogen, and of azote; but almost constantly a certain quantity of sulphur and of phosphorus must be joined to them. These quaternary compounds are the organic substances by excellence; we seek in vain their analogues in the mineral world. They form themselves spontaneously in the texture of living beings; whereas the ternary compounds spoken of above can be brought into relation with the carburets of hydrogen which connect them with the inorganic world.

The azotised vegetal substances form two principal groups,—the group of the alkaloids, and that of the albuminoids. The alkaloids are very complex compounds, capable of combining as bases with an acid. These bodies, unimportant as to quantity, are very important as to their physiological or toxical properties; they are quinine, strychnine, morphine, and so on.

But the substances which, without question, hold the first rank, the compounds essential to vegetal life as well as to animal life, are those which form the group of the albuminoids. We shall see that these substances constitute the nucleus of the vegetal cells, constitute their internal membrane, that they are also found in the liquid filling the cells, in the protoplasm. Among the most important of these substances we must name gluten or *vegetal fibrine*, so abundant in the seeds of the cereals. To it is given as formula, according to the theory of Mülder, $10(C^{40}H^{31}O^{12}Az^{5}) + S$.

In relation to gluten we have to view *glutine*, an analogous albuminoidal substance; it is the coagulable principle of the sap of plants. It is likewise called *vegetal albumine.*

Finally there is extracted from the seeds of the leguminous plants a third albuminoidal substance, containing, like gluten, sulphur, and which is called *vegetal caseine.*

The last substance which we have to mention is the green matter of plants, *chlorophyll.* Its physiological agency is extremely curious and interesting; we shall therefore describe it in detail in the course of our expositions. Here it suffices to observe that chlorophyll cannot be placed in the group of the preceding substances, called *proteïcal.* Neither phosphorus nor sulphur is found therein. It is composed only of carbon, of hydrogen, of oxygen, of azote, and, what is altogether characteristic, of iron. Its formula, still however requiring consideration, would be $C^{18}H^{9}AzO^{8}$ + Fe (in indeterminate quantity).

2. *Chemical Composition of Animals.*

In a preceding chapter we have enumerated the fourteen simple bodies entering into the composition of the most complex of organisms, the human organism.

A glance thrown at this list suffices to show that if the elementary composition is held in view, and the quality of the elements is alone considered, there is almost identity between the vegetal organisms and the animal organisms. But in both kinds of organism these elementary bodies are aggregated in various combinations, with the exception of azote and oxygen, of which a part is in a state of liberty alike in the animal and the vegetal organisms.

In every animal organism also we encounter, in a state of intense blending, immediate principles of the three classes.

The immediate principles of the first class, or mineral principles, penetrate, entirely formed, into the animal economy; and entirely formed they come forth from it: this is the case with water, azote, certain salts, and so on.

The principles of the second class are in general hydrocarbonised ternary compounds such as lactic acid and the lactates, uric acid and the urates, fat bodies (oleïne, margarine, stearine), animal starch or glycogenous matter of the liver, the glycose of the same gland, chitine. They comprehend quaternary azotised products, the result of the disassimilation of the organic elements, such as urea ($C^2Az^2H^4O^2$), creatine ($C^8H^9Az^3O^4$), creatinine ($C^8H^7Az^3O^7$), cholesterine ($C^{52}H^{44}O^2$), and so on. While the principles of the first class pass merely into the organism by coming from the exterior world, those of the second form themselves in the animal organism, but do not sojourn there.

The immediate principles of the third class are numerous neither in animals nor plants, but they play in the first a more important part than in the last. They are the albuminoidal substances, all likewise colloids, and insatiable in their thirst for water. These bodies are very unstable compounds, much inclined to isomeric modifications. They are formed in the animal economy, never leave it when it is in a healthy state, are renewed therein molecule by molecule through the nutritive movement, and from their quantity and from the dominant part they play, they constitute the very essence of the living organism. Their formula, as we have already stated, is still undecided. There has been a disposition to consider them as all formed of the same radical, proteïne, united to atoms of sulphur and phosphorus.

In boiling the epidermic productions, the cartilages, the organic framework of the bones, the cellular tissue, the tendons, and so on, we obtain quaternary azotised substances, chondrine, gelatine, containing less carbon and more azote than the other albuminoidal substances: moreover, containing no sulphur.

The most important animal albuminoidal substances are fibrine, albumine, caseïne, the analogues of which we have signalised in plants. In the same way that in plants we have found a special quaternary substance, chlorophyll, containing a metal, iron, we find also in the superior animals a matter analogous to albumine, but coagulating much less easily when it is dissolved in water. This matter is the substance of the globules of the blood, globuline. Like chlorophyll, it contains iron in its composition, and, like it, also exerts a special action on one of the gases of the atmosphere.

How summary soever may be the short enumeration which precedes, it suffices to establish from a thorough knowledge of the matter a parallel between the composition of animals and that of plants, and to give saliency to the analogues and the differences.

3. *The Organic Substances of the two Kingdoms.*

A supreme fact is evolved from the preceding examination, namely, that there is in the ternary and quaternary substances a dominant element common to them all, carbon. Of all organic substance, carbon is the base. In weight it forms the principal element thereof. The albumine of the blood contains about fifty per cent. of carbon. But in organic substances carbon plays a much more important part still. It is the bond of all the various atoms, which compose the complex molecules of organised bodies. We have already seen that carbon is a tetratomic body, that is to say, capable of fixing, of keeping wedded to one of its atoms four atoms of a monoatomic body, such as hydrogen, or two atoms of a diatomic body, such as oxygen; and so on. We have besides remarked that the atoms of carbon could unite with each other in neutralising reciprocally one only of their affinities, the others remaining free and fit to satisfy themselves, in attracting and fixing either atoms of other elements or even aggregates more or less complex, radicals comporting themselves as a single atom. But these atoms, these radicals, are often only aggregated to the atom of carbon which attracts them by one of their affinities, while the others remain active, exciting the aggregation of new atoms. Let us take, for instance, the iodide of methyl, that is to say, of carbonised hydrogen, an atom of iodine taking the place of an atom of hydrogen :—

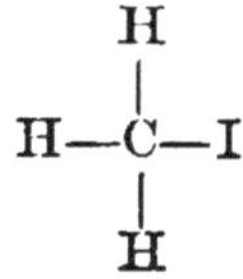

Heating in suitable conditions this body with potash or hydrate of potassium, we determine the displacement of the atom of iode, which combines with the potassium and is succeeded by the oxygen of the potash. But this oxygen is diatomic: the half only of its affinity is satisfied or neutralised by this displacement: the rest still remains free. This is why, without ceasing to form part of the carburet, the atom of oxygen unites itself on its own account with a molecule of hydrogen likewise taken from the hydrate of potassium, and we have thus *wood spirit:*

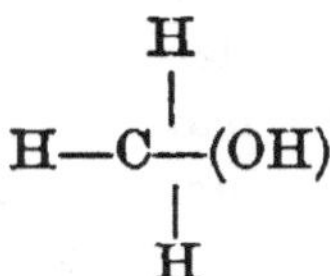

We have taken as example a body in which one atom only of carbon figures. But if we represent to ourselves a poly-carbonised compound we at once see to what a degree of complexity and mobility such a body can attain; therefrom we gain a general idea of what the chemistry of organic bodies is; we recognise that modern chemists have the right to call this branch of their science the chemistry of the compounds of carbon; and we willingly subscribe to this proposition of Haeckel: "It is only in the special chemico-physical properties of carbon, and especially in the semi-fluidity and instability of the carbonised albuminoidal compounds, that we must seek the mechanical causes of the phenomena of particular movements by which organisms and inorganisms are differentiated, and which is called in a more restricted sense *Life.*"[1]

The general statements given above apply equally to organic vegetal substances, and to organic animal substances, forasmuch as we have seen that as regards quality, as regards general

[1] E. Haeckel, *Histoire de la Création Naturelle.* Paris, 1874.

chemical composition, the two classes of substances are manifestly identical. Consequently, there is no radical difference between the organic substances of the vegetal kingdom and those of the animal world. Nevertheless these are notable dissimilarities; they bear on the relative quantity of the ternary compounds non-azotised, and the quaternary compounds azotised, in both the realms of Nature. In effect, the albuminoidal substances which constitute the chief part of any veritable animal organism are from the quantitative point of view little more than accessories. The great mass of every true plant is especially constituted by the non-azotised carburetted substances. Azote, though forming an essential element of the intracellular vegetal protaplasm and of the alkaloids, represents often in weight less than a hundredth of the dry matters: rarely the proportion rises to three hundredths. To sum up, the vegetal kingdom is, quantitatively considered, the kingdom of ternary carburetted substances, while the animal kingdom is that of carburetted substances azotised or quaternary. Consequently there is in the animal world a greater degree of chemical complexity and instability, that is to say, a superior vital activity.

Nevertheless, there is no radical difference. We must henceforth reject that idea of complete antagonism between the two kingdoms, which has so long prevailed in science. We must no longer consider every plant as an apparatus of reduction specially charged to form, all in a lump, at the expense of the mineral world, ternary and quaternary compounds for the nourishment of animals. We must cease to see in every animal an apparatus of combustion whose mission is to destroy those compounds without being able to form any. Cl. Bernard has demonstrated that the cells of the liver fabricate at the expense of the blood an amylaceous matter possessing, according to the analysis of M. Pelouze, the same composition as vegetal starch, and, like it, transforming itself into sugar. Finally, M. Rouget has found this amylaceous matter, glycogen or zooamyline, in the muscular tissue, in the lung, in the cells of the liver, in the placenta, in the amniotic cells, the epithelial cells, the cartilages, &c., of the vertebrates.[1]

For a long time cellulose was considered a substance exclusively vegetal; but after a while, under the name of *chitine*, or *tunicine*, it was found in the tegumentary envelope of the tunicates, in the exterior skeleton of the anthropods, and so on; and M. Berthelot has succeeded in transforming into sugar this tunicine, this animal cellulose, for ebullition and acids metamorphose it into glycose.[2]

Even chlorophyll, that vegetal substance by excellence, has been found in certain rudimentary animals.

Therefore, once more we declare that there is no radical difference, no chasm between the two living kingdoms, from the point of view of the composition and formation of the organic substances. In this respect there is no reason why the two kingdoms should not be included under the denomination of *Organic Empire*, as Blainville proposed.

NOTE.—Both as a substantive and as an adjective, *vegetal* is a good old English word which is often for obvious reasons preferable to *vegetable* or *plant*.—*Translator*.

[1] J. Gasarrat, *Phénomènes Physiques de la Vie*, p. 196.

[2] J. Gasarrat, *loc. cit.*

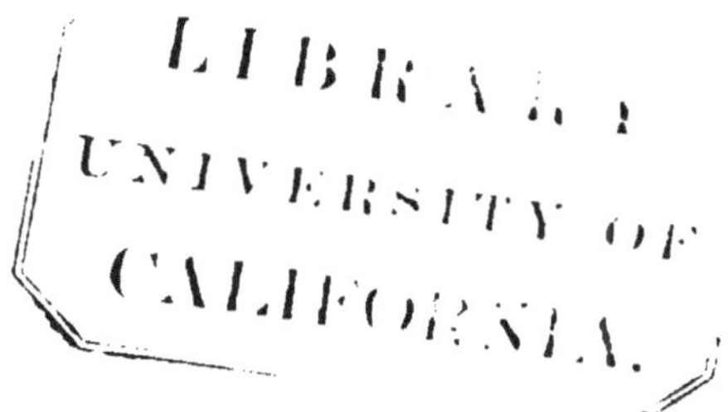

CHAPTER IV.

OF LIFE.

LIFE has long been the mystery of mysteries; and in modern times it has been the last refuge, the citadel of supernaturalism. In fact, so long as there were no clear ideas regarding the constitution of bodies, or the composition of chemical aggregates, so long as so-called organic substances appeared radically different from mineral substances, it was impossible to unravel the mystery of life. We now know that organised bodies do not contain a material atom which was not first derived from, and afterwards restored to, the exterior medium. We have made an enumeration of the immediate principles which constitute living bodies; we have been able to reproduce a certain number of these in our chemical laboratories. We know in what physical state, under what blended conditions, they are found within organised and living bodies. We know, moreover, that the entire universe contains an always active matter, that what is called *force* cannot sever itself from what is called *matter*, that consequently there can no longer be any question of a *vital principle*, of an *archeus*, superadded to living beings, and regulating their phenomena. Even these simple general facts authorise us to affirm that vital phenomena are simply the result of the properties of living matter. To give a just idea of life, it remains to us then to determine what are its properties, and also what are the principal conditions of their manifestation.

We prove then, first of all, that life depends strictly upon the exterior medium, that an alteration in the composition of the aërian or aquatic medium determines the cessation or suspension of the vital movement. We can even at will suspend and reanimate life in certain organised beings.

M. Vilmorin succeeded in reviving, by means of moisture, a dried fern sent from America. By drying and then moistening certain infusoria we may arrest and revive the course of life in them. In America and Northern Russia frozen fishes, brought from great distances, are revivified by being plunged into water of the ordinary temperature.

In Iceland, in 1828-9, Gaymard in ten minutes revivified frozen toads in tepid water.

In the case of dried organisms, the organic substances have been deprived, by evaporation, of their water of gelatinisation, and thereby of their molecular mobility, the instability indispensable to the realisation of atomic changes; in fact, they have been separated from the exterior world, yet without decomposition; whence their easy revival.

In congealing organisms, an analogous result is obtained. By the solidification of water substances lose their colloidal state. They are, in some degree, chemically paralysed, but can nevertheless revive, if congelation has produced neither chemical decomposition of the substances, nor morphological destruction of the tissues and of their anatomical elements.

These facts suffice by themselves to prove that the principal condition of life is the interchange of materials between the living body and the exterior world; but, fortunately, we are not limited to such commonplace demonstrations. Vital activities have been minutely scrutinised, watched, and followed step by step, as we shall see further on. We have been enabled to note the incessant amalgamation with the organism of substances derived from the exterior world, to observe the modifications and transformations which these substances undergo and promote in the midst of living matter; the results of all these biological operations have been summed up, and establish approximately the balance of gain and loss. In short, it is now known that the principal vital phenomenon, that which serves as a support to all the others, is a double movement of assimilation and of dis-assimilation, of renovation and of destruction, in the midst of living matter; that this matter may be either in a semi-solid state, and without structure, as in certain inferior organisms; or that it may be in a liquid state more or less viscous, like the blood and lymph of the superior animals; or finally, that it may be modelled into anatomical elements, into cells and fibres bathed with liquids and gases, as in the bodies of all the superior animal and vegetal organisms.

The living substance is thus a chemical laboratory in constant action. It is the physical or chemical properties of this substance, diversely modified, which underlie all the vital properties, *nutrition*, *growth*, *reproduction*, the *chlorophyllian attribute*, *motility*, and *innervation*.

Now the six properties which we have just enumerated are the six principal modes of living activity, the six categories under which all biological phenomena group and class themselves. The chlorophyllian property is almost exclusively vegetal; but the five other fundamental properties represent, when united, the highest, the most complete expression of life. But they are far from being always united; they are also far from having the same importance. Some of them are primordial, some secondary.

The most important of all is evidently nutrition, the double and perpetual movement of molecular renovation of the living substance. Without nutrition there can be no growth, no reproduction, no movement, no conscious sensitiveness, no thought.

In truth, life can be conceived of as reduced to its most simple expression, to mere nutrition. A being capable of nourishing itself, and destitute of every other property or function, lives still; but if it has not the faculty of reproduction, which, as we shall see, is only a simple extension of the nutritive property, its life will be only an individual life; a moment will come when the nutritive exchanges will slacken, when the nutritive residue, incompletely expulsed, will impregnate the living tissues and liquids, obstructing them, so to speak; then the colloidal plasmatic substances will cease to restore themselves, to regenerate themselves. Soon the retardment will end in complete arrest; then the organised being will have ceased to live; the complex elements which composed it will change, will break asunder, and the groups of their molecules and of their atoms will re-enter the exterior medium, the mineral world.

If, on the contrary, the nutritive property of a living being is sufficiently energetic to rise, as it were, to excess, even to growth and reproduction, the being is sure of living in its offspring; it fills its place in the innumerable crowd of living beings, and can even, if the doctrine of evolution is as true as it is probable, become the source of a superior organised type, can ascend in the hierarchy of life.

In fact, many of the inferior organisms are endowed only with the properties of nutrition, growth, and reproduction. At a greater degree of complication and perfection a new property appears, motility, subordinated likewise to nutrition, when it concerns the individual, to reproduction when it concerns the series. No one is ignorant that large numbers of animals are endowed solely with these four properties, nutrition, growth, reproduction, and motility, which are possessed by a number of plants also, as we shall see hereafter.

Nutrition, growth, and reproduction are truly fundamental properties. They belong to the entire organic world, to everything which lives and lasts. Above these properties must rank three others, all naturally subordinate to the primordial property, nutrition. These three are, the chlorophyllian property, motility, and innervation.

The chlorophyllian property is, with rare exceptions, confined to plants. Motility is, in a measure, common to animals and vegetals. Finally, the last vital property, innervation, is limited to the superior animals. It is also the most delicate, the most subordinated to, the most closely connected with, the integrity of nutrition, the most dependent, directly or indirectly, upon the other vital properties. Let but the nutritive liquids impregnated with oxygen cease to reach the nervous cells, to bathe them, to excite them, to renew them, immediately motility, sensibility, and thought vanish; the animal re-descends, for a time, or for ever, to the level of unconscious organised beings.

From this physico-chemical point of view we can now, without much difficulty, form an idea of the totality of the molecular movements which form the essential basis of life. Every living being is constituted, in a general manner, of colloidal substances more or less fluid, more or less solid, holding in solution salts, gases, and so on. A portion of these salts and gases has been introduced from without, and is ready to combine itself with the unstable colloidal substances; some are the result of combinations already effected; but this process of combination and separation cannot stop; for the atoms of atmospheric oxygen mingle themselves ceaselessly with the organic molecules, separate them, disaggregate them by virtue of their powerful affinities for certain elements which form part of their complex molecules. After a time more or less short, the oxygen, by a slow oxydation, would have thus destroyed the living substance, if food had not likewise been introduced from without into the texture of the living being. These renovating substances, after having often undergone preparatory chemical changes, after having become *nutriments*, that is to say, after having acquired a chemical composition and a physical state which assimilate them to the living substance, identify themselves with it. One by one their molecules take the place of those which have been destroyed. The living being, thus incessantly restored, lasts, continues to live, and would live indefinitely, if this molecular movement never slackened.

But we now know, through the magnificent generalisations of modern chemistry and physics, that in the world there are only atoms in some degree animated, that these atoms transmit to each other mutually the movement which impels them, or which they engender, and that this movement, without ever being annihilated, transforms itself in a thousand ways. These transmutations of movement take place also naturally in living beings, and the impulsions, so complex and varied, of the molecules transmit themselves to the different organic apparatus, producing, here the generation of new anatomical elements, there, the movements of totality of the living substance, elsewhere the nervous phenomena of consciousness, everywhere a certain elevation of temperature and, doubtless, electric phenomena.

It has been said, and may be admitted as a general principle, that the animal world lives at the expense of the stores of matter and of movement accumulated by the vegetal world. We shall have to show, at a future time, what amount of truth there is in this generalisation. We content ourselves, at present, with remarking that these vegetal accumulations are formed under the influence of solar radiation, that is to say, of the vibrations radiated by the central star of our planetary system, and that, consequently, the dynamic solar source is the great reservoir of force, the great motive power, which gives the impulse to the vital movement, and sustains the impulse given.

And now can we define life? For that purpose it will evidently be sufficient if we summarise the preceding facts into as clear, and at the same time as brief, a formula as possible; for it is not our intention to pass in review the very numerous definitions which have been given of life, long before its phenomena were scientifically analysed.

The definition now most commonly adopted in France is that given by Blainville: "Life is a twofold movement, at once general and continuous, of composition and decomposition." This definition, as H. Spencer judiciously points out,[1] is at the same time too comprehensive, and not comprehensive enough. It is too comprehensive, because it is applicable to that which occurs in an electric pile or in the flame of a wax taper, as well as in the primordial nutritive phenomena; it is too restricted, because it leaves out the highest, the most delicate vital activities, the

[1] *Principles of Biology*, vol. i.

cerebral or psychical activities. Lewes says: "Life is a series of definite and successive changes of structure and of composition, which act upon an individual without destroying his identity." In speaking of structure, this definition excludes the activities of purely mineral chemistry, which the first does not, but it also forgets the cerebral activities, and besides it does not embrace the vital acts that take place in the plasmatic liquids, such as the blood, the lymph, which, though destitute of structure, are endowed with life, as we shall presently see.

The definition of H. Spencer, "The continual agreement between interior and exterior relations," has the fault of being too abstract, and of soaring so high above facts, that it ceases to recall them. Besides, just by reason of its vague generality, it might also be applied to certain continuous chemical phenomena.

It would be better to descend nearer to the earth, and to limit ourselves to giving a short summary of the principal vital facts which have been observed. Doubtless life depends upon a twofold movement of decomposition and renovation, simultaneous and continuous; but this movement produces itself in the midst of substances having a physical state, and most frequently a morphological state quite peculiar to them. Finally, this movement brings into play diverse functions in relation with this morphological state of the living tissues, habitually composed of cells and fibres endowed with special properties.

Let us say then that "life is a twofold movement of simultaneous and continual composition and decomposition, in the midst of plasmatic substances, or of figurate anatomical elements, which, under the influence of this in-dwelling movement, perform their functions in conformity to their structure."

CHAPTER V.

ANATOMICAL CONSTITUTION OF ORGANISED BODIES.

EVERYWHERE and always, as we have already expounded, the living or organised bodies are constituted by complex substances, in part albuminoidal, and in that special physical state which is called *colloidal*. The fundamental matter of these living bodies is uncrystallisable. "To live and to crystallise," says Ch. Robin, "are two properties which are never united" (*Éléments Anatomiques*, p. 17). It is enough in effect for a body to be endowed with the humblest of vital properties, nutrition, not to be crystallisable. At the same time living substances are impregnated with crystalloidal solutions and with gases: this is a general attribute; but in the form this attribute is extremely diversified. At the lowest degrees of the organic world we find beings without structure, amorphous: for instance, the genus *Amœba* and the genus *Monas;* they are small contractile albuminoidal masses whose form is modified incessantly. Such are also the simplest rhizopods, living masses rather more considerable, but without definite form; we see them emitting and reabsorbing tentaculiform prolongations of varying length. But if even in a small degree we study by the help of the microscope the structure of beings more elevated in the living hierarchy, we instantly see that the fundamental mass has lost its homogeneousness, that it has fractionised itself into corpuscles generally invisible to the naked eye. These small bodies, these living bricks which by their aggregation constitute every organic edifice a little complex, have been called *anatomical elements* or *histological elements.*

Finally, these anatomical elements float more or less directly in living liquids, which are called *blastemas*. For instance, the freshwater polypus, celebrated on account of the curious experiments of regeneration to which it has given occasion, is solely composed of corpuscles living, spherical, of cells swimming in an intercellular liquid, which is a blastema. This is also the texture of certain infusoria, for example, of the Paramæcia, and likewise of a number of plants.

Besides in plants, and especially in superior animals, exist systems of canals serving for the circulation of liquids as living as the figurate anatomical elements. These liquids, which, like the blastemas, to distinguish them from which there has been a wrong attempt, are both receptacles of disassimilated products and reservoirs of assimilable products, have been called *plasmatic liquids*, or plasmas.

We have successively to describe living substance under the two general forms which it assumes, namely, the *histological* form and the *blastematic* and *plasmatic* form.

1. *Of the Figurate Elements in General.*

The science of the figurate elements of living bodies, whose real origin only remounts to the end of the last century, has long borne the name of *General Anatomy.* It was not till 1819 that Mayer published a treatise of General Anatomy under the title of *Treatise on Histology, and a New Division of the Body of Man.* The word *Histology* has had eager acceptance, no doubt because it is derived from the Greek, and it is now in general use.

The first elementary histological form which organised matter assumes is the cellular form. We must understand by *cell* a microscopical corpuscle, having a sort of independence, an individual life, assimilating and disassimilating on its own account. The cell has generally a form more or less spherical. It is constituted by a substance more or less soft. When it is complete it contains another cellular element which is smaller, a nucleus in which the living activity of the cell usually attains its maximum of power. Moreover, it often happens, especially in plants, that the exterior surface of the cellular corpuscle hardens. This hardened surface then constitutes what is called the *cellular membrane.*

The observations and the inductions of palæontology, of embryology, of the systematic natural history of organised beings, authorise us in considering the organic cell as the corner-stone of the living world, the common mother of all other histological elements. In effect the first figurate living beings have been monocellular, or composed of cells resembling each other, and simply juxtaposed. At the origin of nearly the whole of living beings, animals or plants, we find a simple cell. Finally, when we hierarchically class the innumerable organised beings which people our globe, we encounter, at the lowest, the humblest degree, beings composed of a single cell, or of a small number of identical and juxtaposed cells.

The cellular theory which we have just in summary fashion sketched, is one of the grandest views of Biology. Bichat was the first to attempt the anatomical analysis of living beings, by trying to resolve each organised being into tissues anatomically and physiologically special. Schwann, carrying analysis further, decomposed the tissues themselves into microscopical elements, and was the first to formulate the cellular theory in his work entitled *Microscopical Researches on the Conformity of Structure and of Growth of Animals and Plants.* 1838.[1]

The cellular theory contested at present, or rather differently interpreted on certain points by M. Ch. Robin and his school, nevertheless keeps its ground as a whole. It is not easy to understand without it the genesis and the evolution of organised beings. Finally, this theory has led Physiology to scrutinise

[1] *Mikroskopische Untersuchungen über die Übereinstimmung in der Struktur und dem Wachsthum der Thiere und Pflanzen.*—De Mirbel had already shown that the tissue of plants is composed of utricles and cells. 1831-1832.

more profoundly the mechanism of the vital acts ; it has taught it to refer them to their ultimate agents, that is, to the histological elements themselves, which vary in function and in form in complex beings, and which we must consider as playing a part in the mechanism of organised beings, analogous to that of atoms in chemical aggregates. As Schwann has said, "Forasmuch as the primary elementary forms of all organisms are cells, the fundamental force of all organisms reduces itself to the fundamental force of cells." (*Mikroskopische Untersuchungen*, 1838.)

The cell, properly so called, of which we have given above a succinct description, is a sort of schematic type, scarcely existing in anything except rudimentary beings and tissues. If we study the cell either as a complex organism, or in the hierarchical series of organisms, we see it in effect modifying itself, putting on different forms when assuming diversified functions. Finally, another type of histological element appears: it is the fibre, a microscopical element likewise, springing evidently from the cell in certain cases, where the cells have merely been elongated by juxtaposing and cementing themselves end to end. These derivative fibres exist manifestly in plants, in which they often hollow a passage for themselves as canals. According to Ch. Robin, there is a different process in animals. Here the fibres, with all their essential attributes, would seem to be present at the very dawn of the embryonic life, forming themselves spontaneously by genesis, at the expense of the blastematic liquids secreted by cells. As there are various species of cells, there are also various species of fibres; but the true typical fibres, well specialised, are found scarcely anywhere except in the animal kingdom. We purpose speaking further on at greater length of certain species of fibres, muscular fibres, nervous fibres, and so on.

In sum, passing by some amorphous organised types, points of union of a sort between the living world and the non-organised world, we must consider every complex organism as being constituted by a great number of individuals, living, microscopic, having each special activity and special functions. These anatomical elements are conformed in accordance with a small number of types, and in the superior organised beings they are grouped in tribes, and thus form *tissues*, charged each to fulfil such and such great physiological function, which is the total of all the elementary activities (muscular tissue, nervous tissue, osseous tissue, chlorophyllian tissue of plants).

As a matter of course the degree of differentiation in plants is very variable. It is an organic law that this differentiation of the anatomical elements is carried the further the more the organised individual is perfect. In other words, the great law of the division of labour reigns everywhere in the organised world. Besides, the elements themselves have a more complicated structure the more their function is complex (muscular fibre, nervous fibre). Finally, the more the organisation of an animal, taken as a whole, is simple, the simpler is also the structure of each of the orders of anatomical elements. Thus the muscular fibres of the radiata, the annulata, the mollusca, the nervous tubes, the ganglionic cells of lampreys, are simpler than the same elements in the crab.[1]

But in every superior organism there is a differentiated blending of anatomical elements, having varied functions and varied degrees of structure. We could therefore, in every individual, group the elements in series, according to their degree of perfection, of complication, and we should have a complete scale going from the elements, confused and even amorphous, of the inferior beings up to the elements with complex structure of the superior beings.[2]

At the foot of the organic scale we find monocellular infusoria (polytoma, difflugia, enchelys, monas, amœba) formed of a single homogeneous substance. Some of them are constituted of a sub-

[1] Ch. Robin, *Éléments Anatomiques.*

[2] Ch. Bernard, *Rapport sur les progrès et la marche de la physiologie générale en France.* Paris, 1867.

stance slowly contractile, which seems to be the rudiment, still undivided, of the muscular fibre; it is a sort of non-figurate muscular matter. This matter is called *sarcode*.

There seems to be in inferior beings a confusion of organic materials and functions. Many of the infusoria are endowed with motility and sensibility, with a sort of instinct, and yet they are destitute of muscular elements and nervous elements.

We can place in a degree immediately superior the plants and the animals simply polycellular, that is to say, constituted of a certain number of cells similar to each other and grouped. They are beings formed of a single tissue.

On the other hand, at the outset of their embryological existence, the beings the most complex, the superior animals, man not excepted, commence by being monocellular, then pass through the polycellular state, the most rudimentary; finally, in a last period, their histological elements differentiate.

This gradual histological differentiation, which is observed in the embryological development of superior beings, can also be demonstrated in the palæontological succession of the organised beings on our globe. In fine, it is easy to encounter it anew by grouping living beings hierarchically, from the simplest to the most complex. It is in this triple coincidence that the grand doctrine of evolution, founded by Lamarck and Darwin, finds its most brilliant confirmation.

In the animal kingdom the figurate elements can be classed in two great groups: the group of the constituent elements, which forms the basis, the framework of every organised being, and that of the produced elements, which plays a part more or less secondary, and has an existence more or less provisional. It has been observed that the constituent elements were generally situated in the interior of the body, and the produced elements on the surface. But this division, to which M. Charles Robin first of all, and Mr. H. Spencer afterwards, accorded a supreme importance, is only, like most classifications, a commodious arrangement for grouping the elements. If it were literally accepted—and indeed it is so accepted by M. C. Robin—it would be necessary to class among the constituent elements the globules suspended in the blood, the *haematia*, which yet are evidently elements produced, and of brief duration.

From the point of view of ultimate physical constitution, of the mode of molecular collocation, we must consider every living element as being formed by a blending, molecule by molecule, of immediate principles, belonging to the three classes already indicated. All these immediate principles are dissolved in one of them, in water, which in weight is by far the most important body. In effect, living elements need a certain minimum of constituent water without which they can neither get nutriment nor as a result perform their functions.

In the vegetal elements, as Sachs remarks,[1] we can prove this intimate blending of the immediate principles, by extracting from those elements, by the aid of certain solvents, substances chemically determined, without thereby changing the form of the histological skeleton.

There exists between the anatomical vegetal elements and the animal elements an important difference in the degree of chemical stability. The animal elements are much more easily alterable by physical and chemical agents. In plants there is a certain degree of mineral fixity manifestly in relation with their smaller degree of vital perfection and activity. MM. Naegeli and Schwendener, studying carefully the play of polarised light in the vegetal cellular membranes, the particles of starch, and also in the vegetal crystalloidal bodies, have found that in these vegetal tissues and elements there must be crystallised molecules birefringent and with double optical axes. These facts are perfectly in accordance with the difference of chemical composition of tissues in the two organised kingdoms. We shall see in effect that the most characteristic chemical element of organised substances, azote, enters in relatively feeble proportion into the composition of plants. Now the presence of azote

[1] J. Sachs, *Traité de Botanique*, p. 768. Paris, 1874.

coincides always in living beings, with a more elevated degree of vitality, a greater molecular mobility.

The action of certain chemical and physical agents on the anatomical elements is in manifest relation with their constitution. In effect brought into contact with solutions of bichlorure of mercury, of perchlorure of iron, of chromate of potash, of alcohol, and of other substances eager in their thirst for water, the anatomical elements lose their form and condense; for they then lose their constitutive water.[1] It is for this reason that alcohol definitively arrests the movements of the most resistant of the animal elements, of the vibratile cells, of which we have presently to speak, and that it kills in like manner the vibrions and the spermatozoaries.

Heat, on the contrary, first of all accelerates the vital phenomena; under its influence the mobile cells move with more rapidity, the functions of plants are accomplished with a greater energy; for a certain elevation of temperature facilitates the chemical reactions and renders the osmosis more rapid. In like fashion diffusion increases with temperature. For chlorohydric acid we have in effect the following gradation:—

Diffusion at	15° 5^{c}	= 1
,, ,, ...	27°	= 1·3545
,, ,, ...	38°	= 1·7732
,, ,, ...	49°	= 2·1812

But if the temperature continues to rise, the functional excitation promptly reaches a maximum point, beyond which it first of all decreases and soon is annihilated. Because the heat diminishes by evaporation the constitutive water of the elements, and alters the composition of the albuminoidal substances when it does not coagulate them; a result which is irremediable. Subjected to a temperature too elevated, the anatomical element soon dies; while cold, which likewise slackens and stops the nutritive phenomena, does not always destroy them, sometimes merely

[1] Ch. Robin, *Éléments Anatomiques*, p. 20.

suspends them. The reason is that the vital activities and properties are directly and solely derived from the physico-chemical properties in the midst of the anatomical elements. Consequently we see them grow stronger, or languish, vanish, reappear, or hasten to final extinction, from the sway of the molecular movements and mutations of which they are the expression.

2. *Histology of Plants.*

The vital functions are less numerous, less specialised in the plant than in the animal, it being understood of course that we except the most inferior organisms in the two kingdoms. We are therefore justified in supposing, *à priori*, a less sharp specialisation in the form of the elements. This is what is actually the case. While in the superior animal we find varied histological types very clearly distinguished from each other; in the plant, on the contrary, the elementary forms are less decided, less dissimilar, and sometimes they can be supplemented physiologically.

It is from the microscopical anatomy of plants that has sprung the cellular theory, so contested at present in France, but generally admitted in Germany, and according to which every anatomical element, vegetal or animal, has as direct origin a simple cell. In effect, in the vegetal world, the utricular, the cellular type greatly predominates. Every plant, from the simplest to the most complex, is formed by an aggregation of cells, or of fibres manifestly originating in cells.

Every *complete* anatomical vegetal element is a cell formed of a double wall, of a content, and of one or more nuclei.

The external cellular envelopment is constituted, chemically, by a ternary substance united to certain salts; this is the *cellulose*, composed of carbon, of hydrogen, and of oxygen. The chemical formula of the cellulose is analogous to that of sugars. It is $C^{12} H^{10} O^{10}$. When the histological element is complete, this external membrane is interiorly lined with another very thin vesicle; but the second contains azote: it is albuminoidal. This azotised membrane englobes a semi-liquid substance and one or two small spherical or ovoidal bodies, likewise azotised. These are the nuclei, in which are often included one or two nucleoles. These azotised portions of the cell appear to be the seat of a nutritive movement more intense than the others. They appear also to be bound up with the period of development; for when the cell has lost its fluid content or protoplasm, it becomes incapable of growth and of multiplication.

The contents of the vegetal cells are normally liquid or solid. Liquid they can be formed of oil or of water, holding in suspension either molecular azotised granulations, or particles of fecula, or drops of oil or of resin, or finally, small green bodies, very interesting, called *chlorophyllian bodies*. It is to these last corpuscles that the green parts of plants owe their colour.

The liquid contents, non-oleose, are generally called protoplasm by the botanists. According to certain botanists, this protoplasm is the really important part of the cell; it secretes the enveloping membranes, and the nuclei come forth from it by differentiation.

In any case, this protoplasm is assuredly albuminoidal, for it precipitates through the chemical agents which precipitate albumine, and iodine communicates to it the yellow coloration which it gives to azotised organic substances.

The internal vesicle, the protoplasm, and the nuclei form an albuminoidal whole, which Dutrochet was the first to succeed in isolating, by destroying the external membrane with nitric acid or dilute caustic alkalis.

To constitute the diverse vegetal tissues, the cells assume various forms. For instance, the vegetal vessels by which, especially in plants, the liquids and the gases circulate, are at the outset formed of cells juxtaposed longitudinally. After taking this linear arrangement the cells are cemented, and their walls are reabsorbed at the points where there is contact. The communication once established, the contents of the cells in their turn disappear, and the canal is formed. If the vascular bundle is perfect, every trace of cellular cementation completely disappears; but in the case where the fusion of the elements has been less complete, the vessel remains nodose, being fashioned in the likeness of a chapelet.[1] It is then called *utriculous*. Sometimes, instead of being juxtaposed in linear series, the formative cells assume the shape of a vascular network.

According to M. Ch. Robin,[2] we are able in vegetal tissues to distinguish as clearly as in animal tissues a small number of histological types. These types have as origin a cell; but from their advent they bear a distinct physiognomy, and they are never seen undergoing or accomplishing mutual transformation.

The first of these types is that of *cells* properly so called, offering moreover a certain number of varieties according as they are spheroidal, ovoidal, fibroidal, stellated, or cylindroidal. It would be needful to range in this class the cells of the epidermis of plants, those of the cork-tree, of cambium, and of marrow. We might naturally add the monocellular plants, such as the red utricles which sometimes give a red tinge to the snow of the Alps (*protococcus nivalis* of Saussure), and the diatomous plants. The second type is that of the *filamentous cells*, all more or less cylindrical, and eight or ten times longer than they are broad. We may cite as an example of filamentous tissues the cells forming the mycelium of the cryptogams, and also certain monocellular infusoria, such as the bacteria and the vibrions, if indeed we admit that botany can claim as belonging to its domain these dubiously-defined organisms.

Every plant solely composed of the two histological types spoken of above, is a *cellular plant*.

The *fibrous cells* represent the third type. These are they which, juxtaposed lineally, form the ligneous fibres of wood and of liber.

Finally, the *vascular cells* form the fourth type. These are they which by their linear juxtaposition and the partial reabsorption of their walls, constitute canalicules—vessels. To the

[1] In the sense of *rosary* or *string of beads*.—*Translator*.

[2] *Éléments Anatomiques.*

type of vascular cells belong the trachean cells with spiral filament, the punctuated cells, and the laticiferous cells.

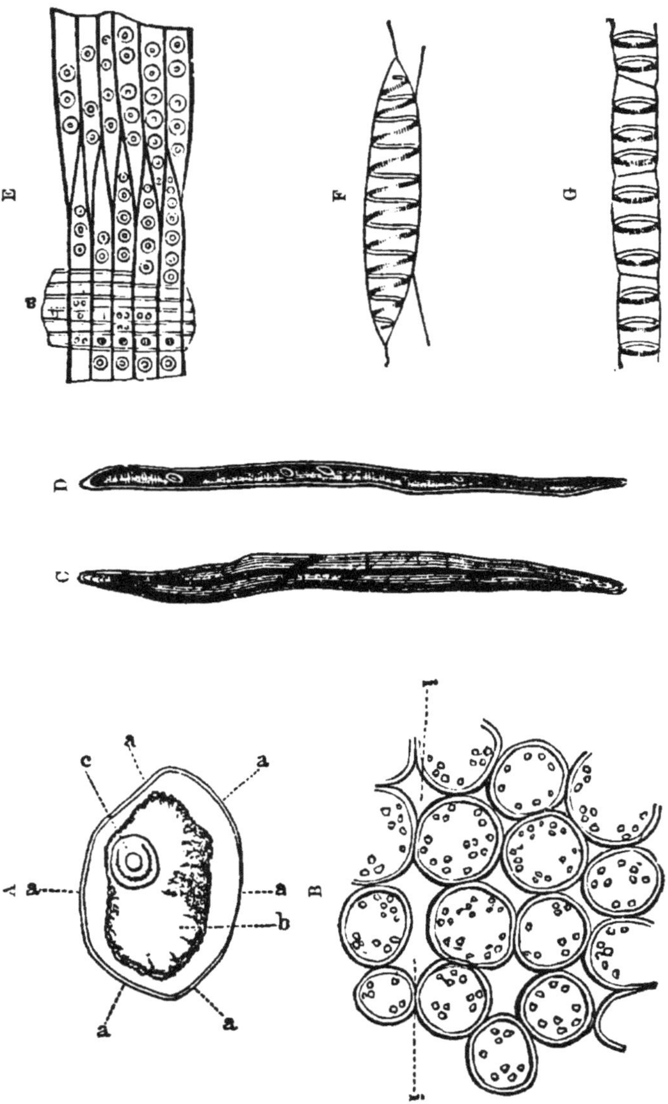

Fig. 1.

A, vegetal cell treated with alcohol. The protoplasm has concreted and separated from the cellular wall. *a*, cellular membrane; *b*, protoplasm; *c*, nucleus with nucleole.—B, spherical cells of the loose vegetal tissues (parenchyma of fruits, pith of the elder tree, and so on).—C, elongated or ligneous fibro-cell of the pine.—D, ligneous fibro-cell of the larch tree.—E, punctuated cells of the conifers.—F, cell like thread wound spirally.—G, another cell like wound thread.

We are disinclined to admit that this classification, so sharp, so decided, can be admitted in all its inflexible rigour; but we may accept it as giving a good general view, as grouping under a small number of heads a great variety of vegetal and histological elements.

3. *Histology of Animals.*

The anatomical animal elements differ in general from the vegetal elements in the threefold point of view of chemical composition, of structure, and of form.

In effect, while in plants the anatomical element is in chief part constituted by a non-azotised ternary substance, the cellulose; the animal elements, on the contrary, are formed especially by the quaternary albuminoidal substances. No doubt we encounter in animals ternary substances analogous to the starch and the cellulose of plants, but partially, secondarily, and in small quantity. It is thus that we find in the tegument of the arthropods, and even in all classes of the invertebrates, a matter very analogous to the vegetal cellulose, chitine, which can transform itself into glucose. But these points of detail do not weaken the value of the grand general fact enounced above.

The general differences of structure are perhaps in proportion to the differences of chemical composition. The albuminoidal substances are in effect essentially colloidal, and consequently must tend to a more unfixed morphology. Thus, while in the vegetal cell we generally find an enveloping membrane with the exactest limits, this membrane is often lacking in the anatomical animal elements. In the latter case the element is a small figurate glomerule, approaching more or less the type fibre or the type cell, and usually furnished with one or more nuclei and nucleoles.

From the point of view of form, the difference between the histological animal elements and the histological vegetal elements is more marked still. The vegetal histological types are few, and are all related directly and visibly to the cell. But the anatomical animal elements have forms much more varied. In those of them which merit the name of *fibres*, there is often no longer any trace of the cellular form ; and the animal fibres are not even derived from original cellular elements, if we admit with M. Ch. Robin the spontaneous genesis of the anatomical elements in the blastemas.

This theory of the spontaneous apparition of the anatomical elements in the living liquids has hitherto been rejected in Germany, where is adopted in all its rigour the axiom of M. Virchow—*Omnis cellula e cellula.*

In accordance with the terms of the German cellular theory, every anatomical element, whatever it may be, has as origin a cell; it comes forth from it by gemmation or segmentation; and every element which departs from the cellular form is simply a metamorphosed cell. It is not easy to understand how assertions so directly opposed should be passionately maintained on both sides by observers equally skilful. We are compelled to admit that on each side there is a portion of truth. We shall see when treating of generation that the French school wishes to reject reproduction by division and gemmation in the vegetal kingdom, in the initial period of embryonic animal life, and in certain *produced* elements. According to this school, most of the elements called constituent, that is to say, forming really the framework of the animal organism, spontaneously arise, by synthesis, by genesis in the living liquids, alike in the embryon and the adult.

Let this be as it may, the cellular theory is convenient for classifying the anatomical elements. In effect, while the partisans of spontaneous genesis admit no bond of direct kinship among most of the anatomical elements, strive especially to note dissemblances and to multiply species, the partisans of the cellular theory, preoccupied with the idea of a common origin, dwell especially on resemblances, and thus arrive at forming a very small number of elementary histological groups. According to them there are only four types of anatomical elements and of animal tissues, namely: 1. The elements of the cellular or connective tissue; 2. The cells remaining autonomous, that is to say, the epitheliums and the glandular cells, to which might be added the globules of the blood; 3. The elements of the muscular tissue; 4. The elements of the nervous tissue.

The first of these tissues forms the general gangue, the support of all the other tissues and elements. It is essentially composed of cells called *stellated cells*, having a diameter of from 0^{mm}, 050 to 0^{mm}, 060. These cells enclose a nucleus containing a nucleole. They emit fibrillary prolongations concurring to form the fibres of the cellular tissue, and which often seem to be anastomotically connected with each other. Other fibres called *laminous fibres*, because they are slightly flattened, form themselves in that same cellular tissue round elongated nuclei, called *embryoplastic nuclei*. The whole resembles a long wire-drawn spindle. The laminous fibres emanating from those cells form the chief part of the cellular tissue. They are very long, grouped in bundles, and with an average breadth of 0^{mm}, 001.

According to this theory, we regard as appertaining to cellular tissue the cartilaginous cells and the osseous cells. All these cells are nucleated. The first are round or ovoid, the second are irregular, and emit in every direction filiform prolongations, anastomotically intertwined. These last cells, which form the living mechanism of the osseous skeleton, have been called stellated osseous corpuscles.

The cells called *autonomous* comprehend the globules of the blood, which we shall describe further on, and the epitheliums. Of these last elements, some serve to line, while protecting, the animal membranes, the skin and the mucous membranes, while the others play in the secretions an extremely important part, to which we shall have occasion to return.

The epitheliums have as their first division the *pavimentous epitheliums*, large cells flattened, and usually polyhedrical, because they are subject to reciprocal compression. They contain a nucleus and a nucleole, and their whole aspect vividly recalls a pavement of hexagonal bricks. In animals they clothe the mucous membrane with many cônduits, and play especially the part of a protecting varnish. As related to them may be regarded the epidermic cells. Other epithelial cells are cylindrical or cylindro-conical. Sometimes the free surface of these cells is furnished with fine and mobile cilia. In that case the epithelium is called *vibratile epithelium.*

We have besides to mention the globulous, the spherical epithelium. We find it especially in the glands where its function is to form, at the expense of the plasmas and the blastemas, the special bodies destined to be secreted.

Finally, we signalise here for the sake of remembrance the muscular elements, distinguished into fibre-cells and into fibres properly so called; then the elements of the nervose tissue, fibres, and cells. We shall have elsewhere to describe in detail the form, the structure, the functions of these elements, which may be characterised as *aristocratic.*

CHAPTER VI.

OF LIVING LIQUIDS.

For centuries the humourists and the solidists have filled the world, we mean the small physiological and medical world, with their discords, with their furious strifes. As in all long wars, there has been in this many a peripetia. Sometimes the triumphant humours submerged their adversaries; sometimes these, offering resolute front with their serried ranks, seemed to have fixed victory for ever to their banners. Observation and experience have ended by imposing on the belligerents their sovereign arbitrament. It is at present demonstrated that between the solids and the liquids of every organism there is less difference than had long been believed. The solids come forth from the liquids; they come forth from them incessantly and return unto them. Finally, between the solids or anatomical elements and the living liquids, that is to say, the blastemas [blastemata] and the plasmas [plasmata], there is a great analogy of composition.

The name of *blastema* is given to every living liquid, that is to say, endowed with nutritive mollecular movement and interposed between anatomical elements. The *plasmas* are also living liquids, but circulating in canals, the blood and the lymph of animals thus circulating. Blastemas and plasmas have a great analogy with each other if we look at them in a general manner. They both are living; they both equally contain materials of assimilation destined for the anatomical elements, and materials of disassimilation which come forth therefrom, crystalloidal substances in process of transforming themselves to become assimilable or organisable albuminoidal substances, and albuminoidal substances tending to become crystalloidal and to be expelled from the organism. The vitality of the blastemas is naturally greater than that of the plasmas ; they may almost be considered as elements which have lost their forms ; physically, they are viscous liquids in which, generally, granulations are interspersed. They are the organisable liquids by excellence. It is in the midst of the blastematic liquids that the anatomical elements of the embryons, or of the tissues constituted in process of development or regeneration, have their origin, for instance, after a wound.

1.—*Vegetal Blastemas.*

The plants, even the most perfect, the dicotyledons, have no special circulatory system. In the dicotyledonous tree, the terrestrial sap mounts through the tissues of the wood, filling the intercellulary spaces, the canals, and passing by endosmosis from cell to cell, from fibre to fibre. Arriving at the leaves it undergoes an important modification, with which we shall have to occupy ourselves ; then it redescends by the more superficial tissues, and especially by the young intermediary tissues between the bark and the wood. The descending sap, that which has been elaborated in the leaves, must be considered as a living liquid, as a blastema ; also, as it goes along we see it organising itself either in a direct manner or through the agency of pre-existent anatomical elements. In truth it is a liquid holding a middle position between the plasmas and the blastemas.

The blastematic liquid, whence spring the buds, is thoroughly comparable with the animal blastemas. Like these last it is in part exuded by the anatomical elements, that is to say by the vegetal cells which it drains off and dissevers. This mode of formation recalls that of the vegetal and animal embryonary blastemas.

The true vegetal blastemas resemble much the semi-liquid azotised content of the vegetal cells, namely, that intracellular substance endowed with spontaneous movements, having its special molecular affinities, refusing, for instance, as long as it is living, to let itself be imbibed by colouring matters. Now we know that this intracellular protoplasma offers all the chemical reactions of the true albuminoidal matters (albumine, fibrine, caseine). Iodine gives it a yellow colour; alcohol, the mineral acids, and heat coagulates it.

Chemically, the vegetal blastemas are constituted by water, by albuminoidal substances, and by some salts.

2.—*Animal Blastemas.*

There has been a wish to limit the name of *blastemas*, in the animal economy, to the interstitial liquids alone in which new anatomical elements are formed, that is to say, to the intercellular liquids of the animal embryons, before the formation of the vessels, to the organisable liquids which are produced in a wound in process of cicatrisation, finally to the liquids of the serous cavities. But in biology, as in all other natural sciences, if it is useful to divide, to classify, it is wise not to accord to divisions and classifications an absolute value. The frontiers which we are obliged here and there to mark out in the vast field of the living world to aid the feebleness of our memory have only a relative value. In effect, in the organic world, even taken generally, all is gradual modification, gradual transition. If this is true, as incontestably it is, in the classifications of natural history, properly so-called, how much more must it be so when the aim is to ascertain the constituent parts of one and the same organism? If we reserve the name of *blastemas* to the small number of liquids living, interstitial, organisable, and generative, what are we to do with the other interstitial liquids, with those which bathe the elements of the tissue called *conjunctive* [connective], of that tissue comparatively coarse in its morphology which serves as gangue and as support to all the others? And in most tissues, wherever the elements do not touch each other at every point of their surface, is there not an interstitial liquid, coming on the one hand from the tissues, and on the other from the exterior medium? But these tissues constantly grow larger from birth to the adult state or age, that is to say, that incessantly during this lapse of time new elements arise and gain place in the midst of the old. Evidently these elements are formed at the expense of the interstitial liquid, which consequently then becomes a true formative blastema.

Let a freshwater polypus be sectionised into several fragments: immediately each of these fragments strives to complete itself, strives to remake a complete individual; but this new individual once formed is not more voluminous than the fragment whence it took its birth. It has therefore modelled and constituted itself at the expense of the interstitial liquid, bathing the cellular tissue of the hydra; therefore this liquid is formative; therefore it is a blastema. But blastema is likewise the intercellulary liquid of the grey cerebral substance, that of the umbilical cord, that of the marrow of the bones, and so on.

The chemical composition of the animal blastemas is a little better known than that of the vegetal blastemas. Like these last, they are composed of albuminoidal substances, of salts, and of regressive crystalloidal substances; lastly, of a great quantity of water. But besides these general characteristics it has been successfully demonstrated that the blastemas of the superior animals have a composition different from that of the blood and the lymph, from which chiefly they are derived. They possess fibrine and albumine in smaller quantity. Largely albumine takes in them its chemical, soluble, and assimilable form; it becomes *albuminose*, no longer coagulates from heat, and coagulates imperfectly and with difficulty through the acids. Fibrine is no longer found therein, and nothing is more natural, for we know that to become assimilable, nutritive, fibrine needs to be transformed isomerically into albuminose.

There has been a desire to make of the chemical instability of the blastemas a distinctive characteristic. No doubt the blastemas are in a state of perpetual mutation; they are never identical for two moments of duration. But we can say quite as much of the blood, of the lymph, of the histological elements themselves, forasmuch as the movement of continuous renovation is the primordial condition of life.

On the whole, between the anatomical elements, the blastemas, and the plasmas, there is from the chemical point of view, merely the difference in the proportion and the nature of constituent immediate principles, but a difference graduated from each to each. They are three forms of living matter, analogous to each other and engendering each other.

3.—*Of the Plasmas.*

In the plant, the imperfect division of physiological labour, and the confusion of functions, are so great that we find it difficult to classify the organic liquids. We have signalised the liquids evidently blastematic of the vegetal embryon, of the buds, the liquid semi-blastematic and semi-plasmatic which is called *sap*. Further on we shall say some words in reference to the secretion of the vegetal glands and of their secreted products; and thus we shall have passed in review all the vegetal organic liquids. The thing is less simple in animals, or at least in the superior animals. The division of labour is more advanced in the tissues and the liquids, and we must carefully classify both of them.

In the superior animal the organic liquids can first of all be divided into two great groups corresponding to the two great divisions of the solid elements; they are the group of the *constituent humours*, and that of the *produced humours*.

The chemical composition of each group, and even of the humours of each group, is very various; but in a general manner we can say, that they all contain immediate principles of the three orders:[1]

1. Principles of mineral origin (water and dissolved salts);

2. Principles of organic origin, some crystallisable, others coagulable (urea, creatine, lactates, choleates, and so on);

[1] Ch. Robin, *Des humeurs*, p. 20, 21, 8vo.

3. Principles non-crystallisable, but coagulable, met with in all the humours, except the bile, the urine, and the sudor. These last immediate principles are the albuminoidal substances properly so called, and the saccharine substances, both of them having the property of dissolving certain mineral or mineralised compounds little or not at all soluble in water. It is thus that albumine fixes silica, phosphate of lime, urates, and so on.

The humours produced or humours of secretion, are formed in the economy at the expense of the constituent humours, and generally by special organs called *glands*. By and by we shall have occasion to study their process of formation. They comprehend the extremely aqueous liquids, produced on the surface of the membranes, called *serous*, which cover certain viscera (brain, heart, lungs, intestines, and so on); the liquids bathing the articular surfaces or synovial liquids; lastly the sperm, the milk, the muci, the salivæ, the bile, the intestinal juice, and so on.

In connection with the produced humours we may view the liquids simply excreted, that is to say, separated from the constituent humours without chemical modifications (sudor, urine, amniotic liquid, allantoïdian liquid, pulmonary exhalation).

All these liquids, save three, are alkaline. The three liquids habitually acid are the gastric juice, the sudor, and the urine. However, this last liquid is sometimes alkaline, sometimes acid, sometimes neutral, in man, at the different stages of digestion. In the herbivorous mammifers it is normally alkaline. But here also the alkalinity depends on the digestion. In effect the urine of the herbivora becomes acid if we feed them on animal aliments, or, which comes to the same thing, if dieting them, subjecting them to inanition, we force them to live at the expense of their own substance. The alkalinity of the animal humours is usually due to salts of bibasic or tribasic soda; but free soda we never meet with in them.

In connection with secretion and excretion we shall speak in detail of the produced humours, contenting ourselves here with indicating their principal distinctive characteristics.

We have now to describe the humours of the first order, the humours constituent, living, those which may be regarded in some sort as liquefied anatomical elements. These liquids, which exist with supreme distinctness only in animals with complex structure, furnished with circulating apparatus well defined, are only two in number, the *blood* and the *lymph*; for we must consider the *chyle*, that is to say, that ultimate product of digestion circulating in the lymphatic vessels, a dependency of the lymph.

The blood and the lymph are contained in circulatory systems, ramified and inclosed in every direction. These circulatory systems imprison the liquids which are formed in them, without, normally, permitting them to break forth in mass. But across their walls they leave an easy passage to many materials coming either from the tissues, or from the ambient medium and to many others which escape from the blood and the lymph, either to nourish the tissues, or to be expelled as unworthy from the frontiers of the organism.

This double movement of coming and going, of exchange of materials, is effected simultaneously, like the nutritive assimilation and disassimilation in the solid elements. In truth the blood and the lymph are living liquids, in process of perpetual renovation. Incessantly they are formed in the system of the canals where they circulate without being destroyed.

The vascular walls which contain the constituent liquids do not seem notably to modify the chemical composition of the substances which traverse them. The part they play is especially physical, osmotic. Therefrom it results that the blood and the lymph borrow their constituent materials already formed from an ambient medium, either from the grand cosmic medium, the air for example, or from the organic medium, the tissues, the myriads of anatomical elements which compose the body of the superior animals.

Considered as organised liquids, the constituent humours necessarily contain immediate principles of the three classes; but those of the third class, the albuminoids of organic origin, not crystallisable or coagulable, predominate therein. They are *albumine* or rather *serine*,[1] and *plasmine*, which severed from the living organism, evolves itself into fibrine spontaneously coagulable and into fibrine called *liquid*. To these albuminoidal substances we must add variable proportions of *peptones* and *albuminose*, that is to say, of albuminoidal alimentary substances liquefied and absorbable.

The blood and the lymph are not homogeneous, physically simple liquids. We meet with, in them, floating bodies which are true free anatomical elements. These bodies have been called globules. The liquids in which these globules swim in immense numbers have received the special name of *plasmas*.

These plasmas isolatedly considered are endowed with nutrition; they are consequently living. They contain in proportions almost equal, immediate principles of the three orders; nevertheless the coagulable principles predominate. If nutrition ceases in a plasma, that is to say, if this liquid dies, its composition immediately changes; the albuminoidal principles which it contains evolve into a liquid portion and another portion spontaneously coagulable. It is this which, in animals provided with a circulatory system, produces the cadaveric rigidity.

The office of the plasmas is of supreme importance; but to obtain a complete notion thereof we must figure to ourselves what every complete organism is from the point of view of texture. At the lowest degree of life and organisation, we find monocellular beings, free anatomical elements, simple infusoria, living habitually in the water; for directly or indirectly, the anatomical elements are generally aquatic entities. These the monocellular organism absorbs and assimilates, disassimilates, and secretes, directly borrowing materials from the ambient medium, or restoring them to it. In beings a little more complex composed merely of cells identical with each other, and simply juxtaposed, the nutritive process is scarcely more complicated. In effect, the polycellular organism, with cells which resemble each other, is definitively nothing more than a collection of juxta-

[1] Ch. Robin, *Des Tissus*, p. 21.

posed monocellular organisms. However there is usually a general enveloping membrane and afterwards an interstitial liquid, a sort of blastema, playing, in respect to the anatomical elements of the polycellular organism, the part of an artificial medium. The cells now no longer plunge direct into the cosmic medium; they are protected and isolated by an organic, an elaborated liquid. In the extremely complex organisms, for instance, in the superior animals, the intrication of the texture is much greater. Here the anatomical elements are not formed in one mould only; they are differentiated according to multiple types, and each type assumes a diverse function. One, for instance the epithelial type, supplies to the living membranes a protecting varnish; to the glands a special agent of secretion; another, for example, the tissue called cellular, serves as gangue, as bond, as support to all the tissues, apparatus and organs, while opening besides a passage to blastematic liquids. A third, exemplified in the osseous element, furnishes to the organism a solid framework. Lastly, the muscular fibre and cell impress on the pieces of the living apparatus the necessary movements, while the nervose fibre and cell endow the organism with sensibility, with will, with thought, and are the sentient soul of the entire being, of which they assume the conscious direction, and so on. But in order that these anatomical elements so diverse, so numerous, grouped into tissues, into organs, into apparatus, into special systems, may live, may perform their functions, may co-operate, it is needful for them to be almost completely withdrawn from the rough and capricious influences of the exterior world; they must perform their functions in an artificial medium, alive like themselves, in which they find a temperature little variable, a magazine always well provisioned with substances elaborated and assimilable, suited to repair their losses; lastly, a place of discharge, into which they throw their used materials, that have become unfit to figure in the vital movement. These nutritive media, artificial, liquid, consequently appropriate for the aquatic life of the anatomical elements, are the plasmas, always in movement, always renewed, yielding incessantly materials to the exterior world and to the tissues, and incessantly taking them back, having a temperature nearly equal, as long as that of the exterior medium does not suffer very great oscillations. For the anatomical elements, the plasmas are a veritable living atmosphere.

4.—*Of the Blood.*

In the vertebrates the most important of the plasmas, the blood, is a red sap unceasingly circulating with greater or lesser activity. In the invertebrates the blood is animated by a much slower movement of translation. It is contained in apparatus not so well constructed, and is generally colourless and transparent like lymph. Often even the blood and the lymph of the invertebrates are confounded; where there are distinct sanguineous cavities and lymphatic cavities, as happens in some annelate worms, the blood is tinted with a special colour—it is sometimes red, sometimes yellow, green, violet, or bluish. The sanguineous cells which it sometimes in that case conveys, are almost always colourless. However, there are sanguineous, coloured globules in the *terebella* and the cephalopods.[1]

The blood of the vertebrates, with which we have especially to occupy ourselves, is, according to a felicitous comparison of Cl. Bernard,[2] an interior medium in which live the anatomical elements as the fishes live in the water. These anatomical elements, moreover, retain in the blood their physiological independence, and though drawing their nutritive materials from the sanguineous plasma, they do not allow themselves to be imbibed by it, which is an essential condition of endosmotic exchanges. Consequently we see, for example in the blood of the vertebrates, potash dominating in the globules, and soda in the plasma.

The physical qualities of the blood of the superior vertebrates

[1] R. Wagner, quoted by Fr. Leydig (*Traité d'Histologie de l'Homme et des Animaux*, p. 509, Paris, 1866.)

[2] Cl. Bernard, *Leçons sur les Propriétés des Tissus vivants*, pp. 55-58. Paris, 1866, 8vo.

are well known. In these perfected organisms the blood is a liquid slightly viscous, of a purple red in the arteries; that is to say, when it is freshly impregnated with aërian oxygen, but of a blackish tint, more or less deep in the veins; that is to say, when from contact with the tissues it has exchanged its oxygen for carbonic acid. It is well known that the temperature of these two bloods differs, that of the venous blood being a little more elevated in the right ventricle of the heart and in the deeper veins—an elevation which we must evidently attribute to the chemical reactions of nutrition, the residua of which the venous blood collects.

It is also known that the blood, as soon as it is drawn from the vessels, separates into two parts—a red coagulum containing the globules, and a liquid part of a lemon yellow, which is called *serum.*

The blood, we have said, is a medium from which all the anatomical elements of the organism derive the materials needful to their life, and into which they pour all their nutritive residua; its chemical composition must therefore be very complex. We find therein in effect immediate principles of the three classes, and we find therein in great number. We must content ourselves, therefore, with signalising the chief of them.

The principles of the first class, wholly mineral, are first of all water, which quantitatively is the most important element, as, moreover, the figure for the density of the blood shows, which on an average is only 1,050. In quantity, water in man represents from 905 to 910 thousandths of the blood. The proportion, however, notably varies; it is more considerable in the infant, the young man, the pregnant woman; in short, wherever the formation of new anatomical elements necessitates the fixation of many solid materials. In this liquid mass all the other immediate principles are dissolved, and the globules travel along.

The immediate gaseous principles of the first class are oxygen, azote, and hydrogen. The first of these gases, which is by far the most important, comes from the exterior air from which it is borrowed by the respiratory organs, as we shall see when speaking of respiration. It is oxygen which, combining with the globules or hæmatia, gives them the vermilion tint which they have in the arterial blood. But the globules do not keep their oxygen long. Elaborated in the fine circulatory vessels, where they are in almost immediate contact with the anatomical elements, they surrender to them their vivifying oxygen, indispensable to the chemical reactions of nutrition. In exchange they take back the gaseous residuum of the oxydation of the tissues, the carbonic acid, which gives them a blackish tinge, that of the venous blood. It must be observed that the sanguineous globules never completely despoil themselves of one of these gases to impregnate themselves completely with the other. They retain them simultaneously, oxygen predominating in the arterial blood, carbonic acid in the venous blood. It suffices, besides, for the gases to be in equal quantity in the blood for the globules to become blackish. The change in the proportion of the two gases is not effected suddenly, but by degrees—in proportion as the arterial blood goes away from the lungs and the heart to approach the tissues, the oxygen gradually yields the place to the carbonic acid.

The water of the blood is not normally free; it is found combined with albuminoidal matters. This is why it cannot filtrate mechanically through the vascular walls. Almost in totality it comes from the aliments, for it is doubtful whether any notable quantity thereof is formed in the organism.

The immediate saline principles of the first class which are contained in the blood are the chlorures, the chlorohydrates, the sulphates, the carbonates, the phosphates, and so on. Of all the salts, the chlorure of sodium is by far the most abundant in the blood of man. The proportion of the salts varies, besides, according to the animal species.

The phosphates predominate in the blood of the carnivora, but yield the superiority to the carbonates of soda and of potash in the herbivora. This is, after all, as we have remarked in reference to urine, a pure affair of alimentation. It is to the basic phosphate of soda and to the carbonate of soda that the blood owes its alkaline reaction.

M. Ch. Robin [1] remarks that the different salts of the blood serve as mutual solvents, and that the phosphates and carbonates of soda permit the sanguineous liquid to dissolve a great quantity of carbonic acid. In the economy, in effect, the blood is never saturated with carbonic acid. Venous blood brought into contact with carbonic acid, still suffices to dissolve thereof 0$^{mg.}$ 48 in 100. Thus it always seeks this gas eagerly, and is ready to disengage therefrom the anatomical elements.

We must cite, by way of remembrance, traces of silica, of manganese of lead, of copper, fortuitous mineral elements, little or not at all useful to nutrition, but drawn along with the others into the living organisms. It is otherwise with iron, which seems to play an important part, to form a really constituent portion of the sanguineous globules, though it exists in a very small quantity, for the total quantity of iron in the blood of an adult man is reckoned to be not more than one gramme.[2]

Other salts, the salts of soda, of potash, of lime, belong like the preceding to the immediate principles of the second class. They are organic salts, nutritive wastes. Let us mention the urates and inosates of soda, of potash, and so on, which probably result from the disassimilation of the muscular tissue, &c.

But the disassimilation of the anatomical elements gives birth to many other principles more complex, more organical, to sorts of alkaloids. These crystallisable principles, always in a state of liberty in the blood, are urea, creatinine, creatine, inosite. There has been an attempt to determine the place of origin of these diverse products. It is said that creatine and creatinine come from the muscles, urea from the tissues, fibrous, laminous, serous.[3]

[1] Ch. Robin, *Des Humeurs.*

[2] A French *gramme* is nearly equivalent to nineteen grains English.—*Translator.*

[3] G. Sée, *Du Sang et des anémies.*

We may view in connection with these substances cholesterine and seroline, formed probably in the nervous tissue.

All these bodies pass by osmosis through the fine capillary vessels, and sojourn in the blood till they are secreted or excreted therefrom.

The other azotised substances of the blood belong to the third class of immediate principles. They are the albuminoidal substances, properly so called. Urea, creatine, creatinine, and so on, are azotised regressive principles, residua of disassimilation: they form part of the material current coming forth from the organism. On the contrary, the other albuminoidal substances are the residuum of the alimentary elaboration. They are destined to repair the waste of the tissues, and form part of the material current entering the economy.

When the blood of a mammifer is drawn from the vessels, and allowed to rest, it separates into a red clot and a yellowish liquid. The clot is composed of an azotised complex substance, which has been called *fibrine*, because it has then a fibrillary aspect, and this coagulated substance retains in its meshes the red globules. The ambient liquid contains in solution another azotised substance, denominated *albumine*, because it has affinity with the albumen of the egg, though containing a half less sulphur. For a long time there was an erroneous belief that fibrine and albumine have as distinct existence in the economy as in the blood when drawn. Fibrine and albumine are, however, isomeric, albumine merely containing a little more water. From important analyses of the blood made by Denis, of Commercy,[1] it results that in effect the blood contains albumine, which Denis prefers to call *serine*; but instead of fibrine there is in the blood another analogous azotised substance, which he calls *plasmine*. This plasmine, he thinks, evolves itself in the blood when drawn, and thanks to the intervention of the globules, into a coagulable part called *fibrine*, and into another soluble albuminoidal substance,

[1] Denis, *Comptes rendus des Séances de l'Académie des Sciences*. Paris, 1856 and 1858.—*Mémoire sur le Sang*, Paris, 1859, 8vo.

which remains in the serum with the albumine or serine and can be separated therefrom. That all these determinations of contemporaneous chemistry are destined to be maintained intact in the future we do not believe.

In effect, the albuminoidal substances are extremely unstable; those of the economy are probably isomeric, or nearly so. Their chemical formula has not even yet been determined. They are not naturally absorbable, except on condition of being soluble; the fibrine and the albumine of the aliments are transformed by digestion into isomeric substances called *peptones*, *albuminose*, whose degree of relationship with the plasmine and the serine of Denis has not yet been determined. In short, these are questions the precise solution of which must be reserved for the chemistry of the future.

To conclude what we have to say on the principal organic materials of the blood, let me mention the fine guttulæ of fat matter floating in the blood after digestion, in the state of emulsion. These fat bodies are absorbed by the tissues, which yield them afterwards to the sanguineous liquid, in the state of combination, saline, saponaceous, and soluble (butyrates, phosphorised fat matters). Let me mention, lastly, a certain quantity of glucose or sugar of grape.

The mineral substances dissolved in the blood are not all in the state of simple blending. If we inject into the blood first of all a solution of a salt of iron, then a solution of prussiate of potash, we do not obtain the characteristic reaction of the salts of iron, the formation of Prussian blue, because the albuminoidal substances of the blood have at the very outset fixed the salt of iron. On the contrary, we obtain Prussian blue if we inject first of all the prussiate of potash, because the albuminoidal substances leave this salt free (Claude Bernard).

5.—*Of the Red Sanguineous Globules.*

If we examine in the microscope a guttula of blood, we find small floating bodies, extremely numerous, pressed against each other, and every one of which seems to be an isolated cell. A more attentive examination soon shows that these bodies are rather glomerules than real cells, for they have neither nuclei nor enveloping membranes. They are small flattened disks, depressed in the centre on their two faces, an arrangement which often simulates a nucleus. Their diameter is from 0^{mm}, 006 to 0^{mm}, 907, in the adult man. But their form varies in the series of the vertebrates, and also in man and the mammifers if we remount to the first stage of the embryological life. In effect, at the epoch when the embryon has only a length of 0^{m}, 02 to 0^{m}, 03, where the globules begin to appear, they are white, and have a veritable nucleus. According to some authors they then multiply by segmentation. They are now much longer, and their diameter attains 0^{mm}, 010, to 0^{mm}, 011.

In the animal series we also observe very notable differences in the form, the volume, and the structure of the sanguineous globules. As long as we are simply considering the class of the mammifers, these differences are slight. However, the sanguineous globules of the camel and of the llama have an ovalar ellipsoidal form, and resemble those of birds and reptiles. In the elephant and in the *didactylus* sloth we also find sanguineous globules much longer than in the other mammifers.

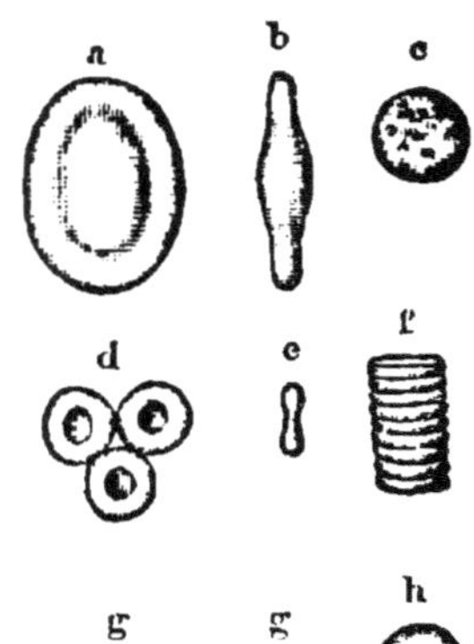

Fig. 2.
Frog.—*a* and *b* front view and profile of a red globule; *c*, white globule, 500 diameters. *Man.*—*d* and *e*, front view and profile of a sanguineous globule; *f*, pile of globules; *g*, white globules; *h*, fatty vesicle of the chyle, 800 diameters.

In the other classes of vertebrates the differences grow more striking. In all the vertebrates the globules are, as in man, of a red colour by reflection, and it is to them that the blood owes its rutilant tint. In general the red globules of the mammifers are less voluminous than those of the other classes of vertebrates. They are elliptical and nucleated in the classes of birds, of amphibia, and of fishes: elliptical only in the class of reptiles. There is no absolute rule, however. The fact of the existence of globules in the blood of an animal is very important; but their form is much less so. We have seen that certain mammifers have elliptical sanguineous globules like those of birds. On the other hand, the humblest of the vertebrates, fishes of an inferior order, such as the genera *myxine* and *petromyxon*, having affinity with the celebrated *Amphioxus lanceolatus*, have like man sanguineous globules, circular and with double depression. Lastly, as a second link of the chain between the *branchiostoma* without globules,[1] as an invertebrate, and the fishes with globular blood, is found a fish with white, colourless globules, the *leptocephalus*. But, on the whole, in spite of some exceptions, the red globules, or *hæmatia*, are found in the blood of nearly all the vertebrates; they are the anatomical and physiological sign of a complete organisation, of a more active respiration, of a higher vitality.

In the blood of man and of the mammifers, the sanguineous globules are in immense number. According to Schumann, Andral, and Gavaret, the red globules in the humid state form in volume the half of the mass of the blood.[2] Furthermore, Vierordt has counted, in a cubic millimetre of blood, from 4,180,000 to 5,551,000 globules.

The younger the individual the greater is the proportion of hæmatia. The blood of the adult man contains 302 thousandths according to Ch. Robin.[3] There is much more in younger persons, and in the new-born child the proportion rises to 600, to 680, and even to 700 thousandths. A German anthropologist, Dr. Welcker, has demonstrated that from the point of view of the form and the proportions of the cranium and of the face, woman holds an intermediate position between the adult man and the infant. In regard to the hæmatia, the relative proportion is the same, rising in women to 320 and to 400 thousandths.

[1] Retzius, J. Müller, De Quatrefages. [2] G. Sée, *Du Sang et des Anémies*, p. 15.
[3] Ch. Robin, *Des Humeurs*.

The hæmatia are veritable histological elements floating in the sanguineous plasma. Like everything which lives, they assimilate and disassimilate incessantly. Each of them has probably only a brief duration. According to the German cellular doctrine, they spring in the embryon, from pre-existent cells, and multiply afterwards by segmentation, by cellular division : but the point is not one which direct observation has yet elucidated. It is certain, however, that the hæmatia are remade in the blood, for a few weeks or a month suffice to cure the anæmia caused by too copious blood-letting or by excessive hæmorrhage. In an animal subjected to abstinence, the globules diminish in number, lose their shape, and shrink. It is probable that incessantly the more aged of the hæmatia dissolve in the blood, and are replaced by hæmatia of new formations. These fresh growths have their birth either in the lymphatic glands, or in the special glands (thyroid body, spleen, and so on).

The physiological characteristic of the hæmatia is the property they possess of absorbing liquids with a great energy. This property is inherent in their very substance, independently of their form. In effect, a solution of this substance grows red in contact with oxygen, and becomes less rutilant from contact with carbonic acid. The affinity of the substance of the globules for oxygen is quite comparable with that of the green matter of leaves, chlorophyll, for the carbon of aërian carbonic acid. It is by reason of this powerful affinity for oxygen that in the sanguineous transfusion practised in men and the mammifers the injection of mere globules suffices to provoke real resurrections. The blood extracted from the vessels continues to appropriate oxygen and to exhale carbonic acid. From contact with oxygen the hæmatia swell, and tend to lose their double depression. Carbonic acid, on the contrary, makes them shrink.

In like fashion in the vertebrated organisms, the function of the hæmatia is to imbibe many volumes of oxygen during their passage through the respiratory organs. Once impregnated with oxygen, the globules give to the blood a tint rutilant, vermilion.

The blood is then called *arterial:* thereupon the globules are conveyed with their provision of oxygen into the circulatory apparatus. Soon they find their way to the finest vessels of this system, where they are almost in contact with the anatomical elements of the tissues. Between the globules and the anatomical elements an exchange of gas is there accomplished which is one of the primordial acts of nutrition.

In effect, the vital condition by excellence for every anatomical element is to be oxydized, more or less slowly. But this process of oxydation produces, along with other chemical compounds, carbonic acid gas, which, if it was not climinated in the degree of its formation, would soon bring death to the anatomical element. The function of the red globules of the blood is precisely to take back that carbonic acid, and to furnish in exchange their vivifying oxygen.

The visible sign of this gaseous exchange is the change of coloration of the globule, which becomes blackish when it has parted with its oxygen to charge itself with carbonic acid. This black or venous blood contains much less oxygen than the arterial blood. According to Magnus, there is in the arterial blood 38 of oxygen to 100 of carbonic acid, and the proportion is only 22 to 100 in the venous blood. The venous blood is blood impoverished by nutrition; it contains fewer globules, less fibrine, and on the contrary more salts, a certain number of which are nutritive residua.[1]

Every anatomical element transforms oxygen into carbonic acid by the mere agency of nutrition; but it consumes a much greater quantity when it exercises a special function, and then the absorption of oxygen is rigorously in proportion to the degree of activity of the chemical element. For instance, if we cut all the veins which are distributed to a muscle, all voluntary contraction becoming for it now impossible, the arterial blood traverses it, losing only a small part of its oxygen, and it is red when it passes into the veins (Ch. Bernard). But if

[1] Longet, *Traité de Physiologie,* t. I., p. 581.

afterwards, exciting one of its nerves by an electric current, we contract the muscle, immediately a certain consumption of oxygen responds to the contraction; in the passage the globules are charged with carbonic acid; they grow black; they become venous. Analogous phenomena are observed for the same reason during the hibernal sleep, during syncope; and the case must be the same for the blood traversing the veins during dreamless slumber.

Claude Bernard has shown that we can produce at will the changes of coloration in the blood by the section and the excitation of certain nerves.

Essentially in all these special cases of organic atony, there is a general phenomenon: the inaction or the diminished action of the tissues, and of the organs; hence the superabundance of oxygen in the blood. We see therefore that in a certain sense we only need to determine the absorption of oxygen by a tissue to measure the degree of its functional activity.

The exchange of gas between the tissues and the globules is probably accomplished by osmosis and diffusion. We must nevertheless remark that in all likelihood the oxygen is in the state of combination with the substance of the globule and the hæmatoglobuline. In effect, an organic acid, the pyrogallic acid, which is greedy of oxygen, and absorbs it with special ease and eagerness when it is in solution in the alkaline liquids, succeeds not however in despoiling thereof the globules of the blood. These globules indeed have for oxygen such affinity, that in the arterial blood they absorb it almost in totality, and rob thereof almost completely the plasma. Like all chemical phenomena, the combination of the globules and of the oxygen is influenced by the temperature. At a low temperature it ceases to be accomplished:—for instance in the body grown cold of a mammifer in the state of hibernal sleep. On the contrary, when the temperature rises, the fixation of the oxygen becomes easy, proceeding as far as 40 to 45 degrees. [Centigrade scale.] Beyond that point the oxydation of the globules tends to exaggeration: there is stable combination, something analogous to what takes place when the globules are in contact with the oxyde of carbon, their special poison: the globule then loses its precious osmotic qualities; it is killed.[1]

6.—*Of the White Globules of the Blood.*

The white globules of the blood, or *leucocytes*, are pale spherical globules, sarcodic, amœboidal, having a diameter of 0^{mm}, 008 to 0^{mm}, 014.

Water and acetic acid pale them and enable us to perceive from one to four granulous masses or nuclei. In the fœtus the substance of the leucocyte is less dense, and not granulous; we see therein one or two granulous nuclei. We find also in the blood free or globuline nuclei from 0^{mm}, 003, to 0^{mm} 005 in diameter, granulous, without nucleoles.

In the blood of Man they are in the proportion of about 1 to 300 red globules: but we meet with them in greater number in woman (1 to 250). They are the more numerous the younger the person is.

From one to two years old	1 in 100
Newborn children	1 in 100 to 130
Human embryon	1 in 80 to 100

We thus see that in relation to the number of leucocytes the woman takes position anew between the man and the infant.[2]

[1] The proteïc matters of the globule differ completely from the surrounding fibrine. They have even their special inorganic compounds. The alkaline phosphates predominate in the globules, and have especially a potassic base. In the serum, soda and lime hold sway. The globule contains ten times more phosphates, two times less chlorure, ten times more potash, and three times less soda, lime, and magnesia than serum.

We likewise behold the predominance therein of fats, especially of phosphorised fats, analogous to those of the nervose substance.

Lastly, we know that the iron contained in the blood is entirely confined to the globules. Besides, the substance of the globule, though albuminoidal, is crystallisable. (G. Sée, *Du Sang et des Anémies.*)

[2] Ch. Robin, *Des Humeurs.*

Sometimes the leucocytes are in considerable number in the blood, to which they communicate a greyish tint or the tint of wine lees. Their number in such cases attains and even surpasses the infantine and embryonary proportion. It is important to remark that in such instances of *leucocythæmia*, there is generally a swelling either of the liver, or of the spleen, or of the lymphatic glands.

The leucocytes are not met with merely in the blood: they are also found in the living plasma, which we have yet to study, in the lymph, and then it is observable that they are more numerous in that liquid when it is examined beneath the lymphatic glands.

Lastly the leucocytes float in variable numbers in most of the humours of the economy. We can view them in relation to the granulous corpuscles existing so numerously in pus. However, in these last globules the amœboidal movements are less evident, and the nuclei not so easily seen.

The leucocytes are met with in the blood of all the mammifers and also in that of birds, of reptiles, and of fishes. However, while the red globule is peculiar to the vertebrates, the white globules, on the contrary, exist also in the invertebrates.

We have signalised the presence of the leucocytes in pus; but we also find them in the blastema of cicatrices, in what is called the plastic lymph.

We can only make conjectures more or less plausible on the office of the white globules. Peradventure we may regard them as a transitory, primitive, state of the red globules.

We know in effect that the first embryonary globules, those which we cannot help regarding as the first pattern of the red globules, are nearly colourless, like the white globules, that like them they have a nucleus, and lastly, that certain inferior vertebrates have only colourless globules. The abundance of the leucocytes in the blood of the mammiferous embryon comes also as a confirmation of this hypothesis. If this manner of regarding the subject had a solid foundation, then by bringing into fellowship with it the presence of the leucocytes in the lymph, their greater number after that this lymph has passed through the ganglions, the swelling of these ganglions, either of the liver or of the spleen in leucocythæmia, we should be tempted to consider these glands and these ganglions as the original sources, as the centres of creation of the sanguineous globules, which before taking the red tint, before growing retractile, to assume the form of hæmatia, begin by being leucocytes.

In leucocythæmia there would seem to be a superabundance of globular formation, while these elements are more voluminous, blended with a great number of free globulines. Hence the difficulty of their transformation into red globules.

7.—*Of the Lymph.*

It is a familiar fact that besides the grand circulatory system, composed of arteries, of veins, in which a propulsive organ, the heart, drives incessantly the blood on, there exists in the superior vertebrates a second circulatory system, without central organ of propulsion. This system, composed of an immense and fine network, which is bestrewn with special glands, called *ganglions*, has its origin partly in the mechanism of the tissues, and especially on the surface of the membranes, partly around the thinnest sanguineous vessels, the capillaries. In this lymphatic system circulates slowly a yellowish, transparent plasma, conveying white globules, and containing, like the blood, immediate principles of the three classes: it is the *lymph*. The materials of the lymph come, in chief part, like those of the blood, from anatomical elements, and they come from them likewise by osmotic process. The portion of the network clothing the stomachal and intestinal mucous membrane absorbs direct the nutriments, and especially the emulsionized fats, which, during digestion, give it a lacteous tint.

The lymph is not borne on in an endless circuit like the blood. In effect the whole system ends, in man, in two principal trunks, throwing itself into two huge venous vessels, the right and the left subclavian veins.

The lymphatic plasma, so analogous to the sanguineous plasma, acts exactly like it when drawn from the vessels. There is evolvement of plasmine and the formation of a fibrinous clot.

Like the blood, the lymph is alkaline. It contains nearly the same immediate principles, but in smaller quantity, and more diluted. As its course is very slow, a result is that it has not everywhere, like the blood, a composition observably uniform. On the contrary, this composition varies with every region, with the hour of digestion, and so on. In general the lymph is the more charged with substances, the nearer our examination extends to its chief trunks of communication with the sanguineous system.

In the mammifers the lymph is a liquid essential, indispensable to the duration of life. If in these animals we form a fistula in the largest lymphatic trunk, the *thoracic canal*, the lymph being no longer able to blend with the blood in sufficient quantity, we see the patients rapidly grow lean and die, even while continuing to take food.

What is the special province of the lymph and of the lymphatic circulation? This is still a very obscure point of physiology. Evidently the lymphatic is an adjuvant of the circulatory system; it connects the immediate principles in the digestive system and in the mechanism of the tissues, then pours them into the whole grand circulation. Its special function the most probable is to form white globules and to convey them to the great sanguineous current. The fact that a fistula of the thoracic canal is followed by death, proves conclusively that the lymphatic system is not a mere ornamental apparatus, and that it plays in the economy one of the most important parts.

BOOK II.

OF THE PRIMORDIAL PHENOMENA OF LIFE.

CHAPTER I.

OF NUTRITION.

We cannot form a precise idea of the mechanism of nutrition, unless we have thoroughly present to our mind the general laws of diffusion and those of osmosis, especially between crystalloids and colloids.

We know that two solutions of different density, put separately into a diffusion vessel and consequently in contact only on a small surface, mingle by degrees intensely, to such a point indeed that after a given lapse of time, the blending has everywhere an identical composition.

We also know that every substance has a degree of special diffusibility, and that generally the colloidal substances have a diffusibility infinitely inferior to that of the crystalloidal substances. A glance at the following table furnishes a complete idea of this difference :—

QUANTITIES DIFFUSED IN EQUAL TIMES.

Chlorure of Sodium	58,68
Sulphate of Magnesia	27,42
Nitrate of Soda	51,56

Sulphuric Acid	69,32
Sugar Candy	26,74
Barley Sugar	26,21
Molasses of Cane Sugar	32,55
Sugar of Starch	25,94
Gum Arabic	13,24
Albumine	3,08

Furthermore, these colloids, so slow to blend and to be diffused, are easily penetrated by the crystalloids, while their analogues cannot traverse them except by taking, through isomeric modification, an altogether special state of solubility. For instance, we shall see when speaking of animal digestion that the albuminoidal substances of the elements, in order to pass into the circulatory system, need previously to be transformed into soluble albuminose.

We must also remember that two substances, incapable of chemical combination, and possessing different degrees of diffusibility, separate to a certain point, when they are put in a blended state, in a diffusion vessel, for then the more diffusible of the two passes out more rapidly than the other.

The application of the preceding data to nutrition is achieved almost of itself, so to speak. In effect, from the simply physical point of view, organised beings are merely masses of colloidal substances, holding in solution crystalloidal substances. This definition is strictly true for a number of rudimentary organisms, for instance the amœbæ and most of the infusoria; in general, for all those beings, neither vegetal nor animal, of which Haeckel has made his group of monera. It applies even to a number of zoophytes, and also to every histological element of the superior organisms, isolatedly considered. Let us take the small amorphous mass of contractile albuminoid substance which constitutes an amœba, a rhizopod, or monocellular beings such as the *Protococcus nivalis*, many infusoria, lastly the globules of the blood of the mammifers, and even the cells and fibres, grouped into tissues to form the body of plants and of the superior animals; we see that in some, every element living, ultimate, isolated, or associated to other analogous elements is only a small mass of colloidal substance. But like all colloidal bodies, this substance is capable of imbibing water and aqueous solutions; it may be said greedily to seek them; thus there is established in the midst of its molecules an incessant aqueous current, which conveys to them soluble substances, modified crystalloids or albuminoids, and which at the same time seizes back other substances, usually crystalloidal, which have become unfitted to form a part of the living body.

The phenomena of diffusion however are not by any means peculiar to the liquid state. In a gaseous medium the diffusion of liquids is simply replaced by gaseous diffusion direct or indirect. We have seen besides that analogous phenomena are produced even in the midst of living liquids, in the blood and the lymph of the superior animals.

But if the physical condition of nutrition is simple diffusion in amorphous beings not yet composed of cells or of fibres, it is a little more complex in the others, if we proceed from the monocellular organisms to the superior mammifers, constituted by fibres and cells cemented or grouped in tissues. Here the diffusion is accompanied by the passage of the liquids through a membrane, that is to say, that there is *osmosis*, and naturally osmosis with a double current from without to within, and from within to without, *endosmosis* and *exosmosis*.

We must needs do for osmosis what we have done for diffusion, that is to say, briefly recall the principal facts appertaining thereto. Osmosis, discovered by Dutrochet,[1] then studied especially by Graham, who gave it the name of *dialysis*, is, as is well known, the blending of two unequal densities, separated by a membrane. Definitively it is diffusion in special mechanical conditions, which permit the superposition of the liquids, whereby

[1] J.-B. Dutrochet, *De l'Endosmose*, in *Mémoires pour servir à l'Histoire anatomique des Végétaux et des Animaux*, t. I. Paris, 1837.

for instance the more dense can be placed above the less dense. As essentially there is a very great analogy between diffusion and osmosis, it follows as a matter of course that the substances which osmosis finds the most sluggish must be the colloidal substances; and that is exactly what experience confirms. In osmosis, the membranous partition, usually organic, which separates the liquids, is traversed simultaneously by a double current; and commonly the stronger current goes from the less dense liquid to the more dense liquid. There are however exceptions. For instance, if water and alcohol are separated by a fragment of bladder, the water passes in greater quantity towards the alcohol. It has been supposed that in this case the direction of the current depends on the inequality of the capillary attractions between the liquids and the two faces of the membrane. The water moistening the membrane better than the alcohol, rises by capillarity in its pores, while on the contrary if we substitute for the bladder a layer of collodium, which is better moistened by the alcohol than by the water, the direction of the osmosis is reversed. All the membranes with which osmotic experiments have been made are in effect really and thoroughly perforated (bladder, collodium, paper, parchment, and so on). But this explanation does not suit all cases, and especially the cases of osmosis through living liquids. In effect, so far as our most powerful microscopes permit us to assure ourselves thereof, the surfaces of the animal and vegetal cells and fibres are absolutely homogeneous. We find no trace of pores. No doubt we are compelled to admit molecular and atomic intervals across which the passage of the solutions must be effected; but in this case the osmosis is accompanied by a chemical action exercised on the dialysing membrane. The liquids, gases, or vapours, which traverse the cellular or fibrillary walls unite themselves, as they pass along, molecule by molecule, to the chemical elements constituting this wall; then, as on the other side of the membrane they find themselves in contact with a new fluid, they forsake forthwith the elements of the wall to combine with those of the fluid which they encounter. This explanation, proposed by M. Ch. Robin, would, if well founded, furnish an explanation of an extremely important fact. It would make us understand why in the organisms themselves the composition of the fluid absorbed is no longer on the hither side of the membrane or living wall what it was on the thither side; a phenomenon peculiar to the biological osmosis, and never produced in the endosmometers and the dialysing apparatus.

But, well founded or not, this explanation seems to me by no means needful to furnish a reason for the changes occasioned in the composition of the fluids by physiological absorption. It suffices to explain this metamorphosis to take into account chemical phenomena at the same time as physical phenomena. Hitherto almost all the osmotic experiments effected in the laboratories have borne upon liquids miscible indeed, but exercising on each other no chemical action. Obviously this is not what happens in the living tissues. There the fluids which have traversed a living wall hold in solution unstable substances, which are by the very fact of the osmosis in contact with other fluids composed of substances of analogous complexity and of different composition. There are evidently at the time of this conflict, exchanges of molecules, chemical reactions; the new substances arising repair the waste of the old, and for that purpose are forced to enter into alliance and combination with them. The residuum of these functions and that of the waste of the substances previously organised are a blending of diverse crystalloidal bodies, which is promptly dragged away from the histological elements, the fibres, and the cells, to be afterwards definitively expulsed from the organism. We have seen that nothing is easier than to separate with a dialyser a crystalloidal substance from a colloidal substance. It is very evident nevertheless that the cellular wall is not inert in all this labour of molecular mutation. It is as living as that which it contains, and must consequently in like fashion participate in the phenomena of transformation.

We could surely in osmotic experiments draw much nearer to what comes to pass in the organised bodies by making to react on colloidal substances oxydant bodies, capable of giving birth thus to crystalloidal bodies, and so on.

The curious experiments of Traube on artificial cells have taught us that it is possible to imitate in a certain measure the physical and chemical phenomena of life.[1] Assuredly we hitherto are far from having imitated within the domain and the range of the possible, the phenomena of animal physics, and of vegetal physics, which form the essence, the support of the vital acts. No doubt the lack of initiative, which, in regard to this matter, experimentalists have displayed, must in a large degree be attributed to the metaphysical and mystical ideas which have been conceived of life. As long as the vital phenomena were considered as of an order altogether apart, as having no relation with the physical or chemical phenomena; as long as there was a belief that to explain what was called "the miracle of life" there had to be invoked directing entities, independent of the bodies, a kind of immaterial gods set over the physiological government of every organism, an archeus, a vital principle, and so on, it was naturally almost impossible that the idea of reproducing artificially the principal physico-chemical acts of life should occur to experimentalists. In our days there is, fortunately, a complete change, and we see men of science venturing on paths which they would never have dreamed of entering half a century ago.

M. Traube has based his experiments on two principal facts. The first of these facts has been established by Graham; it is, that the colloids dissolved are incapable of penetrating by diffusion through the colloidal membranes. The second fact is, that the precipitates of the colloidal substances are themselves colloidal. Starting from these facts, M. Traube has been able, artificially, to make cells, the wall of which was formed of tannate of gelatine. He takes a drop of gelatine, which, by an

[1] *Experimente zur Theorie der Zellbildung und Endosmose* (*Archiv für Anatomie*, &c., von Reichert und Dubois-Reymond, 1867, p. 87).

ebullition of thirty-six hours has lost its coagulability. He lets it dry in the air for several hours, and by the help of a rod fixed in the cork of a vessel half filled with a solution of tannin, he plunges it into this liquid. Then the small quantity of gelatine which dissolves on the surface of the drop combines with the tannin, and the result is a closed cellular membrane. But this membrane is homogeneous, unperforated, as the organic membranes are. Also the diffusion which is established between its contents and the exterior liquid must be effected osmotically across the molecular interstices. The osmosis is produced very energetically. The membrane distends more and more; as a consequence the constituent molecules sever from each other: at a given moment when the molecules of the two liquids brought face to face can easily be introduced into the molecular interstices and blend, they form anew molecules of tannate of gelatine. Consequently, *the membrane grows by intussusception.* In effect, it suffices to arrest all increase, to substitute water for the solution of tannin.

M. Traube forms also, in the same manner, endosmotic membranes, very curious, impermeable by certain substances, very permeable by others; in a word, exercising on the substances in contact with them an elective action, such as the living membranes exercise.

According to M. Traube, every precipitate whose molecular interstices are smaller than the molecules of its components must take the form of a membrane.

Lastly, the endosmosis across the membranes depends solely on the attraction of the body which is dissolved for its solvent.

These experiments are infinitely interesting; nevertheless, they imitate very imperfectly what takes place in the living cells. It is something to have obtained by simple chemical processes a membrane which grows by intussusception, forasmuch as, from time immemorial, this mode of growth has been considered peculiar to living bodies. But in the living cell there is something more; the contents are as little inert as the enveloping membrane; they are modified and renewed unceasingly, molecule by molecule, without being destroyed.

In the primordial phenomena of nutrition there are, in effect, two acts, or rather two principal aspects of the same phenomenon, assimilation and disassimilation. To assimilation relate the facts of absorption and endosmosis; with disassimilation are connected the facts of secretion and exosmosis. We must remember that disassimilation has, as result, the transformation of the colloidal substances of living bodies into crystallisable substances, occupying, after a fashion, a middle position between organic substances and mineral substances.

It is now invincibly demonstrated that these primordial facts of nutrition are identical in all the living universe, as well in the animal world as in the vegetal world. It is also known, moreover, that the principal agent of all these transformations, of all these exchanges, is the oxygen of the air.

In the most rudimentary beings, amorphous or monocellular, oxygen is diffused direct in the midst of the molecules of the living substance; it oxydises this substance by a sort of slow combustion, and determines the formation of diverse organic crystallisable bodies, and of a gas—carbonic acid; the whole is afterwards expulsed.

In the being whose structure is more complex, where there is an aggregation of cells, of fibres, in a word, of diverse histological elements, each having its special form and special functions, while the whole are, moreover, grouped in particular tissues, the oxygen of the air, and generally all the substances which penetrate into the organism and come forth from it, have to undergo a sort of gradatory process before being assimilated or excreted. In the simplest cases when the organism is merely constituted by histological elements of kindred nature, more or less straitly joined together, and bathing in an interstitial liquid, a circulatory system, a respiratory system, a digestive system being wholly lacking, the nutritive substances and the disassimilated substances dissociate themselves in the intercellular blastema. It is in this liquid, impregnated as moreover it is with oxgyen and carbonic acid gas dissolved, that the histological elements choose the materials which suit them and reject those which are no longer suitable. At a more exalted degree of structure are superadded special apparatus, systems more or less ramified with canals, in which circulate the liquids and the gases. But even then the interstitial, intercellular liquid ceases not to exist; it merely renews, revivifies, purifies itself without pause, seeking sustenance in the circulatory fluids, and ridding itself there, in its turn, of the substances destined to elimination.

In sum, the intercellular liquid acts toward the circulatory fluids as the histological elements act toward itself.

We see that it is by a peculiarity of construction that almost all organised beings live in the air. In truth, all the histological elements constituting the complex organisms are aquatic; they bathe in a special liquid, in a living medium, which is alike their essential cause and the result of their nutritive activity. Cl. Bernard has much contributed in these last years to propagate this felicitous idea of the interior media, such as the blood of animals, the sap of plants, and so on: "that *ensemble* of all the interstitial liquids, that expression of all the local nutritions, that source and confluent alike of all elementary changes."[1] We may admit, with the eminent physiologist, though with restrictions and exceptions for the living beings that are wholly rudimentary, that there is no direct nutrition, and that, for instance, the fragment of a fresh-water polypus, when it is reconstituted and completed to the point of re-becoming an entire polypus, avails itself principally of the nutritive interstitial fluid which impregnated at the outset the separated fragment.

Thus, for every complex organised being there are three superposed media, the cosmic medium. aërian, or aquatic—but in this last case holding the air in solution; the sanguineous, or sap-filled medium; lastly, the interstitial, intercellular medium. Naturally, the internal media need, like the external media, to

[1] *Revue Scientifique*, 1874.

maintain themselves in what we call *a suitable state of purity*, that is to say, in a state of composition sufficiently equilibrated for the histological elements to find therein at every instant their food. We shall see in the course of this exposition that in the complex organisms special apparatus of exhalation, of secretion, and of excretion are charged to keep up incessantly across these media renovating currents, exactly as other apparatus, for instance, the digestive apparatus and certain glands, pour in suitable nutriments.

These general data accepted, we can now analyse the acts, the phases of nutrition. We know that this biological property exercises itself in all living substances, figurate or not, as well in the plasmas and the blastemas as in the figurate elements. In both it depends in part on physical conditions of endosmosis, exosmosis, and diffusion, and in part on the physical affinities of substances brought into relation. In all this there is not the smallest place for a metaphysical agent. We have simply to do with physical and chemical phenomena producing themselves in conditions of complexity and simultaneousness entirely special, yet narrowly bound to the variations of the ambient medium. We see in effect these phenomena intensified or enfeebled according as the air is more or less oxygenised, according as the temperature is higher or lower, and so on.

Though the nutritive phenomena are simultaneous and uninterupted, we can, for the convenience of exposition, divide them into phenomena of assimilation and phenomena of disassimilation.

It is by endosmosis that the immediate principles reach the substance of the anatomical elements, reach the living liquids. The principles of the first class, that is to say, the mineral substances, often arrive without modification by simple dissolution, and it is thus, for example, that the chlorures and the alkaline sulphates arrive. Certain of these substances combine with the organic matters, as, for instance, the phosphate of lime combines with osseïne in the bones; but then, in opposition to the laws of non-living chemistry, the combination does not seem to be accomplished in definite proportions. It is a sort of alloy.

In plants the power of assimilation seems more energetic. It is in the mineral medium, in effect, that the plant must seek its aliments direct; consequently we see the green parts of plants assimilate at once the carbon of aërian carbonic acid, and incorporate it immediately with complex organic substances, ternary and quaternary. The same synthetical power is exercised in certain circumstances on the azote of the air, and, normally, on the azote of the ammoniac salts drawn by the roots of plants from the soil.

In the animal, the true phenomena of assimilation are generally exercised at the expense of albuminoidal substances already elaborated. It is an important and remarkable fact that the organic assimilated substances have never, previously, the same composition as those which form the assimilative anatomical elements. The musculine, the elasticine, and the like, peculiar to every species of cell, of fibre, and so on, are, in effect, met with nowhere apart from the elements which they constitute and reconstitute incessantly; they are formed in the animal organism at the expense of the living liquids, by isomeric catalyses.[1]

The anatomical elements can assimilate a great number of substances, but they have necessarily their own special affinities, entirely analogous to those of the bodies of mineral chemistry. Thence result a choice and a selection, which, for a long time, appeared intelligent, though here intelligence no more enters than in the affinity of chlore for hydrogen, of anhydrous sulphuric acid for water, and so on. These chemical combinations formed in the substance of the anatomical elements have excessive instability, and they are the more unstable the more life rises to a superior degree, the more it is animalised. Thus chemical instability is much greater in the anatomical animal elements than in the vegetal elements.[2] In these last we cannot dissever

[1] Ch. Robin, *Anatomie Microscopique des Éléments Anatomiques.* 8vo. Paris, 1868.

[2] Ch. Robin, *loc. cit.* p. 65.

the chemical combinations but by the aid of energetic chemical agents. Moreover, it is much less easy to interrupt the vital movement in the plant than in the animal. But in them both, chemical instability, in various degrees, is the very condition of life. Every combination too stable is the equivalent of death.

Nutritive assimilation has naturally, as a condition, a corresponding disassimilation. In order that new substances may incorporate themselves with an anatomical element it is absolutely necessary that other substances yield their place to them. In effect, incessantly a portion of the substances which formed part of the anatomical element ceases to resemble the fundamental substances, and severs itself from them. The substances thus have not, by reason of the severance, ceased to be complex albuminoidal substances, but generally they have become more oxydised, and have passed into the state of crystallisable matters, —they have taken a step to return to the mineral world.

As to the mineral substances expulsed from the anatomical element, certain of them merely pass through without undergoing any change. This is the case with sundry salts, azote, water, and so on. Other mineral compounds, however, are formed therein by direct combination, just as they would have been formed in a retort. In this fashion are produced in animals, the alkaline carbonates, the lactates, the ammoniaco-magnesian phosphates, the phosphates of lime, the urates, carbonic acid, and so on.

The nutritive exchange is not effected in all the tissues with the same energy. In general it is in the cell, properly so called, or in tissues formed by cellular aggregations, that this double current attains its maximum of power. Nutrition can often be effected without the succour of special circulatory apparatus. The exchange of nutritive materials frequently is achieved in this case from step to step with sufficient rapidity. Things take place thus in certain inferior organisms simply polycellular, in the crystalline of the eye of mammifers, and so on.

If a tissue is at the same time constituted by cells and furnished with a rich vascular network, it is in conditions specially favourable, and the nutrition thus is rapid and energetical: this is what takes place in the mammifers, in the medullary tissue of the bones, in the grey substance of the brain, and so on.

After the exposition of the general facts to which we have devoted this chapter, it will be now more easy for us to present successively in the two organic kingdoms the history of nutrition, that is to say, of the vital property which is the support and the essential cause of all the others.

CHAPTER II.

VEGETAL NUTRITION.

In the composition of every plant we find mineral substances, ternary substances non-azotised and proteïc substances. Now plants not eating each other direct, like animals, it is needful for the vegetal organic substances to be habitually created by the plant itself, at the expense of the mineral medium which environs it. Brought back to their primary mineral elements, the complex organic substances yield carbon, oxygen, hydrogen, azote, and a certain quantity of sulphur and of phosphorus. If we add to these elements chlore, calcium, silicium, potassium, sodium, magnesium, lithium, iron, often manganese, and in the marine plants, iodine and brome, we have nearly the whole elementary sources of alimentation in the vegetal kingdom. Naturally the metals which we have just enumerated form bases which combine with the acids which they encounter, to constitute sulphates, silicates, chlorures, often organic salts, for instance, oxalates, and so on.

To form a sufficient idea of vegetal nutrition, we must follow these mineral elements, note how they enter into the plant, indicate the combinations which they form there, finally, leave them only when, having played out their part in the vegetal organism, they are finally expulsed from it.

Of these chemical elements, some, for example, are derived from the air, others from the soil. The mineral elements taken direct from the ambient air by the plant are hydrogen, oxygen, carbon. Hydrogen and oxygen are absorbed and fixed by the plant, either simultaneously in the state of water, or isolatedly. It is probable, in effect, that the green parts of plants have the power of decomposing water, which they draw in a small part from the atmosphere, but in enormous quantity from the soil. The chlorophyllian parts would effect the decomposition of water, whatsoever its source, and would fix direct its elements in the complex combinations of which we have spoken. This, however, is a point which has not yet been well studied. Certain, at least, it is, that the greater part of the oxygen absorbed by the plant is taken from the atmosphere direct, and a little by all parts of the vegetal organism. As to carbon, which forms in weight the chief part of every desiccated plant, it also is derived by the green portions of the plants from the carbonic acid of the air. This is one of the most interesting, one of the best studied points of vegetal physiology. All the other mineral matters, and almost the whole of the water, are absorbed by the roots of the plant, penetrate into the vegetal tissues, there ascend, there meet with, especially in the leaves, the mineral substances derived from the air, and some of them form complex combinations.

Vegetal physiology is still so confused, the division of labour in the plant is so ill-distinguished, that it is not easy to mark out therein functions thoroughly determinate, functions very different from each other. Everything is connected, everything blends, everything forms the link of a chain. Nevertheless, for clearness of exposition, we are obliged to make divisions more or less natural; we must, in effect, speak of phenomena mingled, entangled, proceeding sometimes simultaneously in the same tissues or organs. We have to tell how penetrate into the plant the numerous mineral substances which chemical analysis discovers there, how those substances have infiltrated themselves into the tissues, what compounds they have formed there under the powerful action of the nutritive movement, finally how and in what proportion they were eliminated during the life of the plant after becoming unsuitable to figure in the nutritive process.

1. *Formation and Circulation of the Sap.*

Let us take as type a complete plant, a dicotyledon, plunging its roots into the ground, displaying its branches in the air. In the spring such a plant impregnates itself incessantly with the materials which it appropriates from the exterior medium. It absorbs them by the roots, by the leaves, by the bark. It is by the osmotic process that the roots draw from the soil the first materials of the sap. It is the delicate cells of the extremities of the roots, radicellular spongioles, which are the principal agents of absorption. These cells contain a protoplasm dense and albuminous, coagulable by nitric acid; they are in contact with the soil by means of their cellular membrane, by means of the hairs which garnish them; they are thus in conditions very favourable to osmotic absorption. It is, moreover, easy to prove that is owing to the process entirely physical of endosmosis, that the roots saturate themselves with the humidity of the soil, for all that is needful to arrest their work of absorption is to plunge them in a saccharine and dense solution. In the soil, on the contrary, the water, relatively little charged with dissolved substances, moistens the extremities of the roots, penetrates by endosmosis into their anatomical elements, and mingles there with their protoplasm, bringing with it ammoniac salts, phosphates, salts of potash, and so on.

But to accomplish their office the hairs of the roots need, like all organised cells, nourishment, that is to say, need to be oxydised, to absorb oxygen, and to exhale carbonic acid. Consequently, the penetration of the air into the soil into which the roots plunge is indispensable to the maintenance of the life of plants. The exhalation of the carbonic acid by the roots has also its utility. In effect it is from the presence of this carbonic acid that certain salts become soluble in water and can thus penetrate into the radical cells. The case is the same, for example, with certain phosphates, and no doubt also for silica, and so on. Once introduced into the cells of the spongioles, of the radical hairs, the substances, borne on by the water which holds them in solution, pass from cell to cell, each borrowing from each, in proportion as the nutritive waste goes on. In plants with roots, this ascensional movement of the liquids drawn from the soil is facilitated by the presence of those vessels and vascular bundles we have previously described; there are indeed no roots except in the plants whose cellular tissue is traversed by vessels. In the spring the flow of the sap is so abundant that it invades everything; cells, fibres, vessels, even the interstices of the cells or intercellular meatus. This flow thus ascends from the root to the leaves, but circulates more rapidly in the vessels, where it encounters fewer obstacles, and is, in a certain measure, raised by capillarity. The grand movement of ascension is accomplished through the central part, through the ligneous body, or through its exterior zone, younger and less incrusted if the vegetal is already aged. Very certainly multiple causes, endosmosis, diffusion, capillarity, the nutritive fixation of the alible materials in the buds, evaporation on the surface of the leaves, co-operate in the ascension of the sap; but the most powerful cause is assuredly the absorption exercised by the ascensional cells of the roots. In effect a plant can live and actively live when its radical extremities alone are placed in water.

Besides this grand general movement of the sap there are others more interesting perhaps, those, namely, of the contents of the cell, of the protoplasm. This liquid, which we know to be habitually an albuminoidal liquid, is granulous, and we see in nearly all plants its granulations execute along the walls of the cell or of the fibro-cell a gyratory movement. They mount on one side and re-descend on the other. This protoplasmic movement is a vital movement connected probably with the molecular exchanges and reactions of nutrition. It is accomplished only within determinate thermometrical limits. The minimum limit approaches 0 degree, the maximum limit is from 45 to 47 degrees. It is toward 35 to 37 degrees that the speed of the protoplasmic current attains its maximum. When it is the cold which arrests this gyratory movement, we can by heating the soil put the liquid again into motion.[1]

With respect to this manifest action of heat it is curious to see that light seemingly influences little or not at all the protoplasmic movement which is accomplished without apparent modification even when the plant is kept in darkness.

After gyrating in the cells, travelling in the vessels and the meatus, after taking from or giving to the elements, which it has traversed or passed beside, certain of the matters which it conveys, the sap arrives at the part truly aërian of the vegetal. There it undergoes very important modifications, thanks to the special action of a substance of which we have now to speak. This substance is the green matter of the leaves, the *chlorophyll*.

2.—*Chlorophyllian Property.*

Chlorophyll is the substance to which all the green parts of plants owe their colour. In the cells of certain lichens and of certain algæ, the chlorophyll sometimes presents itself in the amorphous state, colouring all the protoplasm, sometimes in irregular masses; but habitually in all the vascular plants it has a definite form, that of green granulations of from 0^{mm}, 001 to 0^{mm}, 005 in diameter, of homogeneous appearance and without nuclei. We can obtain from this green matter fat crystallisable bodies, stearine, margarine and so on, and an immediate azotised principle, chlorophyll properly so called, the elementary analysis of which gives oxygen, hydrogen, azote, carbon, and iron. By an appropriate chemical treatment Frémy was able to separate chlorophyll into two substances; the one yellow, *phylloxanthine*, and the other blue, *phyllocyanine*.

Chlorophyll rises spontaneously in the cellular protoplasm. First of all are formed colourless or yellow particles, which grow green afterwards, if the cell in which they are contained is exposed to the light. The chlorophyllian particles, born in

[1] Nägeli, quoted by Sachs, *Traité de Botanique*, pp. 855, 856.

the darkness, remain yellow, but under the action of even a feeble light and a somewhat elevated temperature, of from 20 to 30 degrees, they grow green. All the rays of the solar spectrum suffice to make the colourless chlorophyllian particles green, but by far the most active are the yellow rays.

Once formed, the chlorophyllian particles, if they are favoured by good conditions, increase in size, and at a given moment can multiply by binary division. The solar light does not merely influence the formation of the chlorophyllian particles; their whole evolution is subjected thereto. The particles veridified and subjected to an intense light during a long space of time form in their very substance particles of starch, which are most manifestly the result of a nutrition too active, an aliment in reserve. This starch besides re-dissolves and re-forms according as the green cell is withdrawn from the solar light or exposed to it anew. From a long sojourn in darkness the particles of chlorophyll themselves lose their shape, suffer atrophy, and disappear, dissolving into the colourless protoplasm.

We have seen that all the rays of the solar spectrum have the power to render green the chlorophyll; but all are not capable of impressing on it enough of nutritive activity to form the starch in its particles. That is a faculty limited naturally to the rays the most stimulating, the yellow rays.

Light being the determinating agent of the chlorophyllian formation, it is natural for the chlorophyllian particles to accumulate specially on the best illuminated cellular wall; and this, in reality, is what takes place.[1]

In the persistent leaves, heat appears to have a great influence on the position of the particles of chlorophyll. In effect, when the temperature lowers they quit the wall to accumulate intervolved in the centre of the cell. In spring, or whenever the plant is subjected to a certain elevation of temperature, they, whether in darkness or in light, resume their parietal position.

Lastly, toward the end of the vegetative period, the precious

[1] Franck, *Botanische Zeitung*, 1872.

chlorophyllian substance in a great measure escapes destruction in the perennial plants. It re-dissolves along with the starch which it englobes, and the whole, passing through the petiole, and carrying along even the phosphoric acid and the potash, wanders toward the permanent organs of the plant.[1]

Before we speak of the special properties of chlorophyll, it is opportune to signalise the importance of the metallic element which it contains. The atoms of iron which enter into its composition constitute in effect an integrant part of it; without them it is not endowed with its special properties. Another metal, potassium, though not figuring in the complex molecule of the chlorophyll, seems to play an important part in its nutrition. When the plant does not absorb chlorure of potassium, or at least nitrate of potash, the particles of chlorophyll have less vitality and are incapable of forming starch.

We have succinctly described the morphological evolution of chlorophyll. It remains for us now to speak of its function.

Priestley was the first to observe that the green parts of plants exhale oxygen. He put under a receiver in confined air, where mice had died asphyxiated, some mint plants, which lived and flourished energetically there. The chlorophyllian property was thus discovered; but it was Ingenhouz who attributed the disengagement of vital air operated by plants to its true cause, the action of light.[2] The same observer demonstrated also the inverse action of the flowers and of the roots, which night and day exhale carbonic acid and vitiate the atmosphere.

Every terrestrial or aquatic plant furnished with chlorophyllian cells, and exposed to the solar light, borrows from the air carbonic acid, and restores to the air an equivalent volume of oxygen.

For most plants, the activity of the phenomenon is in proportion to the intensity of the light. Though the chlorophyllian function has need in almost all plants of full light, it still exercises itself, feebly indeed, in diminished light, and there are even

[1] J. Sachs, *Traité de Botanique.*

[2] Ingenhouz, *Expériences sur les Végétaux*, 1780.

plants, mosses, for example, living in the shade of the woods, which cannot, without dying, bear an intense light. But in diverse degrees, light is indispensable to the green parts of all plants. At night, or in darkness, chlorophyll ceases to act, and the plant simply exhales carbonic acid.

The property of decomposing the carbonic acid of the air and of absorbing carbon specially appertains to chlorophyll, just as that of fixing a great quantity of oxygen specially appertains to the hæmoglobuline of the hæmatia. Chlorophyll has also, like hæmoglobuline, its poisons. Thus, as Boussingault has demonstrated, mercury introduced into a receiver where a plant is destroys the chlorophyllian property.

It seems to result from the experiments of Dutrochet,[1] that a part of the oxygen put at liberty by the chlorophyllian action is not immediately expulsed, but penetrates previously into the mechanism of the tissues. The oxygen exhaled direct is merely air overplus. The rest is impelled into the aërian cavities, into the globulous vessels, into the punctuated tubes, and especially into the tracheæ. In this way it descends into the petioles of the leaves, into the stem, and serves there probably the real respiratory function, the oxydation of the tissues, and the production of the carbonic acid disengaged by the plant.

It is by the stomata that this expulsion and this absorption of air seem especially to be accomplished ; nevertheless, the mosses and the coniferæ, which have no stomata, exhale carbonic acid.

We might be astonished that the small proportion of carbonic acid contained in the air suffices for the alimentation in carbon of the whole vegetal kingdom, if we did not think on the great density of the atmosphere, and on the considerable restitutions that are made to the atmosphere by the vegetal kingdom on the one hand, and by the entire animal kingdom on the other. This last, in effect, incessantly consumes oxygen, and exhales torrents of carbonic acid. Moreover, we must add to these principal sources of carbonic acid all the combustions, the

[1] Dutrochet, *De l'Endosmose*, p. 357.

volcanic exhalations, and so on. As to the rest, in calculating in accordance with the presumed height of our atmosphere and the proportion of four ten-thousandths of carbonic acid for a given volume of air, we arrive in estimating the quantity of carbon existing in the aërian medium at the enormous figure of 1,500 billions of kilogrammes.

It is also probable that the air is not the only source of the carbonic acid absorbed by the plant. There is assuredly some produced by the oxydation of the sap and of the anatomical elements, and there is no reason for supposing that the plant does not also undergo the decomposing action of the chlorophyll.

As the carbonic acid is, when light acts, incessantly decomposed by the chlorophyll, there results a sort of carbonic void in the portion of the air in contact with the green cells, and consequently by degrees and by diffusion new quantities of acid arrive. The alimentation in carbon is therefore never lacking.

Though the decomposition of acid is effected in all the green cells, nevertheless, the upper surface of the leaves seems to play the predominant part in the chlorophyllian act, for if we turn the leaves so as to expose to the sun their under-surface, the carbonic action diminishes, and in a few days ceases.[1]

Let this be as it may, the exhalation of carbonic acid, little perceptible during the day and relatively active during the night, is far from compensating the absorption. M. Boussingault has calculated that in twelve hours of the night one square decimetre of green surface burns $0^{g},214$ of the carbon of the tissues, while in twelve hours of the day it assimilates $3^{g},416$.

If we allow the light to reach plants, only by allowing it to pass through glass coloured with the colours of the prism, we see that all the visible rays can put chlorophyll into activity, but that the rays capable of exciting its appearance in the protoplasm are also those which stimulate chlorophyll most. These are, in effect, the yellow rays, which determine the most abundant disengagement of carbonic acid.

[1] Dutrochet, *De l'Endosmose*, p. 355.

If we attentively regard the undulatory amplitude of the luminous rays suitable for making the chlorophyll operate, we see that those active rays have as highest limit $0^m,0006886$, and as lowest limit $0^m,00039968$. They are rays feebly refrangible.[1]

The rays most strongly refrangible, the blue and the violet, as well as the ultra-violet invisible rays, influence especially the rapidity of growth, the movements of the protoplasm and of the zoospores, and so on.

We have recently compared, in passing, chlorophyll and the hæmatoglobuline of the blood. The parallel is so curious that we must consecrate a few lines to it.

Chlorophyll and hæmatoglobuline are both quaternary substances. They both exercise an elective action on a mineral gas.

They both are habitually moulded into globules, without nucleus.

The special property, however, which characterises them seems in both to be independent of the form which they assume. We have seen that a solution of hæmatoglobuline absorbs oxygen, and that amorphous chlorophyll, dissolved in the cellular protoplasm of certain plants, continues, nevertheless, to absorb molecules of carbon.

Chlorophyll and hæmatoglobuline equally seem to form only a temporary association with the mineral element which with special avidity they seek. In effect, the sanguineous globules yield to the anatomical elements of animals their provision of oxygen almost as soon as they have taken it, and, in return, they charge themselves greedily with carbonic acid, thus holding kinship with the chlorophyllian globules in this aspect of their physiology.

Chlorophyll, though absorbing, like every living substance, the quantity of oxygen necessary to its nutritive movement, does not seem to have a great affinity for that gas. It is probable that in the night it ceases purely and simply to exercise its special action without assuming another; but it is certain that it does

[1] J. Sachs, *Traité de Botanique*, p. 873.

not fix more than for a moment the carbon derived from the carbonic acid. As this carbon does not accumulate in the chlorophyllian tissues, as the composition of chlorophyll is always perceptibly the same, the molecules of carbon assimilated by it must be instantly surrendered to the sap to which they bring the supply needful to the formation of complex substances, ternary and quaternary. Whence, besides, could the tissues take the carbon which constitutes the half of their weight, if they had not, to provision themselves, this perpetual supply? We shall have some words to say on this living chemistry. Let us occupy ourselves for the moment with the vegetal function, comparable in everything with what is called respiration in animals, that is to say, with the absorption of oxygen.

3. *Absorption of Oxygen or Vegetal Respiration.*

A green phanerogamous plant is asphyxiated in a medium of hydrogen, of azote, and even of carbonic acid, and if its sojourn in this artificial atmosphere is too prolonged it loses for ever the chlorophyllian property.[1]

Besides, M. de Saussure had already remarked, when extracting by the aid of a pneumatic machine the air impregnating the tissues of plants, that this air contained notably less oxygen than the ambient atmosphere. The proportion is very variable; that found by Saussure was 85 of azote and 15 of oxygen.[2] Moreover it has long been known that during the day when in darkness, and consequently during the night, plants disengage carbonic acid. Ingenhouz had already observed that this disengagement of carbonic gas was constantly operated by flowers and roots. Finally, in our own day, Boussingault, Garreau, Sachs have been able to demonstrate that this exhalation is a permanent and general fact, that it is effected even by the leaves exposed to the sun. In truth we have here to deal with an act indispensable to everything which lives. Without this continual

[1] Boussingault, *Annales de Chimie et de Physique.* IVe série, 1868, t. XIII.

[2] Th. de Saussure, *Recherches Chimiques sur la Végétation.* 1804.

labour of slow oxydation organised substances could not accomplish their metamorphoses, operate the exchanges of matter which constitute the fundamental act of life.

As Dutrochet said in reference to this point: "Life is one; the differences are not fundamental, and when we follow the phenomena to their origin the differences disappear."[1]

This absorption of oxygen correlative to a disengagement of carbonic acid, is the act called *respiratory* in animals, and we can give it the same name in plants, forasmuch as the essential fact is identical in the two kingdoms. Animal and plant absorb aërian oxygen: animal and plant, while producing heat, water, and carbonic acid, burn their fat and amylaceous matters. We even find in the vegetal cells a substance analogous to the principal residuum of the combustion of the albuminoids in animals, urea: namely *asparagine*, immediate azotised and crystalloidal principle.

The respiratory property is essential to life: it is indispensable even to the chlorophyllian cells, which become incapable of reducing the carbonic acid when they lack oxygen, and this is why they are asphyxiated in an atmosphere of pure carbonic acid.

Every living element is athirst for oxygen, and to such a degree that sometimes certain organisms steal it even from stable chemical compounds.

Vibrionians, studied by M. Pasteur, decompose the tartrate of lime and transform lactic acid into butyric acid to procure for themselves oxygen. It is besides by an analogous process that in most of the vertebrates the anatomical elements deoxydise the hæmatoglobuline of the haematia.

In the plant oxygen combines in totality with the carbon of the tissue: for in pure oxygen there is perfect equivalence.

Modern researches the most exact have shown that vegetal growth operates only through the aid of the absorption of oxygen by the tissues of plants, and that this absorption

[1] Dutrochet, *De l'Endosmose*, p. 326.

is proportional to the growth. It is, for example, very considerable in germination. Thus seeds and buds absorb, in evolution, many times their weight of oxygen. It is the same with flowers. Flowers, like all the parts not green of plants, manifestly absorb oxygen; but they absorb it in great quantity, and, a curious thing, in the monoïcal plants, the male flowers absorb more oxygen than the female flowers. A fact deserving of remark is, that in all the organic vegetal combinations, oxygen is always in smaller proportion than would be needful for the complete combustion of their carbon and of their hydrogen, for their total transformation into water and into carbonic acid; and the fact is wholly in accordance with the theory of respiration.

It is known well, that one of the constant effects of oxydation is a certain production of heat, and that the oxydation of the tissues is the principal source of animal heat. Though less active, vegetal oxydation produces likewise perceptible calorific effects, especially in germination and florescence. Barley grains heaped up for the preparation of malt heat much. In the spadix of the aroïdeæ, at the time of fecundation, the excess of temperature above the exterior medium may amount to 10 or 12 degrees. The same fact is observed, but in a less degree, in the isolated flowers of the *Cucurbita*, *Bignonia radicans*, *Victoria regia*, and so on.[1]

However, the respiratory oxydation, and the nutritive or assimilative movement which results therefrom, are not effected except within certain limits of temperature. The minimum is variable as the species varies. According to M. Boussingault[2] the leaves of the larch decompose carbonic acid at 0°, 5 to 2°, 5; those of the grass of the meadows between 1°, 5 and 3°, 5. According to MM. Cloetz and Gratiolet, the carbonic assimilation commençes above 6 degrees in the *Vallisneria*, between 10 and 15 degrees in the *Potamogeton*.

This absorption of oxygen is by no means comparable with

[1] J. Sachs, *Traité de Botanique*, p. 847.

[2] Boussingault, *Comptes Rendus de l'Acad. des Sciences*, LXVIII.

that which is produced after the death of the plant, and which has for result the mineralisation more and more complete of the organised tissues, deprived of life. In the last case special chemical combinations are produced: first of all, a ternary compound, ulmine (C^{40} H^{16} O^{14}), which is transformed into a series of derived products more and more oxygenised (ulmic acid, humine, humic acids, and so on). Such of these products as are acid join themselves to the ammonia formed also during the cadaveric decomposition of the plant; they then constitute soluble salts, which can be reabsorbed by the roots, and thus be restored to the vital movement.

To terminate this abridged description of vegetal respiration, it remains for us to indicate by what paths the air introduces itself into the mechanism of the vegetal tissues.

In the inferior plants, the mosses and the confervæ, there is no aërian circulation, no special apparatus. The oxygen is absorbed direct by the cells. In the complex plants, having true roots, true aërian leaves, canals, there is a commencement of functional specialisation. The air is absorbed a little by the bark, but especially by the leaves. Th. de Saussure, when analysing the interstitial air of plants, found that the air the least modified, the least poor in oxygen, is that of the leaves; that of the stems is less oxygenised, and that of the roots still less.

The air probably penetrates into the phanerogamous plant by the stomata of the inferior surface of the leaves. Thus if we plunge leaves into water and if we subject them to the action of the pneumatic pump, we see the air escaping regularly in bubbles by the section of the vessels of the petiole (Dutrochet). Once introduced by the stomata, the air circulates in the lacunae, the intercellular meatus, especially in the tracheæ and the punctuated tubes, when the first flow of sap in spring has forsaken them.

In the aquatic leaves destitute of epidermis and of stomata the aërian water acts direct on the cells, with thin walls, of the parenchyma. When there are stomata the process is the same as in the open air.

In the complex plant, the air is therefore subjected to a sort of circulation, which Dutrochet justly compares with the aërian circulation in the tracheæ of insects which are likewise furnished with stomata. Once introduced into the vegetal tissues, the air is absorbed by the anatomical elements, and by the sap or the blastemas which it contributes to elaborate. Finally it is probable that a part of the oxygen resulting from the reduction of the carbonic acid by the green parts, and which is consequently in the nascent state, that is to say, eager to enter into new combinations, is also forthwith absorbed, or even mechanically driven into the canals and cellular interstices.

4. *Sap Elaborated or Descendent.*

The water of the soil, more or less charged with the substances in dissolution, penetrates by endosmosis into the cells of the spongioles, and thence is impelled into the vessels, the meatus, the anatomical elements, step by step. Finally it is in some sort inhaled by the whole of the tissues, which all have need of nourishment, and more specially by the leaves and the green parts. These chlorophyllian tissues are the special exhaling and assimilative organs, in the midst whereof the sap undergoes a very important elaboration, and in a fashion passes into the state of true blastema. So far in effect the sap was nothing more than a simple mineral solution. In the leaves the sap is vitalised, it becomes an organisable liquid. It is only after this metamorphosis of the sap that new anatomical elements can be formed at its expense, that the organism expands and grows. This sap thus modified has been called *sap descendent,* because its ordinary course is from the leaves to the roots. It is interesting to follow the flow of this elaborated sap, to see how it passes and distributes itself from the leaves to the root of the plant.

It is needful for us here to take anew into account the effect of the endosmosis, which plays moreover such a foremost part in all the nutritive mechanism of the plant. Besides their special chlorophyllian function, and perhaps in consequence of that function, the leaves exercise on the sap a double propulsion: they aspire it from below to above, and they drive it back, after elaboration, from above to below. This is doubtless a kind of circulation, but rudimentary, without regularity, effected little by little, sometimes by one route, sometimes by another. The cells of the leaves, whose contents are dense and albuminoidal, are well arranged for absorbing by endosmosis the ascending lymphatic sap, to determine there, when it has penetrated into their cavity, synthetic chemical phenomena, to carbonise it, that is to say, to enrich it with atoms of carbon, thanks to their special property, then finally to expulse it by exosmosis. At the same time they concentrate it by depriving it of a notable portion of its aqueous vehicle, which is exosmosed, and escapes by evaporation.

Experience has demonstrated that this exhalation is an active phenomenon, the result of a vital act, probably chlorophyllian, and not a passive evaporation. The confirmatory facts are numerous. Hales was the first to discover that light augments the aqueous exhalation on the surface of the leaves. The phenomenon depends on the solar light and very little on heat: for, simple light diffused, slightly calorific, suffices, whilst in the shade a heat equal or even very inferior does not act much, and at night the aqueous exhalation stops.

If of two simple plants one is exposed to the light, and the other is kept in darkness, the first absorbs much more water than the second.

Sennebier has shown that leafy boughs dipped in water at their lower extremity aspire much more water in light than in darkness. He observed also that heat in darkness has little influence on this sort of suction, and that the results vary according to the species of plants. In general the plants which in the light resist the most the desiccating action of the atmosphere are those which in darkness resist it the least, and reciprocally. The intense contact of the air with the cells of the leaves is necessary to this aqueous aspiration, as the following fact proves.

Dutrochet having plunged into water a leaved stalk of *Pisum sativum*, deprived it of its interstitial air by means of a pneumatic pump. The aërian cavities, once purged of air, were promptly invaded by water. The plant was then taken from the water and the extremity of its stalk was alone kept in it. It was, in this state, exposed to the diffused light, but it had become incapable of aspiring water, and was asphyxiated, or rather it was incapable of aqueous aspiration because it was asphyxiated.

From the preceding facts it clearly results that the ascension of the sap is bound in a certain measure to the functionment of the leaves, and that this functionment, which comprehends simple aqueous exhalation, is subordinated, like every physiological act, to the absorption of oxygen by the anatomical elements, to respiration.

We must note a fact, whose explanation is however easy, namely, that the absence of light has not on the green corollæ the action which it exercises on the chlorophyllian leaves. As these corollæ respire simply after the manner of the animal anatomical elements, by absorbing oxygen and without acting on the carbonic acid, the absence of light does not hinder them from living, from aspiring sap, and so on.

Once elaborated and concentrated by the special action of the leaves, the sap comes forth from the cells by exosmosis, and just as the endosmotic action of the radical spongioles has driven the lymphatic sap from below to above, toward the branches and the buds, the exosmotic impulsion drives back the elaborated sap from above to below, from the leaves to the roots. We have seen that the ascension was especially accomplished through the ligneous centre, or at least through its youngest, peripheric part; the descent of the sap, on the contrary, is more habitually achieved through the bark. In ascending it enriches itself more and more, either carrying along with it the substances contained in the cells which it bathed in its passage, or by the special action of the green leaves;[1] in descending it is impoverished, on the contrary, more and more, for in going along, either it relinquishes organisable substances at the expense of which anatomical elements are formed, or it simply deposits in already prepared cavities masses of nutrimentary or organisable matters.

We have just described, in a fashion, the schematic mode of the circulation of the sap; doubtless there are numerous irregularities, and the sap undergoes in its course diverse impulsions, diverse deviations. The exosmotic impulsion is not the sole cause of the descendent propulsion. We must likewise take into account, as for the ascending sap, the relative void, made by the fixation of a duliquid part, by the nutrition and the formation of the tissues, and lastly, mechanical influences. For instance, it has been remarked that the agitation of the stems by the wind favours the growth of plants. Knight made on this subject precise experiments by immobilising certain stems. We are justified in thinking that the waving and the bending of the stems have as a result local pressions which aid the march of the sap in the canals. This descendent march of the elaborated sap must be slower than that of the aqueous sap; for this time we have before us a liquid, denser, more viscous, even sometimes differentiated. In effect there are assuredly diverse species of sap; for we must consider as such the milky liquid called *latex*, existing in great abundance in the cortical system of lettuce, of the fig-tree, of the euphorbiæ, and so on. The gum of trees of the Pinus kind, and so on, the resin of the conifers, are probably also sap residua.

[1] Knight collected in the spring at various heights the sap of the sycamore and of the birch. These are the variations of specific weight which he observed on the sycamore :—

Level of the soil	1,004
At seven feet	1,008
At twelve feet	1,102

In the last case the sap had moreover a sugary savour.

The sap can descend by the aubier or by the ligneous tissue of the central system. Knight having decorticated circularly a stem of *Solanum tuberosum*, saw the subterranean tubes developing themselves indeed less, yet still developing themselves; in which case the sap could only have descended by the bark.

The same observer has demonstrated that the elaborated sap can take an ascending movement when it is dissolved and carried on by the lymphatic sap.[1]

Furthermore the lymphatic sap can take a descendent movement, for example, when it forms in the leaves, absorbing the water which moistens them. When we cut the stem of a plant containing abundance of liquids, for example a milky plant, we see the latex flowing from the two surfaces of section There is here a simple sap movement, determined by the elastic pression of the canals primitively distended, turgid through the action of the endosmosis. In sum, the sap, elaborated or not, passes where it can. A root which had become naked Dutrochet cut, during winter, lower down than a shoot which it had produced; he saw, the following spring, this shoot continue to live. The development of the leaves had not yet taken place; consequently the shoot lived without elaborated sap, by the mere help of the lymphatic sap, driven now from above to below. Knight having cut off from a forward variety of *Solanum tuberosum* the runners which produce the tubers, there resulted in the stem a plethora of elaborated sap. The plant, which usually is not florescent, had flowers, fruits, and even small tubers were developed on many of the aërian parts of the plant.

If in the spring we cut a vine-root we see the lymphatic sap flowing from the superior or central *tronçon* as from an aërial stem.

Dutrochet observed that the trunk of a tree cut down during winter and completely stripped of its branches presents nevertheless in spring an outflow of elaborated sap under its bark. This sap existed therefore in the central part of the vegetal

[1] Knight, *Philosophical Transactions*, 1805.

tissue, forasmuch as it came neither from the leaves nor from the roots. It had been already prepared, and must have reached the periphery by means of the transversal medullary radii.

It is probably because the monocotyledonous plants have none of these transversal medullary radii that there is not in them any flow of sap between the cortical system and the central system, and consequently no growth at the point of junction of these two systems.[1]

In sum, there is no true circulation of sap, forasmuch as the course of the nourishing liquid is at the mercy of a number of accidents. There are only two principal sources of the sap: the spongioles of the roots, which introduce into the plant the aqueous or lymphatic sap, and the leaves, which elaborate this sap already thickened and enriched, to impel it afterwards to the lower part of the plant, by all the paths that are practicable.

5. *Algæ and Mushrooms.*

We have just presented a rapid picture, or rather given the enumeration, of the different phases of nutrition in a complex plant having roots, stems, and leaves. Naturally things come to pass otherwise in the inferior plants, composed almost wholly of cellular tissues, in what are called in these days *Thallophytes*, in the algæ and mushrooms.

Here there are neither true vessels nor true roots. The aliments are absorbed by the cells which are in contact with them, and transmitted by endosmosis from point to point.

Nevertheless the algæ have still fellowship with the higher plants by the presence of chlorophyll, even in the coloured algæ, where it is masked by colouring matters (nostochineæ, and so on). Chlorophyll acts there as it acts everywhere; it absorbs and assimilates carbon, and only operates in the light.

In many thallophytes the elementary cell has taken the

[1] Dutrochet, *De l'Endosmose*, p. 387.

elongated form of a filament, simple in certain intermediary algæ (phycomyceta), divided transversely by partitions in others, and sometimes ramified (mucedineæ).

The nutritive characteristic of the mushrooms is the absence of chlorophyll. Through this characteristic the mushrooms resemble the animal organisms. They are incapable of assimilating direct the aërian carbon, and have no need of light. We can even see some of them living under the soil, as for instance the truffles, and so on. From the absence of chlorophyll it also results that the mushrooms, like animals, have need of aliments wholly prepared, of combinations carbonised, assimilated by other organs. Thus when they are not parasites they have to live at the expense of organic matters in process of decomposition.

Furthermore, and this is a general characteristic, they are incapable of forming a single granule of starch. This fact explains in a certain measure the office of chlorophyll in the synthetic chemistry of the vegetal organism, and serves us as transition to pass to the following chapter.

CHAPTER III.

VEGETAL ASSIMILATION AND DISASSIMILATION.

1. *Organic Substances.*

BEFORE speaking of the chemical transformations of which the vegetal organism is the seat and the agent, it will not be useless to enumerate the various bodies which penetrate into the living plant from the ambient medium.

We have seen the roots, and sometimes the leaves, absorb water, of which the plant has an imperious need. We know that the water absorbed by the roots is loaded with substances in solution. Among these substances, some are terreous, and we shall pass them under review at a future time in this exposition (salts of lime, soda, potash, ammonia, &c.). It is through the roots that the larger part of the azote which is necessary to the plant penetrates into it. This azote is introduced into the plant in the form of salts of ammonia and especially of a series of substances produced from the slow oxydation of the organic detritus. The type of these substances is a compound of C^{40}, H^{16}, O^{14}, *ulmine*, which, according to Mülder, is transformed by a gradual sur-oxygenation into ulmic acid, humine, humic and geïc acids, &c. The acids of this series eagerly absorb ammonia, and compose with it soluble salts. They form also soluble alkaline salts, and terreous salts, which would be insoluble if the ammonia did not form therein an aggregation of double salts. According to Mülder, water co-operates with these formations, being decomposed to oxydize these matters, and a part of its hydrogen combines itself with the azote of the air and forms ammonia. Plants raised in ulmic acid and powdered charcoal, entirely deprived of ammonia, shut up in an atmosphere and watered with water which was also destitute of it, have yielded to analysis a double or treble quantity of azote to that which their seed contained at the commencement of the experiment. In this case, the plant being deprived of ammonia, it is very possible that it had absorbed the azote directly from the air. The following facts render this a very probable eventuality :—

Schroeder sowed cereals in flower of sulphur watered with distilled water, and contained in vessels of glass or porcelain covered with a receiver. The seeds germinated, and produced halms from 2 to 14 inches in length, bearing short ears, which however flowered. When dried, they were five times the weight of those seeds from which they sprang.

Boussingault sowed twenty-nine seeds of clover in sand previously reddened in the fire. The plants which they produced weighed 67 grains at the end of three months. (*Annales de Chimie*, t. lxxvii.)

Peas treated in the same way yielded in the same space of time plants weighing 72 grains, loaded with flowers and perfect seeds, &c. (Quoted by Burdach, *Traité de Physiologie*, t. ix. p. 255.) Finally, as we shall see further on, recent experiments of M. G. Ville put the normal absorption of atmospheric azote by the plant almost beyond doubt.

If the anatomical elements of plants bring about many chemical syntheses, they are also very capable of disaggregating mineral compounds. We know that chlorophyll decomposes the carbonic acid of the air. It has been also observed, that seeds germinating in water decompose it, and absorb a portion of its hydrogen.

The oxygen contained in a plant comes in a large degree from the aërian medium ; a notable portion of it from the decomposition of the carbonic acid by chlorophyll. The plant also procures it in another way. In general, as we have seen, the nutritive substances absorbed by the roots are very rich oxygenised compounds; on the contrary, the assimilated substances, forming a large portion of the dry matter of plants, are either poor in oxygen, or do not contain it at all. Assimilation must then be specially an act of dis-oxygenation; now we know that it takes place particularly in the chlorophyllian cell. We have then here, as Sachs justly remarks, the place, the conditions, and the time of this dis-oxygenation.[1]

In fact, it is in the chlorophyllian cells that assimilation, or rather the vitalization, of various substances introduced into the plant takes place. It is towards this living and active laboratory that they converge; it is there that they enter into conflict with each other. The chlorophyllian cell is a powerful apparatus of synthesis, effecting organic combinations which still defy the power of contemporary chemistry. The molecules of azote, carbon, oxygen, hydrogen, &c., penetrate the cellular cavity, some free, others entangled in combinations more or less complex. There they are mixed together, dragged into the circular current of the cellular protoplasm, subject to the attraction exercised upon them by the chlorophyll, and, on the other hand, shaken by the undulations of the yellow rays of solar light which add their impulsion to the vibrations by which they are already animated. The atoms and the molecules yield to these united influences; those particles which are involved in combinations abandon them, resume their liberty, and all unite to form living organic substances, ternary and quaternary substances.

We have already said a few words upon the advent and aspect of one of these substances, one of the most important in vegetal physiology, starch. Starch is especially formed in the green cells, and even in the interior of the chlorophyllian bodies. We have seen that its formation is in direct dependence upon the chemical operation of the chlorophyll, consequently upon light. It

[1] J. Sachs, *loc. cit.*, p. 821.

appears under the influence of light, and in its absence re-dissolves. We seize here chemical synthesis in some degree in the very act.

Assuredly proteïc substances are also formed in the chlorophyllian cell. Doubtless the phenomenon is less evident here; but we know that the sap reaches the leaves in the state of a fluid still very slightly charged, and that it issues from them as a living liquid in the state of nutriment and blastema.

We have just said that chemical synthesis, exercised upon the chlorophyllian cell, had probably resulted in a subtraction, an emission of oxygen; but there are, on the contrary, organic compounds that result from super-oxydation; these are the vegetal acids. Their molecule contains more oxygen and hydrogen than are required to form water. Oxalic acid, one of the most frequent, contains three atoms of oxygen to two of carbon (C^2O^3). In laboratories, it is obtained by oxydizing sugar and fecula with azotic acid. In the plant, it is probably produced by the direct action of oxygen upon the same substances. Oxalic acid, in fact, as well as the ternary, acetic, citric, and malic acids, and so on, is not formed in the synthetic laboratory of the chlorophyllian cells, but in those parts which are not green, or which are sheltered from the light.

Organic vegetal substances are naturally subject in a high degree to isomerism and instability, like all aggregates of this kind; they also often undergo metamorphoses during the course of the nutritive process. Their molecular mutations succeed and engender each other.

The spores, seeds, bulbs, tubers, rhizomas, the vivacious shoots of ligneous plants are in truth nutritive reservoirs, where the organised sap deposits organic substances, utilizable at a later time, either for germination or for nutrition, when the flow of lymphatic sap reappears in spring. These substances are then drawn away, and furnish materials for the development of the folial and floral buds, for the nutrition of the tissues, which could not otherwise obtain aliment, frondation not yet existing; but before being drawn away and utilized, these substances in reserve often undergo metamorphoses.

The fecula, deposited in autumn in the ligneous body of the trees, is often in spring transformed into sugar, which, in certain plants, the maple, for example, mixes largely with the ascending sap. MM. Payen and Persoz have shown that there is first of all produced a matter called by them *diastasis.* This matter has the property of rendering the fecula soluble, by transforming it first into dextrine, then into sugar. It produces this isomeric transformation by means of very small quantities, for it is capable of rendering soluble five thousand times its own weight of fecula.

As to the albuminoïdal substances in reserve in expectation of the spring sap, such as starch, sugar, inuline, and fat, certain among them, not having lost any of their solubility, are simply seized anew, drawn away, and finally assimilated. Others also undergo their isomeric modifications, specially in germination. Thus, during the germination of the leguminous plants, caseïne is metamorphosed into albumine in the cotyledons. In the gramineous plants, the gluten of the endosperm, which is insoluble in water, is dissolved during germination, and is conveyed into the plantule. The energetic oxydation which always accompanies germination also determines the formation of regressive substances, for example, asparagin, which is perhaps re-assimilated at a later period.

A certain portion of carbonated hydrates is also totally burned, converted into water and carbonic acid by vital oxydation. In this case, the hydrates are burned either in the anatomical elements, or in the sap of the plant, as they are in the blood of the animal, proving once more that there is no antagonism between the vegetal kingdom and the animal kingdom, and that essentially the primordial phenomena of life are identical in the two kingdoms.

As Cl. Bernard has very well observed, if an animal eats the sugar accumulated in beet-root, that does not prove that this sugar was made for him. On the contrary, it was destined (if we may employ this word in speaking of blind organic finality)—it was destined to be burned by the beet-root itself during the second year of vegetation, when florescence and fructification are developing themselves.

However this may be, the nutritive substances, when sufficiently elaborated, are assimilated ; that is to say, their molecules intercalate themselves between the organic molecules already formed, either to replace those which have been destroyed, or to bring about the enlargement of the cell ; others spontaneously organize themselves in the blastemas.

We may form some probable conjectures as to the special office of the various categories of organic substances. It is probable that the hydrates of carbon and the fatty bodies are especially transformed into cellular membranes, these membranes being principally constituted of cellulose, the molecule of which is almost indentical with that of starches, sugars, and fats. Thence come also without doubt the vegetal acids, tannin, and the colouring matters.

The proteïc substances are probably employed in the formation of internal azotized utricles included in the cells, and of the cellular protoplasm and the chlorophyllian particles.

They also furnish, but by incomplete oxydation, by degradation, asparagin, of which we have spoken, and without doubt the vegetal alkaloïds (quinine, morphine, strychnine, &c.), quaternary but not proteïc substances, and of which the molecule, very complex, very rich in carbon, and tolerably rich in hydrogen, contains also a notable proportion of oxygen and very little azote.

2. *Inorganic Substances.*

Besides atmospheric gases, a number of mineral substances are introduced into the plant through the roots. The principal are alkaline and terreous bases : potash, soda, ammonia, magnesia, lime, generally salified by sulphuric, azotic, phosphoric, silicious and carbonic acids. Many of these substances do not contribute anything to the nutrition of plants; they are mechanically driven by the water which penetrates into the plant, and are deposited in the deep tissues, after having formed insoluble salts with the vegetal acids.

Habitually, these mineral particles are uniformly intercalated between the organic molecules; also by incineration, or by treating the vegetal anatomical elements with certain acids, we may succeed in destroying the organic substances which they contain, preserving only the mineral skeleton, but nevertheless keeping the form of the destroyed anatomical element.

Often the mineral salts also form true crystals in the vegetal tissues and even in the cells. Such, for example, are fine granulous incrustations of carbonate of lime, bundles of spars of oxalate of lime, or even tolerably voluminous crystals.

The amount of mineral matters in a plant is proportioned to its age, and also to the quantity of water which flows through it; in short, to the activity of the vegetation. The proportion of soluble substances may undergo certain variations. As to the insoluble substances, they accumulate ceaselessly, mineralizing the plant more and more, and contracting the range of action of life. If most of the mineral substances drawn from the soil are of little use in the development and nutrition of plants, yet some of them are very important, for example, ammonia, the phosphates and the sulphates. Others again appear sometimes necessary to one plant or group of plants, sometimes to another.

Most plants which grow on the sea-shore contain much soda, and this soda is necessary to them; for they grow only upon the sea-coast or near saline deposits inland.

We always find in the tissues of plants of certain very natural families, the same mineral substances. They have thus chosen them in a certain degree. We may mention the gramineous plants, the stems of which, without exception, contain silica, whilst the fruits contain phosphate of magnesia and of ammonia.

But often the same plants contain different salts, according as they grow in various soils.

Let us notice, however, that it is possible, under the conditions of artificial culture, to rear, without silica, plants which habitually contain much of it, maize for example.

Lime seems to be indirectly useful to plants, by serving as a vehicle to the sulphuric and phosphoric acids, and by neutralizing the oxalic acid, hurtful to the plant in which it is formed; but any other base could do as much.

3. *Influences of Light, Heat and Electricity.*

We have already spoken at length upon the influence of solar light upon vegetal nutrition. It is such, that, without this influence, nearly the whole of the vegetal world would cease to exist. In consequence, the animal kingdom, which, in the continental parts of the globe, lives, directly or not, at the expense of the vegetal kingdom, would also become extinct, at least in its higher branches, on the surface of these terrestrial regions. In effect, without solar light, mushrooms alone, of all plants, could still live upon the continents.

But animal life would probably find refuge in the sea. A number of the lower marine animals are nourished and live without the help of plants. Now these rudimentary organisms, to which we must grant the faculty of synthetising complex organic substances, after the manner of plants, could themselves furnish alimentation for higher aquatic animals, as often actually happens. All this aquatic fauna can thus live without light. This is even the case now, as the dredgings and soundings made, during the last few years, at the bottom of the ocean and of large lakes has proved.

Solar light only penetrates the sea to a very small depth. Below 50 fathoms (1 fathom = $1^{m}\cdot 82$,) a very sensitive photographic paper is no longer impressed. At this depth also, the vegetal kingdom is only represented by rare specimens, and below 200 fathoms it absolutely disappears. Now the bottom of the Atlantic ocean, which is as much as 1400 and 1500 metres deep, is covered with organized beings, all belonging to the group of the Protozoa of Hæckel. This bottom is everywhere carpeted with those little sarcodic, gelatinous, contractile organisms, called by Huxley *Bathybius Hæckelii*. Amongst these bathybians live foraminifers, rhizopods, radiolites, sponges.[1]

We therefore must admit that certain of these rudimentary animal organisms can decompose water, carbonic acid, and ammonia, or assimilate the numerous organic detritus in suspension in the marine waters, arising from the dejections of animals, and from the decomposition of their dead bodies. It is only thus that they can live, can multiply, and can aliment more complex animals.

Analogous facts have been observed by MM. de Candolle, Forel, and Dufour, in the Lake of Geneva. There also the most sensitive photographic paper ceases to take an impression beyond the small depth of 50 or 60 metres in summer, and 40 or 50 in winter. Nevertheless, at the bottom of the lake live from thirty-five to forty species of lower animals.

The idea of a necessary solidarity between the two organic kingdoms must then be abandoned. This solidarity only exists, in a very large measure, in the organized terrestrial world. There, we may admit, as a general thesis, that the vegetal kingdom is necessary to the animal kingdom. Now nearly the whole of plants cannot live without light. But the calorific and chemical solar rays are not less necessary to the maintenance of life. Solar irradiation is then one of the grand causes of the production, development, and duration of organized beings. The great theory of the transmutation and correlation of physical forces is therefore applicable to biology. We must guard, however, against making this application with a mathematical and inflexible rigour which the subject does not allow of. We think also that there has been too great an inclination of late years to consider solar irradiation as the unique and universal

[1] *The Depths of the Sea*, by C. Wyville Thomson. 8vo. London, 1873. French Translation: *Les Abîmes de la Mer*, 8vo. Paris, 1875.

cause of life upon the surface of the globe. With these reservations, we cannot deny that solar irradiation is fixed, accumulates in the plant, and that, in the infinitely numerous cases in which the plant serves as animal alimentation, this irradiation is treasured up by it, transforms itself into various series of molecular vibrations, into heat, movement, thought, etc.

We have not to return here to the chief office fulfilled by light in the nutrition of plants; but, besides the phenomena of synthetic chemistry accomplished by chlorophyll, light also produces in plants certain secondary phenomena, which probably depend upon the chlorophyllian property. Thus many plants *sleep* in the night, that is to say, droop their leaves or follicles more or less, or rather gather them up along the stem or principal petiole. This phenomenon assuredly depends upon the light, since De Candolle was able to make sensitive plants sleep in the day in artificial darkness, and, on the other hand, succeeded in waking them at night by the light of lamps.[1] This so-called slumber, this sinking down of the leaves, arises probably from a diminution of the turgescence of the tissues, and this diminution is assuredly the result of the inaction of the chlorophyllian cells, which, no longer exhaling, no longer forming organic syntheses, summon less water into the petiolary canals, the meatus, and so on.

If the luminous undulations are, as it were, the soul of vegetal life, caloric undulations are not less indispensable to the nutrition of plants. Above and below a certain temperature, vegetal life ceases. Speaking generally, the thermometric limits of vegetal life are 0 degree and 50 degrees.[2] It must be noticed that the cellular juices, being liquids very full of substances in solution, do not congeal at 0 degree. M. Uloth has seen seeds of *Acer platanoïdes* and *triticum* germinate between the fragments of ice

[1] *Mémoire sur l'Influence de la Lumière artificielle sur les Plantes* (*Mém. des Savants Étrangers de l'Institut*, t. i.)

[2] Sachs, *Ueber die obere Temperaturgrenze der Vegetation* (*Flora*, 1864).

in a glacier, and send down numerous roots, several inches in length, into ice destitute of crevices.[1] It is then probable that germination can still take place at 0 degree, and, as it always develops a remarkable quantity of heat, the result must have been, in the case mentioned by M. Uloth, the partial fusion of the ice, whence it became possible for the roots to penetrate.

We have given the limits of temperature, but these vary according to species. Most plants succumb after remaining for ten minutes in water at from 45 to 46 degrees. In the air, phanerogams are killed after remaining from ten to thirty minutes in a temperature of from 50 to 51 degrees. To produce a lasting alteration, it is sometimes necessary that the temperature should reach 60 degrees. Death then takes place through the coagulation of the albuminoïdal substances.

Dried vegetal tissues (seeds) can naturally support more extreme temperatures of heat or cold. Thus, tissues full of sap are killed at 50 degrees, whilst dried seeds of *Pisum sativum* may be heated with impunity to 70 degrees for an hour.

In every plant each function has its special limits of temperature. We cannot enter here into the enumeration of facts in detail. Moreover, we have already stated the temperatures necessary for the operation of chlorophyll.

In cells killed either by congelation or by too high a temperature, the cellular membranes become permeable; liquids filter through: consequently turgescence ceases. First of all, the protoplasm becomes immovable, takes a sombre tint, and rapidly loses its water by passive evaporation.

Life is not sufficiently active in plants for them to have, like the higher animals, a temperature of their own, independent in some degree of that of the exterior medium. The small aquatic plants, and the subterraneous parts of terrestrial plants, have generally the temperature of the ambient medium. On account of their volume, the massive stems follow more slowly the

[1] Uloth, *Flora*, 1871.

exterior thermometric variations, and consequently, they may be hotter or colder than the exterior medium.

Electricity seems to exercise very little influence upon the nutritive movement of plants. Feeble constant currents, small sparks of induction, have no apparent effect either upon the movement of the protoplasm, or upon that of the mobile leaves. A too powerful current, a spark too strong, causes either the arrest of the protoplasm or that of the tissue. Thirty Grove elements instantly stop the protoplasmic movement.

As to the electric state of the tissues, it is curious to observe in the plants phenomena very analogous to those which have been observed in the nerves and muscular fibres of animals, and have served as a basis for many theories. In effect, in a cut stem, there is a current from the surface of the stem to the centre of section.[1] The electric currents, which, without doubt, result from the chemical assimilative and dis-assimilative reactions of nutrition, are not peculiar to such and such a vegetal or animal tissue. There again the two organic kingdoms touch and intermingle.

[1] Buff, *Annal. der Chimie und Pharmacie*, 1854, Band 89. Jürgensen, *Studien des Physiologisches Instituts zu Breslau*, 1861, Heft I.—Heidenhain, *ibid.*, 1863, Heft II.

[Note.—A French metre is equal to $39\frac{33}{100}$ English inches, or to rather more than a yard. The small *g* occasionally occurring indicates *gramme;* the small *m*, *metre;* the small *mm* a *millimetre.* —Translator.]

CHAPTER IV.

OF ANIMAL NUTRITION.

After what has been already said concerning the chemical and anatomical constitution of animals and plants, nutrition in general, and the nutrition of plants, it will be possible to abridge considerably the general exposition of the phenomena of nutrition amongst animals. In effect, if nutrition is traced back to its fundamental activities, that is to say, to assimilation and dis-assimilation within a living matter, amorphous or figurate, the principal phenomena are plainly similar in both kingdoms; and, even if we only consider certain rudimentary animal organisms, for example the *gregarine*, we can say that there is an identity of process, not certainly with that which takes place in the superior plants, where there already exists a high degree of differentiation in tissue and of division of labour, but with the nutritive mode of purely cellular plants. In effect, the dissimilarity between plants and animals decreases in proportion as we compare the inferior types of the two kingdoms.

Correctly speaking, we should rather compare the *gregarine*, a parasitical animal without organic differentiation, to a mushroom. In effect, as mushrooms live by assimilating the products of decomposed organisms, and sèem powerless to effect the chemical synthesis of which complete plants are the agents, in the same way the *gregarine* absorbs and assimilates direct in the intestines of articulated animals and worms, where it lives upon the complex organic substances which it would be incapable of forming. The same general remark may be applied to a number of infusoria, the amœbæ, and the rhizopods.

At a later period we shall have to describe the principal modes of anatomical and physiological differentiation, by means of which the complex animal organisms prepare the way for interior nutrition, and we shall see them gradually working, becoming more and more complicated in proportion as we rise in the animal series. For the present, we have only to consider the general nutritive phenomena; the principal exchanges of matter, which are accomplished between a complex animal and the exterior world.

In such an organism, there is no direct interchange between the anatomical elements and the ambient medium; intermediary liquids, which may be compared to the elaborated sap of superior plants, have the office of receiving the different kinds of nutriment fully prepared for assimilation, of conveying them—of presenting them, as it were, to the anatomical elements; the latter, by virtue of their innate constitution, of their affinities, draw from these interior and living mediums the substances which suit them, and reject those of their molecules which are altered by the action of life.

We have already shown what is the composition of those living mediums called plasmas and blastemas. We know that therein are found three orders of immediate principles, mineral substances, ternary matters, amyloïdal, saccharine, fatty, finally and especially substances proteïc or albuminoïdal. Naturally, the diverse chemical compounds of living liquids are so much the more numerous, the more varied, in proportion to the complexity and differentiation of the animal, since in every organism each kind of anatomical element has its special alimentary preferences. Also, whilst amongst the superior plants we find some products of dis-assimilation little varied, as for instance asparagin, and perhaps some vegetal acids; we see among the superior animals the proteïc substances alone furnishing, when the anatomical elements reject them after having oxydized them, urea, urates, creatine, creatinine, inosic acid, inosates, and so on.

A very general fact, and one which it is very important to point out, is that all truly animal organisms cannot live without constantly absorbing complex organic substances, and that they are powerless to create these themselves, at least in sufficient quantity, from the mineral world, whence it is necessary that they should draw a supply, either from other animals or from vegetals.

The blood of the superior animals is the great mart in which the anatomical elements provide themselves with the substances which are necessary to them, and reject those which encumber them. As a consequence of its office in the vital movement, the blood is itself necessarily subject to a perpetual exchange of materials. It gives to, and takes from, the anatomical elements on the one hand, and on the other, to and from the ambient medium. It borrows oxygen from the outer air by means of the respiratory organs; it receives from the digestive organs immediate principles, restorative nutriments liquefied or dissolved. Finally, the totality of the lymph, when it exists, continually diffuses itself in the blood. On the other hand, the tissues yield to it the dissolved or liquid residues of vital combustion.

The expenditure of the blood balances its acquisitions. It exhales, by means of the lungs and skin, water, carbonic acid, and azote; it expels through diverse special channels, water and complex organic substances. Of the latter the most part are purely regressive, and destined to be returned to the mineral kingdom; the others are modified by special secreting organs, and become fit to play a more or less important part in the various physiological functions. Finally, on the other hand, the blood yields to the anatomical elements the three orders of immediate principles indispensable to their conservation, the oxygen which burns and vivifies them, the water which soaks them, diverse mineral salts, hydro-carbonic ternary substances, new proteïc matters which, in each anatomical element, replace, molecule by molecule, the exhausted materials.

This double movement of acquisition and expenditure is simultaneous in the blood, as the twofold movement of assimilation and dis-assimilation is simultaneous in the anatomical elements.

After this general sketch of the nutritive going and coming within the complex animal, we can now occupy ourselves with each of the three groups of immediate principles which enter into the composition of the body of every animal, and indicate in bold outline under what form they enter the organism, and under what form they leave it.

[Note.—A word which frequently occurs in the French original is *intime.* To the Latin *intimus*, the French *intime*, the German *innig,* in their scientific import, there is not any exact English equivalent, though *ultimate* and other words are employed in lack of better.—Translator.]

CHAPTER V.

OF ASSIMILATION AMONGST ANIMALS.

1. *Mineral Principles.*

THE mineral principles which ceaselessly introduce themselves into animal organisms are, gases of the air, a large quantity of water, and dissolved composite solids, of which the principal are phosphate of lime, phosphate of magnesia, bi-carbonate of lime, fluoride of calcium, silicious acid, chlorure of sodium, carbonate of soda, alkaline phosphates, oxyde of iron, etc. Amongst the lower organisms, the two chief gases of the air penetrate direct through the limiting surface of the animal. Amongst others, apparatus exist, varying according to species and medium, but whose office is limited in final analysis to bringing the outer air into contact with the black or venous blood, without other barrier than a very thin organic membrane—that is to say, to realizing in a special part of the organism the conditions of osmotic absorption which exist over the whole surface of the bodies of inferior animals.

Of the two principal gases of the blood, the one, azote, is absorbed in smaller quantity, and seems normally to exercise no function in the animal economy. It is rejected as it was absorbed, and by the same channels. Nevertheless, the animal, in a state of inanition, fixes a certain portion of it in its tissues. Probably there is then a direct synthesis. As a compensation, when the animal enjoys an abundant azotic alimentation it rejects all the aërian azote, and, in addition, a certain portion in excess, arising from the reduction of the aliments.[1]

Oxygen has a very different part to play in the exercise of physiological functions. It penetrates into the blood, is conveyed by it into contact with the tissues which impregnate themselves with it, and burn, thanks to its assistance, the immediate principles which constitute them, particularly the complex organic substances. Finally, the anatomical elements in return restore to the sanguineous medium an equal volume of carbonic acid, of which a large portion is exhaled, particularly through the respiratory surfaces, a smaller part forming alkaline carbonates, by combining itself with the bases previously existing in the tissues.

This oxydation of anatomical elements is one of the primordial phenomena of life. It must take place, under pain of death, in every living substance, figurate or amorphous, vegetal or animal. It is a fundamental, biological phenomenon, and all the reactions of living chemistry depend upon it. Without its intervention, there can no longer be either assimilation or dis-assimilation in any organized substance, and the anatomical elements become incapable of exercising their special functions; the chlorophyllian cell then ceases its chemical synthesis, the muscular animal fibre no longer contracts, the nervous cell becomes incapable of feeling, volition, or thought; there is no longer either maintenance, or development, or generation of the anatomical elements.

Amongst the invertebrated animals, whose blood does not contain globules, it is the sanguineous plasma itself which conveys the dissolved oxygen to the tissues. Amongst almost the whole of the vertebrates, on the contrary, this office principally devolves upon the red sanguineous globules, as we have already indicated. The substance of these globules, which has a great affinity for oxygen, imbibes it, fixes many volumes of it, and becomes vermilion. The oxygen thus flowing on with the

[1] Ch. Robin, *Des Humeurs.*

globules is conveyed even into the finest capillary ramifications of the circulatory system. There the globules meeting the waves of carbonic acid, the expulsed residuum of the oxydation of anatomical elements exchanges with this acid its oxygen, with which it had only made a shifting combination. The oxygen thus yielded goes, by endosmosis, to re-vivify the tissues, whilst the carbonic acid, carried away in its turn by the globules, to which it has imparted a sombre red tinge, is exchanged by them for a fresh provision of oxygen, in the capillary vessels of the special organs called *respiratory*. The respiration of the most complex animals is summed up then in a series of very simple acts: absorption by the red globules of aërian oxygen, cession of this oxygen to the anatomical elements, in exchange for a residuum of carbonic acid, finally the exchange of this carbonic acid for a fresh supply of aërian oxygen. It is a ceaseless circuit, which stops only with death.

Almost the whole of the water contained in such large proportion in the animal organism penetrates it from without, either through the general surface of the body, as in inferior organisms, or by that of the organs called *digestive*, as in the complex animals. Nevertheless, it is probable that a considerable quantity of water forms itself in the interior of the living tissues themselves, by the complete oxydation of certain ternary hydro-carbonated matters. The water besides conveys along with it into the economy important quantities of calcareous, alkaline, and ferruginous salts, and also a certain quantity of air, which it holds in solution. Boussingault has stated that the quantity of mineral substances imbibed each day at the drinking trough by a milch-cow rises to 50 grammes.

In weight, water forms the major part of every animal organism, of every plasma, of every anatomical element. The albuminoïdal substances retain a notable quantity as constituent water, another portion as water of imbibition. Water does not seem to be directly decomposed amongst animals, and its office is specially mechanical, but none the less important. It is water which furnishes to the immediate principles the fluid medium, indispensable to nutritive exchanges, which maintains in the tissues the same principles in a state of solution and emulsion, in a word, of needful molecular division, so that the chemical reactions indispensable to life may be accomplished. Deprived of their water, living substances lose their mobility and their instability, and become in a measure mineral. We know that certain infusoria, rotiferous, tardigrade, and anguillule, lose or recover their physiological functions according as they are dried or moistened.[1]

Water issues from the economy in various ways. It is exhaled through the external cutaneous or mucous surfaces directly in contact with the exterior medium, either in the state of vapour or in a liquid state. But, amongst complex animals, it is especially through certain secretory or excretory organs, particularly through the renal and sudorific glands, that the water of the organisms escapes and returns to the mineral kingdom.

A quantity of mineral salts, oxydes, etc., penetrates into the economy either with the water, or with the aliments, whatever they may be. The principal are phosphates of lime, of magnesia, of soda, of potash, sulphates and carbonates of lime, of silica, of chloride of sodium, alkaline carbonates, salts of iron, etc.

Many of these salts, though insoluble in water, are dissolved in the blood, by means of the albuminoïdal substances, the carbonic acid, and other salts, which act upon them as solvents. Once in the blood, they pass into the substance of the anatomical elements; thus, in the muscular flesh, we find phosphate of lime, phosphate of magnesia, and tribasical phosphate of soda. The alkaline carbonates are also met with in most of the solids and liquids of the economy.

Certain of these salts form themselves within the tissues by the process merely of nutrition; thus carbonic acid, the result

[1] J. Gavarret, *Nouvelles Expériences sur les Rotifères* (*Journal du Progrès*, t. IV., p. 421; t. V. p. 1).

of the oxydation of the anatomical elements, combines itself with alkaline bases; the oxydation of the proteïc substances causes, specially in the nervous tissue, the formation of phosphoric acid, which combines itself also with the alkaline bases, united with less powerful acids, etc.

Tribasic phosphate of lime, which is dissolved in the blood, or combined with the albuminoïds of this humour, is deposited in a solid state in the bones, the teeth, and the hair.

There is a remarkable proportion of chlorure of sodium in the blood, and in all the liquids and solids of the economy. Its office is not yet correctly defined, but its presence seems necessary. It cannot help figuring in the alimentation of men and animals. Livingstone, during his travels across the African continent, having to live for whole months without animal food and milk, relates that he was seized with a violent desire to eat sea-salt; he adds, that after thus abstaining from them, animal food and milk had at first, to him, a very salt taste.

Though there is in the blood a very small quantity of iron, it is a substance absolutely necessary. We have seen that it forms part of the hæmoglobine of the blood. Its curative office in anæmia is as beneficial as it is well known.

2.—*Hydro-carbonized Principles.*

The hydro-carbonized principles, that is to say, specially the amyloïdal and saccharine substances, seem only to penetrate into the circulation, and consequently the anatomical elements, after having been transformed into dextrine or glycose, in the same way that amongst plants the amylaceous masses are saccharified before being made use of for nutrition. In the blood, and also in the anatomical elements, these ternary substances meet the oxygen, which reduces them, when the oxydation is complete, to water and carbonic acid. This transformation of hydro-carbonized substances leaves no organic residuum, but develops heat, and contributes powerfully to facilitate all the reactions, all the exchanges of material which are the very foundation of nutrition. Cl. Bernard has proved that a portion of these ternary compounds form themselves direct, by synthesis, in animal tissues as well as in those of plants. The liver of animals forms in its cells an amyloïdal matter denominated *glycogene*, which is afterwards transformed into sugar. The placenta of the larger number of mammifers is also a producer of glycogene. The placenta of ruminants seems to form an exception, but the glycogenical function is then fulfilled by platings, by special villosities of the amnios. We also find sugar in the muscles of the fœtus, in the larger portion of the epithelial cells covering its mucous membranes, in the bones and the cartilages. In the crab, where development is achieved by sudden bounds, by moultings, the liver only contains glycogenical matter during those periods of rapid growth.

It seems certain that the glycogenical cells of the liver, and all the analogous cells, can synthetically form glycogenical matter at the expense of the aliments, whatever they may be; but it seems also incontestable that they directly utilize substances of a similar composition contained in the aliments. M. W. Pavey has, in effect, shown that in dogs fed exclusively upon animal aliment, the mean proportion of glycogenical matter was 6·97 in the 100, whilst it rose to 17·23 per 100 in dogs fed upon boiled potatoes, barley-meal, and bread. Finally, the proportion was 14·5 in the 100 when a considerable portion of sugar was added to the ration of animal food.[1]

We have said that the combustion of hydro-carbonized matters in the organisms developed heat, which was afterwards utilized in the reactions of living chemistry. We shall see farther on that a portion of this heat can also transform itself direct into muscular movements. In effect, we find sugar in the muscles of the fœtus, in the tissue of muscles immobilized by the section of their motory nerves, and in the muscles of animals in the

[1] *The Influence of the Diet on the Liver.* London. (*Guy's Hospital Reports,* 1859.)

sleep of hibernation. Finally, we find, in the juice extracted from the muscles, lactic acid and inosite, which seem products derived from the oxydation of hydro-carbonized substances.

We must remark also, that during the last stage of intra-uterine life, when the muscles already contract frequently, only lactic acid is met with in their tissue, as in the adult. We know, moreover, that when a muscle is immobile, only a very small part of the arterial blood is transformed therein into venous blood. It seems then that, in effect, muscular action may be one of the principal causes of the oxydation of hydro-carbonized substances contained in the blood and in the tissues. As to the question of the direct transmutation of heat into movement by the muscle, the subject is so complex as to demand a separate chapter.

Substances analogous to the amyloïdal and saccharine substances, fat bodies, also enter the blood in a state of very fine division —a state of emulsion. In the superior animals, this emulsion is accomplished by special liquids, bile and the pancreatic juice; the latter passes with these matters into the blood and decomposes them into glycerine and into fatty acids; for they do not seem to be always utilized in nature. In the blood the fatty bodies evolve themselves by forming fatty acids, oleïc, stearic, etc., which become oleates and stearates of soda and potash; besides, cholic and choleïc acids, whence are formed cholates and choleates; finally, cholesterine, and a fat phosphoric substance, cerebrine. It is probable, moreover, that like the larger portion of amyloïdal matters, they also end by being oxydized, in developing utilizable heat. We know that the alimentation of the inhabitants of the polar regions contains an enormous quantity of fatty matters, and that, under the same conditions, European sailors have found the advantage of adopting the same kind of alimentation as the natives.

It is also certain that animal organisms can produce fat from saccharine matters. MM. Dumas and Milne-Edwards, having fed bees exclusively upon honey, obtained thus an overplus of wax.[1]

Whatever may be their source, the fatty matters of the economy, when in excess, are deposited in the meshes of the cellular tissue, and generally in the larges quantity where the organic activity is the least.

3.—*Albuminoïdal Substances.*

Are proteïc substances completely formed, by chemical synthesis, in the economy? That is possible, but not sufficiently proved. It is most probable that the albuminoïdal substances of the anatomical elements are derived by means of multiple isomeric elaborations and modifications from the compounds of the same kind existing in the aliments. These substances the inferior animals, without differentiated anatomical systems, absorb direct from the ambient medium; the other kinds of animals find them already prepared in other organisms. According to a general formula, which has for a long time found acceptance, it is the vegetal kingdom which is forced to provide for the animal kingdom. We have seen that probably the marine fauna forms an important exception to this formula, which is correct as regards a very large majority of the terrestrial animals.

But, finally, whether the first origin of organic substances be vegetal or animal, it is not the less true that the least complex animal has need to borrow from other organisms non-mineral aliments, without which it cannot live. If, as we have just seen, the fact may in a certain degree be contested as regards the albuminoïdal and saccharine compounds, it is undeniable in regard to the more important materials of living beings, the proteïc substances.

The animal never either absorbs or assimilates directly these substances; it is necessary that they should undergo, before

[1] *Comptes Rendus de l'Académie des Sciences,* 1843, t. xvii. p. 531.

absorption, important isomeric modifications, due, in the superior animal, to the action of fermentescible agents called *pepsine*, *pancreatine*, etc., which are secreted by certain portions of the digestive apparatus. Under the influence of these agents, the albuminoïdal vegetal and animal bodies undergo an isomerical transformation. All become soluble and incoagulable by heat; they are then *peptones*, and it is under this form that they enter the circulation. Amongst the vertebrates, the intestinal capillaries, and consequently the intestinal veins, are the principal channel for the absorption of the albuminoïds. Once in the blood, these substances pass from the state of peptones, or albuminose, to that of albumine, of plasmine or fibrinogenous matter. This complicated elaboration is necessary: thus a solution of albumine of egg injected into the lungs, into the serous vessels, into the cellular tissue of an animal, into one of its veins, is never assimilated, and passes intact into the urine.[1]

The peptones were not coagulable by heat, but the azotised principles which result from them, and which at last form themselves in the blood, become again coagulable. Electively absorbed by the anatomical elements, they incorporate themselves therewith molecule by molecule, after having undergone a final isomeric modification. In effect, justifying the title of *proteïc* which has been given to them, they become, in final result hæmoglobine, keratine, syntonine, osseïne, etc., all these substances having a chemical composition analogous to that which produces them, but being endowed with very different properties.

But the albuminoïdal substances do not alone penetrate the web of living tissues. They bring with them a number of mineral matters, upon which they act as solvents,—for example, calcareous salts, silica, etc.[2]

The peptones, which have the same property, incorporate with themselves in some degree, certain dissolved substances. In

[1] Cl. Bernard, *Progrès et Marche de la Physiologie Générale en France*, p. 197.

[2] Ch. Robin, *Leçons sur les Humeurs*, pp. 103, 104.

effect, artificially digested, and mixed then with much sugar, they nevertheless offer no reaction if treated with the reactive of Trommer.[1]

The complex azotised matters assimilated by the tissues are not the only ones which the blood furnishes. Other azotised substances are formed by means of the coagulable principles of the blood, through special secretory organs called *glands*. These substances are destined to be the specially active agents, the source of various liquids to which we shall have to refer. Let us cite, as examples, the digestive ferments, pepsine and pancreatine, of which we have just spoken.

It will be seen from the preceding remarks, how difficult it is to strike the balance of the nutrition of an animal. The alimentary substances are, in fact, only utilized after having undergone a succession of elaborations, of metamorphoses, and, as we shall see, before being eliminated, they are elaborated anew. For the anatomical element is not a passive element allowing itself to imbibe like a filter: it is a living agent, endowed with special needs, which it satisfies, under penalty of death, by choosing amongst the substances in contact with it, by forming sometimes completely the substances indispensable to its nutrition, then by decomposing and rejecting them; it is an agent at once of synthesis and of chemical reduction.

It is especially within the nucleus of the anatomical element that its power of chemical attraction and repulsion seems to dwell. In the glycogenical cell, according to Cl. Bernard, the sugar only accumulates while the nucleus is persistent; and in the muscular fibres, the nucleus is persistent as a regulator of the nutrition.[2]

On the contrary, the degeneracy and death of the element would be the consequence of the disappearance of the nucleus.

[1] M. Schiff, *Digestion*, t. ii. p. 154.

[2] Cl. Bernard, *Progrès et Marche de la Physiologie*, etc., p. 99.

CHAPTER VI.

OF DIS-ASSIMILATION AMONGST ANIMALS.

When once the nutriments are elaborated and assimilated, the first half of the nutritive movement is accomplished; but in the living substance there is never any repose, and these same materials, employed so far in a work of reparation, of reconstruction, are in their turn worn out; in their turn they become unfit to figure in a living tissue, and are eliminated under various forms.

The substances expulsed from the animal mechanism, the elements of regression, may be divided into three great categories; (1), mineral substances; (2), mixed substances, altogether mineral in their properties, but not formed spontaneously apart from animal organisms; (3), finally, complex, azotized, but crystallizable compounds. No truly proteïc substances are met with. These last are not normally expulsed from the animal organisms.

The mineral substances are either gaseous, or liquid, or solids in solution. The gases in almost all animals are azote, oxygen, and carbonic acid. A small part of the azote is produced by the complete combustion of a portion of the azotized substances; the larger part is simply aërian azote which, after having penetrated the tissues, is expulsed from them without having fulfilled any appreciable function. But the experiments of M. Regnault have shown that animals, even when in a state of complete repose, and particularly when they have a richly azotized diet, exhale a small proportion of azote, the residuum of the complete combustion of a small quantity of albuminoïdal matters.[1]

The exhaled oxygen also no doubt is derived from the outer air. It is an overplus which cannot be utilized. So far, at least, it has not been shown that the chemical reactions of nutrition set at liberty oxygen gas.

As to the carbonic acid abundantly exhaled from the general surface of the bodies of animals, and especially through their respiratory organs, it is almost wholly derived from the slow combustion, the oxydation, of the tissues. In effect, we cannot, in striking the nutritive animal balance, place any value upon the small proportion of carbonic acid contained in the atmosphere. The carbonic acid exhaled by the animal is a true principle of dis-assimilation ; formed in the very substance of the anatomical elements, it is first sent into the circulatory torrent of sanguiferous animals, to be finally restored to the aërian medium. The inmost mode of formation of the carbonic acid is besides far from being well known. Thus the quantity of carbonic acid exhaled by the lungs does not exactly correspond to the quantity of oxygen absorbed. It seems then that we have not here a simple question of combustion.[2] It has been possible to compare, in the various tissues and organs, the power of the formation of carbonic acid, that is to say, to estimate, in a certain degree, the relative energy of oxydation in the superior animals. The parenchymas of certain very active glands, such as the liver and the kidneys, yield a strong proportion of carbonic acid. The muscular tissue furnishes less of it, and the nervose tissue less still.

It is not in the blood that the greater part of the exhaled carbonic acid is produced, but in the histological elements themselves ; also we find that the blood has always a lower temperature than the tissues from which it proceeds.

As a mineral substance, and an exhaled liquid, we have to mention water, excreted specially through the skin and kidneys

[1] J. Gavarret, *Des Phénomènes Physiques de la Vie*, pp. 177, 178.

[2] Cl. Bernard, *Progrès et Marche de la Physiologie Générale en France*, p. 191.

in a liquid form, through the respiratory organs as vapour. Let us note, by the way, that this pulmonary vapour is charged with putrescible, azotized particles.

But water, in its liquid state, is not excreted alone; it carries with it a quantity of mineral substances, mostly saline, some acid. The salts in solution in the water are carbonates, sulphates, phosphates, etc., certain of which have only traversed the organism to leave it as they entered it; some of them, however, are directly formed in the organism. Carbonic acid, for example, so abundant in living solids and liquids, can unite with the bases which it meets there. On the other hand, it is probable that the combustion of proteïc matters in nutrition leads to the formation of sulphuric and phosphoric acids, which, on their side, also combine with the bases, either directly, or by displacing weaker acids. It is certain, in fact, that these salts always come together in the ashes of proteïc matters directly burned. Finally, it has been proved that all prolonged intellectual labour in man corresponds to an abundant excretion of phosphates through the kidneys, and it is known that the nervose tissue is very rich in phosphorus.[1]

To the mineral acids must be added organic acids; some, not azotized, such as the lactic acid ($C^6H^5O^5,HO$) probably result from the oxydation or evolvement of the hydrocarburets of the fats; others are azotized, and proceed from the oxydation of proteïc matters. These acids, which have a fixed formula, like mineral acids, are only formed in the tissues of animals, and once constituted, they act exactly like the ordinary mineral acids, and unite with the bases to form salts, which are also expulsed by the ordinary ways. The principal of these azotized organic acids are uric acid ($C^{10}H^4Az^4O^6$) very abundant in man, the carnivorous mammifers, birds and reptiles; and hippuric acid ($C^{18}H^8O^8Az,HO$), which is met with most in the herbivorous mammifers.

[1] Byasson, *Essai sur le Rapport qui Existe, à l'État Physiologique, entre l'Activité Cérébrale et la Composition des Urines.* (Thèses de la Faculté de Médecine de Paris, 1868, nº 162.)

It is in the anatomical elements themselves that the urates are formed, and thence they afterwards pass into the blood. They are created particularly in the fibrous and muscular tissues. The formation of uric acid seems to respond to an incomplete, imperfect oxydation of the tissues, at least in man, and yet we find an enormous quantity of this acid in the urine of birds, whose respiration is extremely active,—a paradoxical fact, of which there are many in biology, proving that much remains to be done by physiologists.

We must put in the same category with uric acid a substance yet more important, urea ($C^2Az^2H^4O^2$), proceeding, like the uric acid, from an oxydation of the albuminoïdal substances, but a deeper and more complete oxydation. It is necessary to guard against considering urea as an alimentary residuum; this substance is incessantly and regularly formed, and is produced by the dis-assimilation of the anatomical elements themselves. It is produced even during abstinence, since Lassaigne found it in the urine of a man executed after eighteen days of absolute abstinence; and it exists also in the urine of the newly-born. As it is constantly forming, it is also constantly met with in the blood, which appropriates it afresh from the tissues.

Urea and urates are not the only products of the molecular dis-assimilation of the anatomical elements. There is a whole series of analogous compounds; for, each kind of anatomical element, having its own composition, assimilates and dis-assimilates particular substances. The principal azotized quaternary substances proceeding from proteïc dis-assimilation are creatine, creatinine, leucine, inosic acid, and inosates. All these substances are crystallizable. They are more specially produced by such or such a tissue. Thus creatine seems to be almost exclusively the product of the dis-assimilation of the muscular substance. Leucine is found rather in the tissue of the glands, with or without an excretive canal, in the lymphatic ganglions, and also in the grey substance of the brain.[1]

[1] Ch. Robin, *loc. cit.*, pp. 686—688.

Oxygen is certainly the determinating agent in the composition of all these regressive bodies. Nevertheless, according to M. Robin, there is not here simple oxydation, but oxydation with evolution, and often simple evolvent *catalysis*, in technical terms. It would even be necessary, in relation to this, to establish a kind of antagonism between assimilation and dis-assimilation. The one would operate through metamorphic *catalysis*, and the other through evolvent *catalysis*. These are data which can only be admitted with considerable reservation. In fact we can, in a certain degree, note in passing the substances which enter into an animal organism, and those which issue from it; but, from the ingress to the egress, a succession of reactions and transformations has taken place, which we are far from having traced step by step, since we have not yet even succeeded in giving the definitive formula of the albuminoïdal substances.

CHAPTER VII.

ULTIMATE PHENOMENA OF ANIMAL NUTRITION.

M. Cl. Bernard has recently expressed some general views upon the phenomena of nutrition, supported only by a very small number of facts, but certainly true in a degree, and highly interesting.

According to him, we must altogether reject the idea of direct alimentary assimilation. The histological element he thinks acts not upon the complex substances absorbed by the animal organism, but upon the elements of these substances, which it thus decomposes, not uniformly, but according to its needs, by selecting, as a chemist does in his laboratory, and drawing forth the elements which are indispensable to it from complex compounds with very diversified formulas. The following fact, quoted by Burdach, would support this view:—"John has seen seeds perish when sown in powdered silica, with potash and fresh white of egg; they only germinated when this last was already putrified. Here organic matter only acts as an aliment when it is on the point of returning entirely to inorganic nature."[1] M. Cl. Bernard has himself observed that, without doubt, the saccharine aliments introduced into alimentation favoured the production of the glycogenical matter of the liver, but that other bodies altogether different, such as glycerine, chloroform, etc., did not favour it less. All these bodies simply acted as nutritive excitatives. The hepatic cell appears at first

[1] Burdach, *Physiologie*, t. ix. p. 392.

in the liver without determinate characteristics; then we see an amidonised product forming in its cavity, as starch forms in the chlorophyllian cell, and without doubt by an analogous chemical synthesis.

The larva of the fly, the asticot, may be considered as a bag of glycogenical matter; now this larva is developed and nourished especially by animal food in a state of putrefaction, tending to decomposition, to mineralization. Accurate experiments seem to have shown that this larva develops itself, not at the expense of animal food in general, but of a small part of the extractive substance; nevertheless it fabricates glycogenical matter with this azotized substance.

The glycogenical function is diffuse in the lower animals, and specialised in man and the higher mammifers. M. Cl. Bernard, concludes from this that the case must be the same in the formation of albuminoïdal substances, that consequently there must be animal cells, where the synthesis of proteïc matters is specially accomplished at the expense of the dissevered elements of the nutriments.[1]

It is very certain that the various species of histological elements fabricate varied quaternary substances, since each of them has its special chemical composition, and their assimilation is carried on at the expense of the same sanguineous plasma. But where are the special agents of the albuminoïdal fabrication in the complex animal organism? No one yet knows. Besides, it is incontestable that the animal cannot live without azotized aliments, already containing, in a prepared state, substances analogous to those which constitute the web of its tissues. No animal can aliment itself simply with distilled water and air, in which however it finds wherewith to compose, elementarily, ternary and quaternary substances.

The precise amount of our knowledge of this very interesting

[1] Cl. Bernard, *Des Phénomènes de la Vie Communes aux Animaux et aux Végétaux* (Revue Scientifique, 1874).

question it is very easy to show: verily the chlorophyllian cell makes starch, and probably proteïc substances, by combining the elements of the atmosphere and those of the ascending sap. Verily, also, there are in animal organisms cells which fabricate an amyloïdal substance, either by isomeria or by synthesis. All the rest is in a state of hypothesis.

A very certain fact is, that animals cannot aliment themselves without substances already elaborated either by plants or by other animals, and that it will be the same with man, as long as chemistry cannot fabricate completely, synthetically, the immediate organic and complex principles of which he stands in need.

The synthesis of albuminoïdal substances seems as if it would baffle the science of the chemist for some time longer. Nevertheless, these substances are supremely the materials of life; but their mobility, their instability, that is to say the very qualities which render them suitable to form living anatomical elements, are serious obstacles not only to their artificial fibrication, but even to the determination of their formula, and, up to the present time, chemistry has only succeeded in synthetizing hydrocarburates, or regressive azotized bodies, of which urea is the type.

A very important property of the coagulable albuminoïdal substances proceeds from their very mobility; it is the faculty which they have of transmitting to each other in living organisms and by simple contact their molecular conditions. These are only simple isomeric changes, and it is by this process that each anatomical element, whatever may be its composition, fabricates, at the expense of the plasmas, of the common nutritive juices, the special compounds which it must assimilate.[1]

After the same manner also we may explain the action of the different kinds of virus, as well as hereditary transmission and morbid contagion.

We may observe analogous phenomena in proteïc substances

[1] Ch. Robin, *loc. cit.*, Introduction, p. xiii.

deprived of life. It is thus, for example, that muscular flesh and blood putrefy with great rapidity when brought into contact with the organic particles called miasmata, and that we obtain, by condensation, aqueous vapour from the atmosphere of marshes, or from confined air vitiated by agglomerations of men or animals.[1]

We have spoken, in one of our first chapters, of the great affinity shown for water by albuminoïdal substances. We consequently comprehend what extreme importance this property has in the scheme of nutrition. It is, in effect, one of the principal conditions of absorption and assimilation. In reality, all the anatomical elements of animals, even of those which live in the air, are aquatic organisms. They live in the plasmas and blastemas, as marine animals live in the water. They have their inner liquid mediums, which are their safeguards from the destructive action of the exterior aërian medium. It is in these inner mediums that they take their nutriments and throw off their secretions. Also all the animal membranes lose their vital properties by drying, and the aërian animals, as well as the aquatic animals, can only respire through the intermediation of tissues incessantly lubrified.

After what we have already said of the chemical composition of animals and plants, as well as of nutrition in general, it is self-evident that the ultimate phenomena of nutrition, already sc analogous in the two organic kingdoms, are identical in the animals called *herbivorous* and in those called *carnivorous.* There is only a difference in the constitution of the apparatus charged with the preparation of the alimentary materials. We know besides that many animals are herbivorous or carnivorous indifferently; and we shall see further on that the transformation of a herbivorous animal into a carnivorous one is only a pastime for the physiologist. This transformation takes place spontaneously under the influence of abstinence. Then the urine of herbivorous

[1] Ch. Robin, *loc. cit.*, Introduction, pp. 195—199.

animals, normally muddy and alkaline, becomes clear and acid like that of the carnivorous animals. Besides, if we compare each day the quantity of oxygen absorbed by an animal in a state of inanition, by respiration, with the quantity of the same gas which enters into the composition of the carbonic acid exhaled, we see the proportion established first of all as if the animal had been subjected to an alimentation of fat, then as if it had been exclusively nourished on animal food, but in an insufficient quantity.[1]

In effect, assimilable, complex, ternary, and quaternary substances are necessary to the anatomical elements of animals; but it matters very little whether these substances are derived from the vegetal kingdom or the animal kingdom, provided they have undergone a suitable preparation. If these aliments of exterior origin fail, the nutritive movement continues nevertheless for a longer or shorter time in complex animals, but then it is exercised at the expense of the living substance itself; the organism literally eats itself. The blood first yields all that it can yield of assimilable substances, and repairs its losses at the expense of the tissues, and of the anatomical elements, of which it previously furnished the materials. It takes back what it gave. This is called *physiological abstinence.*

This abstinence may be provoked by various causes. Everything which can disturb the internal mutations of the tissues in such a way as to cause the nutritive expenditure to predominate over the receipts is a cause of physiological abstinence; for example, an ill-chosen alimentation, even if it be copious, to which is wanting any one of the constituent principles indispensable to the organism; or an incomplete reparation, whether it be a pathological defect of the organs, or an alteration of the exterior air (encumberment, confined air, etc.). Abstinence bears simply upon the oxygen then; but we have seen that this gas is indispensable to the utilization of alimentary materials.

In the lowest scale of the organized world, in beings formed of

[1] G. Sée, *Le Sang et les Anémies.*

one cell, or of a group of identical cells, abstinence causes the dissolution of the organisms in the ambient medium, the direct return to the mineral world. This is what indeed takes place in the most complex beings; but here the scene is more varied; for the body of the superior animals is an aggregate of various organs and apparatus, constituted by the grouping, the intrication of anatomical elements dissimilar to each other, peculiar in form and function. The body of the higher mammifers must then be considered as a republic hierarchically organised with division of labour. To one tissue is attached secretion, to another movement, to another sensibility, motility, thought, &c., to all life; that is to say, the movement of assimilation and dis-assimilation, but more or less tenacious, rapid, and energetic. The result is that towards abstinence each of these tissues takes a different attitude, that when once the food is cut or furnished in an insufficient quantity, our histological citizens are more or less slowly re-absorbed and perish. Those which are first destroyed are either the least robust, or those which have the most urgent nutritive needs, those which, from their essence, are the seat of the most rapid mutations. They dissolve, and the immediate principles which constituted them are re-absorbed and pass back into the circulatory torrent; for there is circulation in all the very complex animals. The blood gives back to the tissues that which it has just taken from them; but it takes from some to give to others; it first satisfies the most exacting to the detriment of the more patient. In short, the more voracious and active of the anatomical elements devour those which are more feeble, and live at their expense.

According to Chossat,[1] whose work is the authority upon this question, the death of an animal from total or gradual abstinence takes place when the animal has lost about forty-hundredths of its total weight, or sometimes fifty-hundredths, if the animal is fat. The following table will indicate the unequal repartition of loss in the different organs:

[1] Chossat, *Recherches Expérimentales sur l'Inanition*, 4°, Paris, 1843.

Parts losing more than the average.		Parts losing less than the average.	
	0,400		0,400
Fat	0,933	Stomach	0,397
Blood	0,750	Pharynx, œsophagus	0,342
Spleen	0,714	Skin	0,338
Pancreas	0,641	Kidneys	0,319
Liver	0,520	Respiratory apparatus	0,223
Heart	0,448	Osseous system	0,167
Intestines	0,424	Eyes	0,100
Locomotive muscles	0,423	Nervous system	0,019

A very curious fact arising out of this table, is the feeble deperdition undergone by the nervous centres, even at the moment of death. We may say that, before dying, they have really devoured the other organs, and it is to their almost complete integrity that we must trace the relative integrity of the intellectual faculties observed in men in a state of inanition up to the moment of death.

The physiological disturbances proceeding from abstinence are not less interesting to study. The temperature of the body gradually lowers, in proportion as the nutritive movement is retarded. Chossat, experimenting upon various animals, guinea-pigs, rabbits, crows, turtle-doves, pigeons, chickens, has found 3 degrees to be the daily average lowering of the temperature, with a more considerable depression the last day. At the moment of death, the average heat of the body falls to 24°·9.

Normally the nutritive movement is retarded during the night, and the temperature, which strictly depends upon it, falls in hot-blooded animals to about 0°·74; but during abstinence this oscillation is always augmented with regard to amplitude and duration, and the thermometric depression reaches 3°·28.

The secretions and excretions are also closely connected with the nutritive movement, since their principal function is definitively to expulse the products of dis-assimilation or to favour the absorption of the aliments; it is then very natural that during abstinence they should be diminished or suppressed. The liver ceases to form animal starch, which normally contributes to its fabrication of sugar (Cl. Bernard). The urine still contains urea, the principal residuum of the combustion of the proteïc substances of the tissues; but naturally, it contains less and less of it.

The digestive tube shrinks; sometimes even it becomes ulcerated by the local re-absorption of its anatomical elements.

Vital combustion being gradually extinguished, respiration, which furnishes it with the necessary quantity of oxygen, or at least with the major part of this oxygen, performs its functions with gradually decreasing energy. The movements of the muscles are retarded. It is the same with the beatings of the heart, which, in a man in a state of inanition, have been known to fall to 38 per minute. The wearing out and rapid re-absorption of the very tissue of the heart is now added to the general nutritive depression.

The functional disturbances of the nervous centres are still very imperfectly determined. They are sometimes absolutely wanting in man. In animals, however, notice has been taken of insomnia, first a period of agitation, then a period of stupor, finally one of fury; in man, there are often hallucinations.[1]

Abstinence is naturally better supported when the movement of life is less energetic. Burdach[2] has in an interesting manner connected with this subject some facts observed in the most various animals. After six weeks of abstinence, a snail had only lost an eleventh part of its weight; fresh-water polypi can live without food for five and even ten years; toads, two years; Chinese gold-fish, several years; salamanders, six months; a toad shut up in a porous vessel tightly closed, the vessel being surrounded with earth saturated with humidity, was still living, though very thin, at the end of two years. It had been kept in

[1] Collard de Martigny, *Journal de Magendie*, t. viii.

[2] *Traité de Physiologie.*

an almost uniform exterior temperature.[1] On the contrary, birds, whose nutritive movement is intense and rapid, succumb very quickly. One day of abstinence kills a sparrow, three days a thrush, &c. The same law is verified in the same animal organism, according to the various ages. Thus, in middle life, man can resist complete abstinence for one or two weeks. The child succumbs sooner, and the old man much later.

Everything which retards the quickness of the nutritive exchanges helps to support abstinence; for example, repose in bed, sleep, and the substances which provoke it.

Without having for the animal the primordial importance which it has for the plant, solar light is nevertheless a powerful excitative of life in the animal kingdom also. Moreover, abstinence is better supported in darkness. Eight miners shut up for thirty-six hours in a coal-pit without food said they had not suffered from hunger.

Nevertheless, in all animals death is sooner or later the result of abstinence, and in all, it takes place when the organism has lost a part of its weight, always perceptibly the same for all the species. If the animals known as cold-blooded live much longer than others, it is only because their diurnal loss is normally much less. They expend less, and consequently their capital is not so soon exhausted.

[1] Cl. Bernard, *Leçons sur les Propriétés des Tissus Vivants*, p. 49.

CHAPTER VIII.

OF THE MODIFYING AGENTS OF NUTRITION.

VARIOUS influences modify the nutritive movement, more or less, for better or worse. These influences dwell either in the organism itself, or outside it.

A. It might be said that nutrition is the object of life, if life has an object; but it has none, since it is simply the result of a fortuitous concurrence of cosmical, geological, climacteric, and even orological facts. It has not always existed on the surface of our little planet. It will be extinguished there one day. But, if we cannot say that nutrition is the object of life, we are justified in saying that it is the basis thereof. A given organism lasts, prospers, develops and reproduces itself the more surely the better nourished it is, and it is the better nourished the better organised it is, that is to say, in more complete harmony with the exterior medium. Also, as we shall see hereafter, the various organic systems of every complex animal are regulated with regard to nutrition. The result of the special functionment of each of them is to render nutrition possible and easy. Here is a series of organic harmonies, of which we shall give an exposition farther on. We shall see that there are systems of organs adapted, some to render the aliments absorbable, others to convey the nutritive substances to the tissues; whilst certain others have the function of ventilating to some extent the anatomical elements, furnishing them with oxygen, and relieving them of their carbonic acid. Finally, upon others devolves the task of relieving the animal organism of the non-volatile residue of denutrition, &c.

But in order that the action, the utility, of these varied organic mechanisms, their influence on nutrition, may be thoroughly comprehended, it is first of all necessary to describe them, which shall be done in its place. We only wish to point out here, in passing, some interesting facts relative to the influence of the nerves and nervous system upon nutrition; for it would be difficult to place these facts elsewhere.

It is certain that nutrition is, to some extent, independent of the nervous system. In effect, we see it in operation in the plants, in the lower animals, destitute of a nervous system, and in the embryo, which, as yet, has none.

Professor Schiff has seen absorption carried on by the stomachal mucous membrane, after the section of nearly all the nerves which communicate with the stomach[1] (solar plexus destroyed, section of the threads of the pneumo-gastric nerve which are joined to the œsophagus).

Claude Bernard has likewise observed that the section of the nerves in the wing of a pigeon does not prevent the feathers from shooting.

But there are very curious contradictory facts. We shall see farther on that the expenditure, the afflux of blood in the finest sanguineous canalicules, the capillary vessels, is regulated by the influence of certain special nervous threads, called for this reason *vaso-motors.* Under the influence of these nerves, the capillary vessels contract or dilate; now these vessels form a net-work of meshes more or less compact in the web of nearly the whole of the organs and tissues, and the afflux of the nutritive sanguineous plasma is more or less abundant, according as their calibre enlarges or contracts. Here then, the influence of the nerves is necessary. Moreover, if we paralyze, by section, the vaso-motory nerves of an organ, we see that this organ can no longer support abstinence as easily as the others. It seems that it may become

[1] M. Schiff, *Leçons sur la Physiologie de la Digestion*, 1867, t. ii. p. 404.

accustomed to a more abundant nourishment, and if this fails, it becomes inflamed and suppurates.[1]

That paralysis of the capillary vessels, and consequently their persistent dilatation, occasions an excessive afflux of sanguineous plasma, is proved by the following experiments of Professor Schiff. He cut one of the lower maxillary nerves of a mammiferous animal; then, at the end of a certain time, having killed the animal, he discovered an increase of volume in the bone of the jaw, with rarefaction of the osseous tissue.

In a dog two months old, the same physiologist cut the two principal nerves of one of the posterior members, the sciatic and crural nerves. Four months afterwards the bones of the member were hypertrophied, and new osseous productions had formed in the foot. This result is due to the section and consequent paralysis of the nervous vaso-motory fibres intermixed with the other fibres of the nerve.

The same experimentalist has also observed paralysis of the vaso-motory fibres of the ear of a rabbit cause abundance of hairs to grow upon that ear.

B. The exterior medium also powerfully influences nutrition by its principal physical modalities, especially by light and heat.

Doubtless, light has not that primordial importance for animals that it has for chlorophyllian plants. The animal kingdom has representatives at the bottom of the Atlantic, at a depth of nearly two kilometres, whither a solar ray never penetrates, but there it is only represented by inferior animals. Nevertheless, in the fauna of caverns, vertebrated species are found; but they are modified, and modified for the worse, by the absence of light. Very commonly, the organs of sight are atrophied. Besides, very frequently, the skin of the animals of caverns is destitute of the colouring pigment, as has been observed in a blind fish of the cavern of the Karstgebirge, and in the *proteus anguinus* of the Mammoth Cave.[2]

[1] Cl. Bernard, *Rapport sur les Progrès de la Physiologie Générale*, p. 215.

[2] Leydig, *Histiologie Comparée*, p. 94, 95.

Every one knows that a man obliged to live, not in complete darkness, but only in a dim weak light, grows pale, languid, etiolated. In this case, as we have proved above, when speaking of abstinence, the nutritive exchanges are retarded.

To the animal, heat is much more important than light. Without wishing at present to approach the study of animal heat, we can already prove that animal and vegetal life is only possible between certain limits of exterior temperature. Doubtless animals, and especially the superior animals, do not yield with a readiness as perfect as that of plants to the temperature of the exterior medium, but they only resist it up to a certain point. In all animals life is extinguished when the interior temperature, that of the anatomical elements, remains for some time below zero, for then the humours congeal. In the higher mammifers death takes place even sooner. The child in that state of slow asphyxia which has been called *œdema of the newly-born*, dies when its temperature descends to about 20 degrees centigrade.[1] Experimentalists also indicate this temperature of 20 degrees as being the minimum of inner temperature. Nevertheless, the temperature of the blood may descend to two or three degrees with impunity in a hibernating animal, just as this animal can then sojourn without injury in an irrespirable gas; this is because the nutritive exchanges are in this case extremely retarded. The anatomical elements can consequently be satisfied with a small nutritive ration, which, at a higher temperature, would not suffice to keep them alive.[2]

Neither is the superior limit very high. A cold-blooded animal dies when a temperature 30 degrees above zero penetrates its anatomical elements. For the mammifers the highest limit is 45 degrees; it rises to 50 degrees for birds. Without doubt, an animal can live a certain time in these extreme temperatures, but only a very short time, and on condition that the ambient

[1] Ch. Letourneau, *Quelques Observations sur les Nouveau-nés*, 1858. (*Thèses de la Faculté de Médicine de Paris.*)

[2] Cl. Bernard, *Leçons sur les Propriétés des Tissus Vivants*, p. 50—53.

atmosphere is dry. Then, in fact, cutaneous evaporation produces a certain degree of coolness, which makes a momentary compensation. Thus, a mammiferous animal, in a dry hot-house will support, during some minutes, temperatures of 80, 100, and even 120 degrees, but only when its inner temperature does not exceed by more than five degrees the normal temperature. It first pants; then at the end of a very short time, it falls instantaneously dead, without agony. When a temperature of 45 degrees has thus invaded the anatomical elements the muscular sap coagulates, and the animal dies in that state which is known as *cadaveric rigidity*. A thermometer then quickly introduced into the heart marks about 45 degrees.[1]

[1] Cl. Bernard, *loc. cit.*, p. 231.

[NOTE. A *kilomètre*, that is to say, a measure of a thousand *mètres*, is equal to rather more than five hundred fathoms.—TRANSLATOR.]

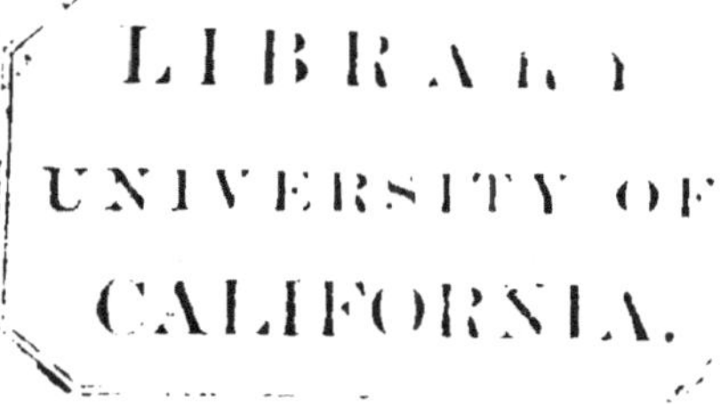

CHAPTER IX.

OF THE MEANS OF ANIMAL NUTRITION.

Every organised being is the seat of an incessant and double movement of assimilation and dis-assimilation. In order that this double movement may be accomplished, and this is the fundamental condition of life, it is necessary that it should be constantly alimented by substances so composed, that they can be incorporated in the organism.

In almost all vegetals, and in a certain number of the lower animals, the work is relatively simple, since they draw food direct, and without preparation from the ambient medium. In the superior animals, the phenomena are complicated. Before us are then brought very diversified organisms, where the division of physiological labour is pushed farther and farther. But in this case, the anatomical elements seem to have lost in vegetative energy what they have gained in fineness. In order to live, they need to absorb highly elaborated organic substances, and, for this purpose, the organism is furnished with a special apparatus, whose office consists in forming a kind of physiological kitchen, to modify the elements, to accomplish the first chemical transformation, which renders them more suitable for assimilation. This apparatus is the *digestive system.*

But this chemical elaboration would be useless if the alimentary substances, once modified and transformed into *nutriments,* did not come in contact with all the anatomical elements, superficial or hidden, in the suitable physical and chemical conditions, if the residua unfit to sustain the vital movement were not given back to the elements in the degree of their formation. A special apparatus performs the task of conveying to each anatomical element, whatever may be the situation and function of that element in the organism, new materials, while at the same time it carries away, sweeps off the nutritive residua. This apparatus is the *circulatory system.*

Moreover, an incessant atmospheric ventilation is an indispensable condition of the ultimate nutritive exchange. Every living substance needs to combine with the oxygen of the air, and to exhale carbonic acid, which is one of the principal products of this combination. The special apparatus which provides an easy entrance and issue to the circulating gases in the complex organism has been called the *respiratory system.*

But the respiratory apparatus can only provide an issue for gaseous products; now denutrition, as we have seen, originates a whole family, yea, even several families, of regressive products, salts, quaternary substances. These bodies are taken afresh from the anatomical elements by the circulatory system; but this system would be incapable, by itself, at least in a very complex organism, of removing them definitively beyond the frontiers of the living histological federation, which constitutes every superior organism. For this there must be special apparatus, *excretory glands.*

For the sake of clearness of exposition, we are obliged thus to cut the complex organism into a certain number of great divisions; but in reality, in the living being, all these functions are bound up together, are indispensable to each other, and are carried on simultaneously, to such a point, indeed, that, to understand the working of one of them, it is almost necessary to know the working of them all. Consequently, in giving an exposition of them successively in the order of our enumeration, we shall be compelled to leave behind us, on the way, many gaps, many obscure points, which will only disappear in the course of the exposition. Finally, when we shall have passed in review digestion, circulation, respiration, and excretion, it will remain for us, in order to complete the description of the adjuvant means of nutrition, to examine the organs of motility and innervation, which, while being more specially consecrated to the life of relation, are, in one sense, not less indispensable to the exercise of the great functions mentioned before. For, in complex organisms, each anatomical element, while having its own existence, and a certain degree of nutritive and functional independence, nevertheless does not cease to be closely and wholly united to the other histological citizens of the confederation. There is unity in diversity.

Before commencing the exposition of the great physiological functions, we have yet a general remark to formulate. In organised beings, from the lowest to the highest, the most differentiated, there is a graduated hierarchy. From the physiological confusion which exists at the lowest step of the ladder, we pass, step by step, through a series of organic models, better and better finished, to the most perfect specialisation. Nothing is more interesting than this seriation of organs, especially from the point of view of the great doctrine of evolution, still so much discussed, but which more and more vivifies all the branches of natural history. Now, however succinct may be the general picture of life which we have undertaken to draw in this book, the reader will nevertheless find there all the great features, the principal outlines of the animated world. We shall try then to describe briefly, but clearly, the various degrees, the successive advances of each great function in the animal world, the only one in which organic specialisation has attained a high degree of perfection. Apart from all application to the doctrine of evolution, the result of this mode of proceeding will be to give a more complete and exact idea of each function.

CHAPTER X.

OF THE DIGESTIVE APPARATUS IN THE ANIMAL SERIES.

Digestion is the introduction in mass of aliments into a special organic cavity where these aliments are analysed, elaborated, and absorbed, leaving usually a residuum which is afterwards expulsed. We see that this function does not exist normally in the vegetal kingdom. At the most, we can make a slight exception for the carnivorous plants. It is very interesting to follow along the animal series, the specialisation more and more perfect of this grand function.

Fig. 3.
Amœba Sphærococcus at the different degrees of its evolution. A, amœba encysted. Protoplasmic mass (*c*) containing nucleus (*b*) and nucleole (*a*); enveloping membrane (*d*); B, amœba come forth from the enveloping membrane; C, amœba commencing to divide; D*a* and D*b*, amœba totally divided into two independent amœbæ.

In certain very inferior organisms, in certain protozoaries, digestion is as completely lacking as in the vegetal kingdom. In the *gregarine*, for example, the alimentary substances are absorbed in the state of solution by all the points of the surface indifferently.

In the amœbæ we behold the very birth of the digestive function. Alimentary particles glue themselves on the viscous surface of the amœba at any point whatever of this surface; then penetrate by degrees into the central parts of this elementary organism, by digging for themselves, in some sort, a temporary digestive canal, which closes behind them. Finally, the aliment dissolves; its molecules incorporate themselves gradually with the substance of the amœba and the residuum not assimilable is expulsed through some point of the surface. The same alimentary process is observable in the rhizopods with this difference, that the viscous prolongations or pseudopods of the animal roll themselves, first of all, round the alimentary particle. Things take place exactly in the same way in the *actinosphærium* which, however, is an organism differentiated into cells. (Fig. 4).

In the *acinetous* infusoria the radiating appendices clasp the prey, which is usually another infusorium whose soft substance lets itself be liquefied and passes along the appendices as along tubes, to come finally to accumulate in droplets in the parenchyma of the animal.

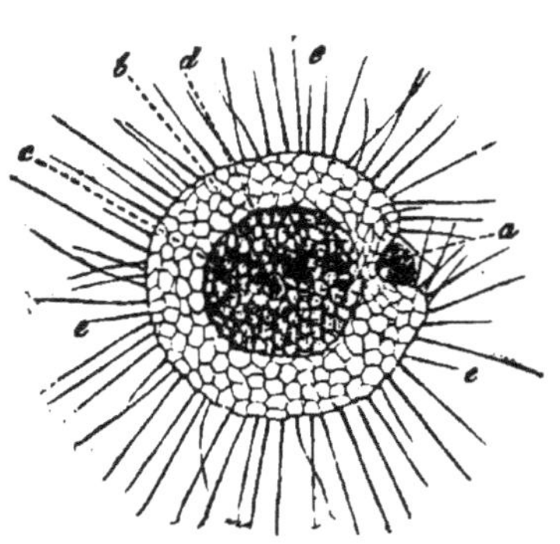

Fig. 4.
Actinosphærium. *a*, alimentary fragment penetrating into the cortical layer; *b*, *c*, central parenchyma; *d*, globules of nourishment; *e*, pseudopods of the cortical layer.

The specialisation which commences already to be discernible in the acinetous infusoria takes increased point and pith in other animals of the same class. There is not yet any digestive apparatus: but there is a special orifice of ingress and sometimes also a special orifice of egress. The aliments penetrate by the orifice consecrated to that use, and which is furnished with vibratile cilia, making for themselves a way through the parenchyma as in the previous case; then the residuum is expulsed.

It is thus that digestion is effected in the infusoria, very common in vegetal infusions,—the *paramaecia*. (Fig. 5.)

In young sponges, there is a permanent cavity which is also the general cavity of the body. This cavity, which at first had no special covering, clothes itself afterwards with vibratile cilia, and ramifies into anastomosed canals by which circulate the alimentary matters. It is, as we see, a rude model of digestive system, and of circulatory system, which are still confounded (nardon). In other sponges there are a mouth, a canal and fine ramifications debouching on the surface of the animal by smaller orifices (sycon, &c).

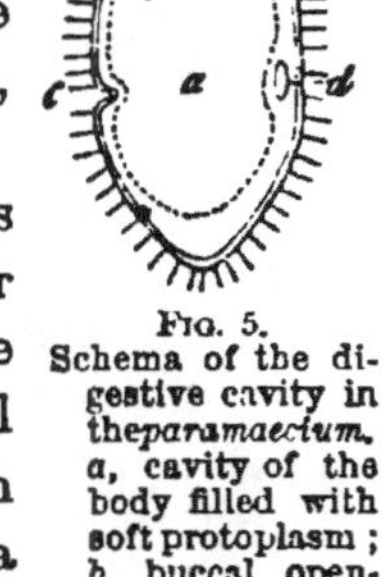

FIG. 5.
Schema of the digestive cavity in the *paramaecium*. *a*, cavity of the body filled with soft protoplasm; *b*, buccal opening; *c*, anus; *d*, *d*, contractile cavities.

Gegenbaur, whom we chiefly follow in this description,[1] places after the sponges the inferior animals, to which he and Haeckel give the name of cælenterates (κοῖλος, ἔντερον). The morphological characteristic of these animals is a body with constant cavity, which may be considered a digestive cavity, sometimes ill differentiated, however; for the freshwater polypus, for example, if we turn it inside-out, after the fashion of a glove-finger, and keep it thus by means of a thread, can digest with what was previously its exterior surface. Nevertheless in many hydrarians we already see developed the division of the digestive tube into three parts, an œsophagian portion, a dilated portion, and a narrowed portion, terminating in a cæcum.

The organised beings live as they can, become what circumstances permit them to become, and all processes are good to what we call nature, provided they attain their object. Thus the hydrarian polypi living in colonies, have an intestinal tube in common, prolonging itself through the whole tribe, and realising thus the most perfect communism.

In the colonies of *Siphonophores* the specialisation has taken a

[1] Gegenbaur, *Manuel d'Anatomie Comparée.*

form more strange still. In them has been established a division into castes, as in our old human communities: but the caste in question is a physiological caste. Certain members of the colony specially adapt themselves for the digestive function. For that purpose they come to bear the form of dilatable sacs, and are in communication interiorly with the digestive cavity common to all the tribe.

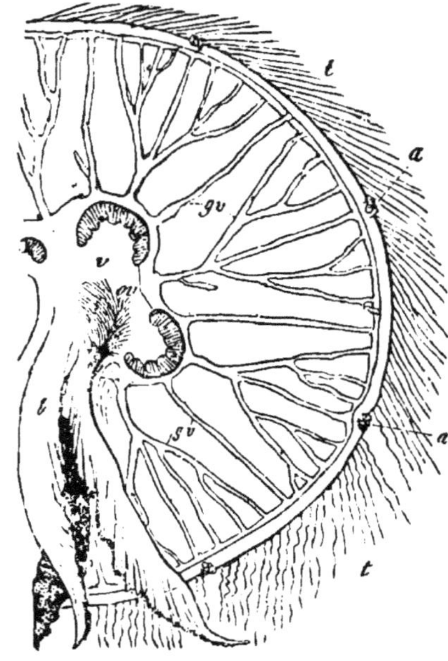

Fig. 6.
Half of *aurelia aurita* seen underneath. *a*, marginal corpuscules; *t*, marginal tentacles; *b*, buccal arms; *v*, stomachal cavity; *gv*, canals of the gastro-vascular system, which ramify toward the edges and throw themselves into the circular canal; *ov*, ovaries.

In the medusæ, and also the actiniæ, the digestive and circulatory systems are still confounded, reducing itself to a median cavity, whence start canals which unite on the edge of the ombrelle to form a circular canal. (Figure 6.)

The bryozoaries have a mouth surrounded with tentacles, an enlarged intestine, sometimes furnished with dentiform projections destined to mastication. Occasionally there exists a sort of stomach with orifices of ingress and egress (cardia and pylorus). In the annelid we find a digestive tube nearly complete with two orifices. The digestive tube of the lumbric has a powerful muscular portion useful to an animal which feeds on the humus of the soil.

In the tunicates we meet with an œsophagus, a portion enlarged, stomachal, and a rectum. The mouth is often at the bottom of a large sac, whose walls serve at the same time for respiration. It is another instance of physiological confounding. We have mentioned above, tribes of animals bearing an intestine in common: in the compound ascidia there is likewise a cloaca in common where meet all the anus of the colony.

Differentiation takes a sufficiently important step forward in the bryozoaries and the tunicates. The elaboration of the aliments has in them become more complex by the existence of glandular sacs though these are still isolated and rudimentary.

In a certain group of *echinoderms* the progress continues. Sometimes the buccal edges, hardened, play the part of masticatory apparatus: sometimes (echinoids) there are complicated apparatus of mastication. Besides a special secretion proceeds on the internal surface of the intestine, where we remark a covering of coloured cells.

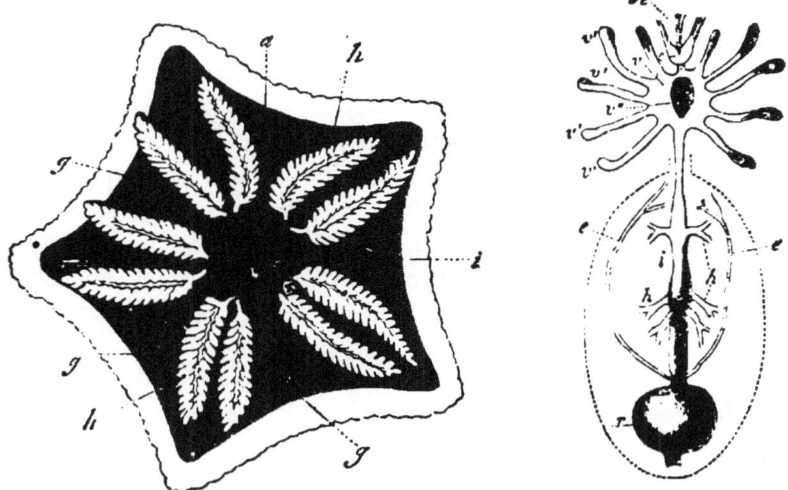

FIG. 7. FIG. 8.

FIG. 7.—*Astericus verruculatus*, open at the dorsal face. *a*, *anus*; *i*, stomach expanded in form of rosette; *h*, radiating and tubular appendices of the intestine; *g*, genital glands.
FIG. 8.—Digestive organs of a spider. *œ*, œsophagus; *c*, superior œsophagian ganglion, cerebroidal; *v*, stomach; *v′*, lateral prolongations; *v″*, appendices directed upwards; *i*, median intestine; *r*, intestinal extremity enlarged into cloaca; *hh′* openings of the liver into the intestine; *e*, urinary canals.

The asteroids, properly so called are without anus, and have a stomach stellated like their body. The stomach is in effect furnished with caecal appendices extending by pairs in each radius. (Fig. 7.)

In the sea-urchins and the holothuria there are a mouth, teeth, jaws, a distinct intestine, an anus.[1]

In the grand class of the arthropods (crustaceans, arachnida, myriapods, insects), there exists a digestive apparatus regularly and definitively constituted. (Fig. 8). Often, especially in the crustaceans and the insects, the intestinal epithelium is clothed with a hard layer of chitin, sometimes emitting projections destined to crush the aliments. In the crustaceans, certain anterior parts become buccal pieces. The glandular system is enriched and diversified. The crustaceans have a voluminous liver, full of a yellow-greenish bitter juice. This liver decomposes into ramose cylinders.[2] Most insects have also œsophagian glands tubulated, enrolled, lobated, ramified, and which are supposed to be salivary. Other glands forming simple or ramified tubular sacs, debouch into the median intestine; it is conjectured that they represent the liver of the superior animals. In insects, the secretion has still a miscellaneous character. In the long slender and flexuous tubes, which are usually inserted at two distinct points of the intestine, and which are called the *vessels of Malpighi*, there is secreted by the side of the urine, another matter probably of a bilious nature. The yellowish canals are believed to represent the biliary canals.[3]

In many arthropods we already remark that the alimentation considerably influences both the form and the dimension of the digestive apparatus. Carnivorous animals have a digestive apparatus conspicuously and comparatively short. The larva of the butterfly, which makes an enormous consumption of aliments, has a very wide intestine, while the butterfly itself, which eats little, and only liquid aliments, has a long and slender digestive tube.

Certain insects (bombyx, ephemerides, œstri), which are very voracious in the state of larvæ, are, in the adult state, destitute

[1] Dugès, *Physiologie Comparée*. [2] Dugès, *loc. cit.*

[3] Leydig, *Histologie Comparée*, p. 336. Dugès, *loc. cit.*, t. III. p. 400.

of organs of manducation. Wholly destined for generation, they cannot take any nutriment; hence the brief duration of their life.[1]

If we except the *brachyopods*, whose intestine terminates in an imperforated cæcum, the *mollusks* have a very complete digestive system, and comparable, in a certain measure, with that of the *vertebrates*. In the *cephalopods*, we find a long œsophagus, an enlarged stomach, an intestine with circumvolutions, and a rectum. In the *littorine* we observe an arrangement of the stomach, which exists in certain vertebrates; there are a cardiac part and a pyloric part, separated by a salient fold. Sometimes the stomach is furnished with triturating hooklets of varied form. But it is especially by the development of the glandular appendices that the stomach of the mollusks is distinguished from that of the animals hierarchically inferior. These organs, in effect, go on perfecting and complicating themselves more and more in the diverse families of mollusks, and especially in the most advanced of the mollusks, the aquatic cephalopods (Cuvier), there are œsophagian salivary glands with short cæcums, a liver developed, compact, divided into lobes, provided each of them with an excretory conduit, and all these conduits open together or separately at the origin of the median intestine, or into the stomach. The liver of the mollusks functionates essentially like that of the vertebrates. That gland is in this class very voluminous, and it fabricates sugar at the expense of the blood though in the mollusks this liquid is destitute of globules.

From the point of view of the general form of the digestive tube, properly so called, the vertebrates have signal fellowship with the mollusks. However in them the degree of organic specialisation is more elevated still. Thus the buccal cavity which is not divided in fishes and the amphibians, commences in reptiles to be sectionised into two divisions, a nasal or aërian division and a buccal or digestive division.

Corneous teeth, analogous to the appendices of the same kind

[1] Dugès, *Physiologie Comparée*, p. 279.

existing in the inferior animals,—are also found on the buccal orifice, in form of a cupping-glass, of the cyclostomes. The amphibians have likewise analogous organs on the edge of their jaws.

In the fishes, the *ganoids* and the *teleostians*, there are teeth on the jaws, the palatal bones, the vomer, the hyoid bone and the bronchial arcs; the *selacians* have none except on the jaws. (Fig. 10.)

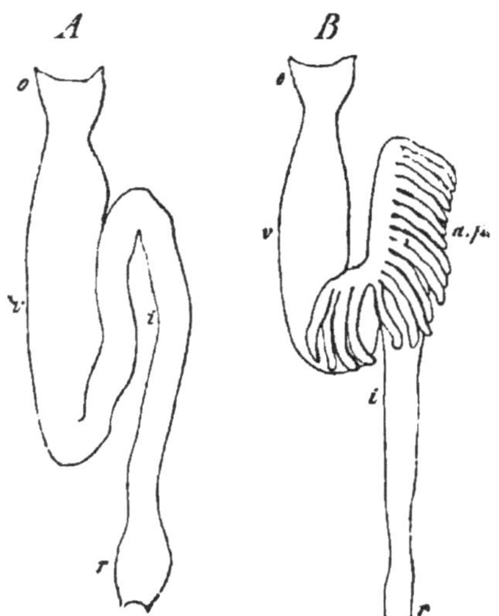

FIG. 9.
Digestive tube of fishes. A, *gobius melanostomus*. B, *salmon*; o, œsophagus; v, stomach; i, median intestine; ap, pyloric appendices; r, rectum.

FIG 10.
Buccal opening of the *petromyzon marinus*, with corneous teeth.

The teeth of fishes, of amphibians, of reptiles fall and are renewed during the whole of life.

The three divisions of the digestive tube, properly so called, grow gradually more perfect in proportion as the animal rises in the hierarchy of the vertebrates.

In fishes the first part of the digestive tube has direct continuity to the stomach without difference of diameter (Fig. 9).

This first portion or œsophagian portion is only distinguished from the stomachal portion by characteristics drawn from the structure of the mucous membrane. Its function, however, is still imperfectly specialised. In effect, at its origin, in the portion which may be called *pharyngian*, the cavity is in part circumscribed by the branchial arcs and has consequently respiratory uses.

The proteus (amphibian) has a conformation more inferior still; there is no trace in the animal of stomachal dilatation.

The first portion of the digestive tube of birds is characterised by diverse dilatations: the first, annexed to the œsophagus, is the crop. Two other divisions are more specially stomachal. They are the *gizzard*, very rich in glands, then another very

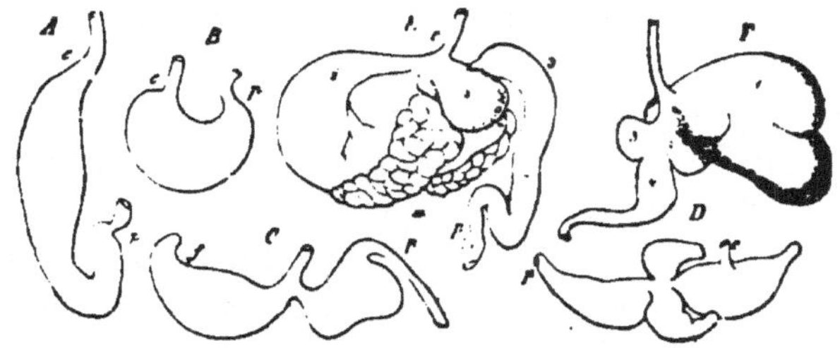

FIG. 11.
Stomachs of different mammifers. A, *seal.* B, *hyæna.* C, *hamster.* D, *lamentine.* E, *camel.* F, *sheep.* 1. paunch. 2. bonnet. 3. manyplies. 4. caillette. *c*, cardia, *p*, pylorus.

muscular dilatation, especially in the granivorous birds, and which is sometimes clothed with a corneous layer. This last pouch fulfils especially a mechanical office.

It is only in the mammifers that the stomach becomes regularly transversal (Fig. 11). At the same time it has enlarged, and its walls have not everywhere the same structure. Therein are distinguished, as in man, a *cardiacal* portion connected with the œsophagus, and a *pyloric* portion debouching into the intestine. But in certain families of mammifers the division is not merely, as in man, characterised by a difference of structure. Every one knows the stomach with four compartments of the ruminants: the *paunch*, the *bonnet*, the *manyplies* and the *caillette*, this last division being more specially charged with the secretion of the gastric juice (Fig. 11).

The stomach of the vertebrates debouches into the intestine properly so called, or rather into the first part of that intestine, the duodenum, and has continuation in the small intestine. It is however distinctly separated from it by a membranous fold called the valvula of the pylorus.

In its region nearest to the stomach the intestine receives the excretory conduits of two of the most important glands, the *liver* and the *pancreas*. The median or small intestine is, with some exceptions, longer in the herbivorous than in the carnivorous vertebrates. There are exceptions to this rule, but they are more apparent than real; for, in effect, the development in width usually comes to compensate for the defect in length. The mole, the cetaceans, though carnivorous, have very long, but at the same time very narrow intestines.

Caterpillars have an intestinal tube, which is not much longer than their body, but it is nearly as wide.

The last portion of the digestive tube, the large intestine, really develops itself in length and breadth only in the *amphibia*. Beginning with reptiles, the large intestine is furnished on its passage with a dilatation called *cæcum*, much more developed in herbivorous than in carnivorous animals.

In the termination of the intestine we likewise remark an unequal differentiation in the diverse vertebrated groups. In the selacians, the amphibians, the reptiles, the birds, and the monotrematous mammifers, the large intestine opens into a cavity called *cloaca*, into which debouch the urinary conduits, and the genital conduits.

We have not here to give a detailed description of the structure of the digestive tube, and we content ourselves with recalling that, in the vertebrates, its walls are composed fundamentally of three tunics encased in each other, and intimately connected. They are, from within to without, a mucous membrane sometimes furnished with tufts or *villosities* clothed besides with cells comparable with the cells of the epidermis, and called *epithelial cells:* a muscular layer of which the contractile elements are in part longitudinal, in part arranged as a ring round the digestive tube: a layer of cellular tissue, which gives a surface of insertion to the contractile elements, and to the canal a sufficient degree of solidity and of resistance. It deserves remark that the villose tufts which garnish the internal face of the stomachal and intestinal mucous membrane, to the number of many thousands the square inch in man, are lacking in the invertebrates. Lastly to the digestive apparatus are annexed glands, some of them voluminous, having their body outside of the digestive wall, which only their secretory conduits traverse, others in the very substance of that wall.

To conclude the very general morphological description of the digestive tube, it remains for us to say a few words on those glands.

Of the Glands of the Digestive Apparatus.

It is especially by the glandular apparatus that the digestive system of the vertebrates in general and of the mammifers in particular is distinguished from that of the inferior animals. The digestive glands, in effect, are more numerous, more differentiated as to function, and of a more complex structure.

The salivary glands are numerous, in clusters granulous and closely pressed. Already in the tortoise we find under the tongue a pair of glands which are probably salivary glands. Analogous glands, but larger still, exist in birds. In the mammifers they are distinguished into three couples, the *maxillary*, the *sublingual*, and the *parotid* glands. It is in the herbivorous mammifers that the three pairs of salivary glands attain their greatest volume, and we shall see that this fact is in relation to the kind of alimentation.

As Claude Bernard has shown, these salivary glands, though

very analogous to each other, secrete salivas of diverse species. In general the salivary glands are wholly lacking in the aquatic animals.

In the salivary glands, the organs of secretion, the glandular elements are agglomerated; but in the stomach of the superior vertebrates these elements are separated and hidden in the substance of the mucous membrane. Their number, moreover, is immense; they are blind glands and separated from each other only by a thin layer of cellular tissue. These small glands are related and proportioned to the numerous capillary vessels, and to the nervose threads. Besides the glandules of the stomach there are many others, scattered over the whole extent of the digestive canal. They assume variable forms: sometimes they are simple depressions, small purses, small tubes: sometimes they ramify.

But the most important glandular appendix of the digestive canal of the vertebrates is assuredly the liver.

In the most inferior vertebrate, the one which in a certain measure connects the vertebrates with the mollusks, the *amphioxus*, the liver is represented only by a cæcum of a greenish tint, reminding us of the simple *culs-de-sac*, short, not ramified, which seem to play the part of liver in the invertebrates. It is also under an analogous form, under the form of a pointed cone, that the liver originates in reptiles, birds, and mammifers.[1] In the superior mammifers, the liver—very voluminous—is, abstraction being made of the cellular tissue, essentially constituted by vessels and nerves, first of all by special cells, charged, as we have seen, to fabricate a glycogenous matter. These cells are irregular, rounded, or polygonal, have a simple or double nucleus with nucleole (Figs. 13 and 14). Their contents, finely granulous, can include accessorily granules of fat. Between these cells pass very fine capillaries, emanating from the great venous trunk, or vena portæ, coming from the intestine (Fig. 12).

[1] Leydig, *Histologie Comparée*, pp. 401—410.—Gegenbaur, *Manuel d'Anatomie Comparée*, p. 752.

Such is, schematically, the elementary structure of the glycogenous hepatic organ, forming in itself the major part of the liver. But by the side of this organ, the former of sugar, and which we must consider as a gland without excretory conduit, a sanguineous gland, there is another organ, the secreter of bile, constituting, schematically also, a gland of clustering form. It consists of small groups of glandular cæcums, agglomerated in the fashion of fern leaves, and dispersed along the branches of a greatly ramified excretory conduit, into which they open and secrete. The wall of the glands and of the conduits is clothed with epithelial cells of divers form.

FIG. 12.

Capillary network of the liver, injected from the super-hepatic veins.

Between the glycogenous elements and the biliary elements there is only a relation of juxtaposition. The superior vertebrates offer us, therefore, here an example of organic intermingling such as we find so many of in the inferior animals. In them the liver has separated itself from the renal glands, with which it is confounded in certain invertebrates; but it remains still joined to the glycogenical gland. If, as some partisans of the doctrine of evolution believe, the present superior vertebrate is not yet the last term of organic progress and differentiation on the earth, the two glandular apparatus may separate in the future in the more perfect being destined to arrive as successor.

The hepatic biliary conduit or *choledochal canal*, debouches into the first portion of the small intestine or *duodenum.* In this same region and very near the origin of the choledochal canal opens, in man, the principal excretory canal of another and very important gland of cluster-shape, formed also of small clusters or *acini*, each diverticule whereof is terminated in a *cul-de-sac.* It is also clothed with epithelial cells. We shall have to speak of the digestive office of this gland called *pancreatic.*

In certain vertebrates the pancreatic canal opens at a certain

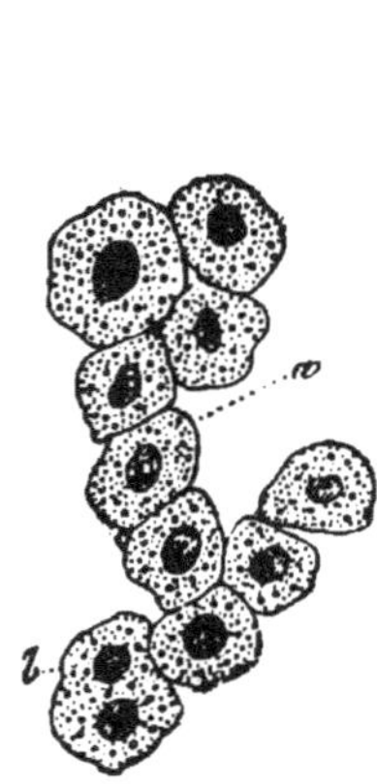

Fig 13.

Isolated cells of the liver; *a*, with simple nucleus, *b*, with double nucleus.

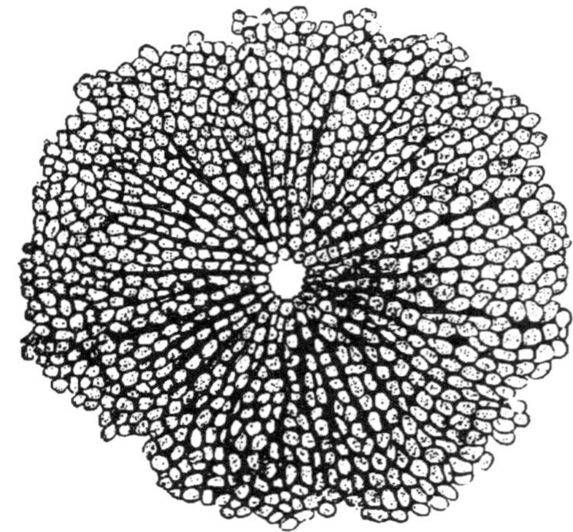

Fig. 14.

Arrangement of the cells of the liver in a lobule cut transversally, with the section of the hepatic vein in the centre.

distance from the hepatic canal, but generally below, about from 30 to 50 centimetres (rabbit, hare, beaver, porcupine, ostrich).

The pancreatic gland already exists in the most inferior vertebrates—the fishes: but in them it is generally smaller and of a simpler structure. Its *culs-de-sac* [blind ends] are wider, more or less independent. The pancreas of divers vertebrates (lizard, and so on) is joined to the spleen as in the superior mammifers, the biliary gland is joined to the glycogenic gland. A degree of glandular entanglement is observable in the *chimaera monstruosa,* in which the pancreas adheres alike to the spleen and the liver.

After the pancreas no large gland figures in the rest of the digestive tube. There are merely small mucous glands scattered here and there. We have already said a few words about them, and, when treating of digestion, we shall have to indicate the part they play.

CHAPTER XI.

OF ALIMENTATION IN GENERAL.

WHETHER animals absorb the complex organic substances, in their cruder state, after simple isomeric modifications, whether, in final analysis, they resolve them into their ultimate elements, certain it is that the most complex substances are at the same time the most easy to assimilate, and also those which have the greatest alible value.

The most important constituent principles of the animal anatomical elements are chemically formed by azotized quaternary compounds, and it is indispensable that analogous principles should be found in a considerable quantity in their aliments; now we know that these quaternary substances are relatively weak in vegetals, that, moreover, they are therein combined with mineral substances. It is then from the animal kingdom itself that animals must especially derive their nourishment. "Eat each other" is for them one of the most imperious nutritive rules. Nevertheless, some of them are herbivorous, but then they must, like the ruminants, have a complex, differentiated digestive tube, a perfected nutritive alembic.

Most of the lower animals, whose rudimentary digestive tube we have briefly described, are limited almost exclusively to animal nourishment. Many of them, moreover, are marine animals, that is to say, living in the midst of a vast organic dilution, full of detritus, dissolved, diluted, or fragmentised. Those among them which are fixed, the polypi, the cirrhopods, even the bivalvous mollusks, are forced to content themselves with sucking in the dissevered organic molecules as these pass along, or sometimes the animalcula, which the aqueous medium brings them. Some aid chance by causing a current of water to pass incessantly into their digestive cavities.

In short, most invertebrated animals live upon animal substances, excepting always the majority of insects; but these latter have moreover a complete digestive system, already differentiated, at the adult age. In the state of larvæ, on the contrary, they often feed upon animal matters. Fish, and even reptiles, are also, for the most part, carnivorous. We must come to birds and to mammifers to find *important groups of animal species*, living habitually upon vegetals.

A general fact, strange at first sight, arises out of the comparative examination of the digestive system in marine herbivora, whatever may be their place in the animal hierarchy. In effect, in the terrestrial animals, the herbivorous digestive tube is more differentiated, more complex, furnished with gastric pouches, with cæcums vaster and more numerous than the carnivorous digestive tube possesses. The porcupine has as many as fourteen stomachic cavities.[1] Now we observe precisely the contrary in aquatic animals, in which we see herbivorous alimentation generally coinciding with a simplified digestive system. Thus in the herbivorous cyprinus there is not even any stomachal distention; and it is the same in the tadpole. The cetaceous herbivora (the lamentine, dugong, &c.) have only one stomach with simple or double dilatation, while the cetaceous carnivora, (dolphin, whale, &c.) have three, four, five stomachs, and the *squalus peregrinus* many stomachal cavities.

The explanation of this anatomical paradox seems to us easy to give. The general principle is, not that every herbivorous animal must have a more complex stomach, but that every animal must possess a digestive system more differentiated in proportion as its alimentation is more varied; now, in the

[1] J. W. Draper, *Human Physiology*, &c., p. 59.

aquatic mediums, the flora is little diversified. It is almost entirely composed of sea-weeds, plants of very simple structure, of a soft consistency, and of a chemical composition everywhere almost identical. The digestive apparatus of the marine herbivora has then only a uniform work, of relative simplicity, to accomplish, while, on the contrary, the aquatic fauna being extremely varied, the alimentation which it furnishes resembles it; and consequently necessitates a digestive system adapted to render absorbable, aliments very dissimilar from each other.

Alimentary variety seems even to be a necessity for certain herbivora, and Magendie has seen rabbits only live fifteen days, that is die of inanition, when compelled to live solely upon one of the vegetals which constituted their ordinary alimentation (carrots, cabbages, barley, &c.).

Besides, we must guard against attaching an absolute value to the denominations *herbivora* and *carnivora*. As M. Schiff points out in his excellent *Traité de la Digestion*,[1] there is no essential difference between the gastric juice of herbivora and that of carnivora. Both disaggregate the vegetal or animal aliments, both dissolve the albuminoidal substances, and everything that is soluble in acidulated water. The herbivorous gastric juice is only a little less active, and, though of equal weight, it digests fewer albuminoidal matters than the carnivorous gastric juice. But, by causing a rabbit to absorb, either through the blood or through the stomach, various soluble substances (peptogens) which have the property of rendering the gastric juice more active, we succeed in making the stomach of a rabbit functionate like the stomach of a dog or a cat. At all events the peptones prepared at the expense of the albumine by the stomach of a herbivorous animal are directly assimilable by the tissues of a carnivorous animal, and if we inject them either into its stomach, or direct into its veins, they are perfectly absorbed, and we do not find them again in the urines, as happens with every non-assimilated substance.

[1] T. II., pp. 183, 184.

The identity of the process and the result of digestion in herbivora and carnivora being demonstrated, there is no cause for astonishment in the facility with which a herbivorous animal can become a carnivorous animal, and inversely.

The herbivora especially become accustomed without much difficulty to an animal diet. Organised for the greater, it costs them little to accommodate themselves to the less.

Numerous facts of this kind have in science an almost common-place notoriety. Spallanzani had accustomed a pigeon to eat meat to such a point that it afterwards refused seeds.

The cows and horses of Iceland feed willingly upon dried fish. Horses and oxen accustomed to feed upon fish have been seen to enter the water to fish for themselves.[1] Besides, all the herbivorous mammifers are necessarily carnivorous during the period of lactation. Also, during this period, the *rumen* of ruminants is not yet developed.[2]

Amongst the carnivora, the most robust, the most typical sometimes actually refuse vegetal nourishment. The tiger, the lion, the eagle, habitually allow themselves to die of hunger rather than touch it. Nevertheless, Spallanzani had accustomed an eagle to eat and digest bread.

Native repugnances can be most frequently overcome by calling in the help of the culinary art. It is, moreover, in a great measure to this circumstance that man owes his omnivorous character. Dogs and cats do not eat corn, but they will eat bread. The rabbit refuses large pieces of raw meat, but it willingly accepts and digests meat minced or boiled.

Certain animals are at once carnivorous and herbivorous, as, for instance, a number of birds. Others are frugivorous in winter, and insectivorous in summer. The small frugivorous monkeys eat insects, and seek eagerly for eggs and for little birds scarcely hatched.[3]

[1] Burdach, *Physiologie*, t. IX., p. 241.
[2] Gegenbaur, *Manuel d'Anatomie Comparée*, p. 748.
[3] M. Schiff, *Digestion*, t. II., p. 187.

Moreover, animal species very closely connected in the taxinomy differ in their mode of alimentation. The polypi, like most inferior animals, usually feed exclusively upon animal matters. Even the hydras habitually reject vegetal substances, which their rudimentary organization does not permit them to assimilate; but the *tubularia gelatinosa* nevertheless feeds upon the flowers and seeds of the water lentil.

Among the coleoptera, the plantigrades, and the cetaceans, certain kinds are carnivorous, and others of the same group are herbivorous, &c.

We have already said above that during abstinence every animal becomes carnivorous. It consumes its own tissues, and even its stomach becomes charged, at the expense of these tissues, with a carnivorous gastric juice, destined for the absent aliments (L. Corvisart).

Before concluding this very incomplete sketch, it will not be useless to say a few words on geophagy.

The earth-worms, the naïds, swallow the humus of the soil; this humus is kneaded by a kind of musculous gizzard, probably diluted with a secretion, and the residuum is rejected in pulpy cords. According to Swammerdam, the larva of the ephemeris eats clay only. Its colour even varies with the colour of the clay.

Among the lower beings, geophagy easily explains itself. The vegetal humus, which the earth-worms eat, the moist clay, which the larvæ of the ephemerides also swallow on the banks of the rivers where they live, contain a large quantity of organic detritus and of soluble salts, which alone are absorbed. Geophagy is less easily comprehended in man; nevertheless, independently of pathological cases, there are numerous examples of it.

The Otomacs eat, or rather ate, every day, a pound and a half of unctuous and ferruginous clay.

Spix and Martius[1] have observed analogous cases amongst various tribes on the banks of the Amazon. According to

[1] *Reise in Brasilien*, t. II.

Labillardière, the New Caledonians ate a white and friable steatite, etc., but only in time of famine.

We may certainly admit that the human stomach, like that of the earthworm, can separate certain mineral substances from organic remains and from salts, but in very small quantity, and it is probable that here geophagy only plays a fictitious alimentary part. It simply distends the stomach, neutralizes more or less its gastric juice, mechanically deadens the feeling of hunger, as carbonic acid, for example, or any inert gas does.

CHAPTER XII.

OF DIGESTION.

A. The object of digestion is the preparation of assimilable substances. The processes are the mechanical division followed by a chemical transformation of the aliments. In final analysis absorption is only exercised on liquefied and dissolved substances.

True digestion, therefore, does not exist among the beings rudimentary, not differentiated, that form the first stage of the animal kingdom. In an inferior degree it exists in animals with a simple digestive tube, having neither at the orifice of the alimentary canal, nor at any point of its cavity, organs suitable for tearing, lacerating, and crushing, and consequently not being able to assimilate easily almost anything but aliments liquefied apart from their organism.

Among animals possessing buccal, pharyngian teeth, or apparatus which occupy the place thereof, there exists usually a special glandular apparatus charged to render alimentary trituration more easy by imbibing the aliments with a liquid called *salivary*. In a more general manner we may say that some kind of saliva exists often when there is an apparatus for the prehension of aliments. Thus the fly emits on the particles it is about to draw in with its proboscis a brownish liquid which dilutes them. Naturally the saliva is the more abundantly secreted the harder is the habitual aliment. Thus the saliva is null or scanty in aquatic animals, whatever they may be (crustaceans, fishes, crocodiles, palmipedes, birds, carnivorous cetaceans). It is, on the contrary, very abundant in the granivorous and herbivorous animals. Schultz observed in a horse that a single parotid gland produced in twenty-four hours 1,678 grammes of saliva.

It is especially in the terrestrial vertebrates that the secretory glandular apparatus exists well developed. It furnishes a liquid, transparent, slightly viscous, usually alkaline, sometimes acid or neutral.

In the superior mammifers, and in man, there are three pairs of salivary glands, called *sublingual*, *submaxillary*, and *parotid*. Claude Bernard has shown that these glands do not secrete an identical liquid. In effect the liquid secreted by the glandules of the buccal mucous membrane seems incapable of transforming starch into sugar, when it is blended with the parotidian saliva. But it accomplishes very easily this transformation when it is mixed with submaxillary saliva.[1]

The chemical agent of this isomeric transformation is a sort of special ferment, *ptyaline*. It is formed in the midst of the closed cells which originate in the substances of the acini of the gland; these, breaking, surrender to the purely excretory liquid the product of their elaborations. Thus we can easily obtain by maceration of the glandular tissue an artificial salivation.

The diversity of the products of secretion of glands in appearance identical permits us to range in the category of the salivary glands the venomous glands of serpents, which seem indeed to form part thereof anatomically. The same reason authorises us not to reject with too much disdain the facts, seemingly legendary according to which the human saliva itself can acquire venomous properties under the influence of a violent burst of anger. The substances produced by living chemistry undergo with an extreme facility isomeric metamorphoses: and these metamorphoses bring along with them new properties.

It is especially during mastication, when there is rapid contact

[1] Draper, *loc. cit.* p. 43. —Cl. Bernard, *Leçons sur les Liquides de l'Organisme*, t. II., p. 239.

of the aliments with the mucous membrane, that saliva is secreted abundantly, and assumes all its characteristics. Mitscherlich observed that the parotidian saliva of man was acid in the state of comparative repose of the gland, but became alkaline during mastication. The accidental acidity of the saliva is probably due to a secondary modification of the starch which becomes first of all sugar under the influence of the salivary diastasis, then evolves into lactic acid. In its turn this acid is carried with the aliments into the stomach, where it contributes its share to the gastric digestion.

The alimentary mass, more or less impregnated with saliva, passes from the mouth into the first portion of the digestive canal, into the pharynx when there is one, lined in certain reptiles with a vibratile epithelium destined to aid in the transfer of the alimentary particles, but clothed in man alone with a simple epidermoidal epithelium called *pavimentous*. From the œsophagus the alimentary mass passes into the stomach, where the most important digestive acts are accomplished.

B. *Gastric Digestion.*—The gastric digestion is the principal act of the digestive function, since, as we are about to see, the stomach is the grand laboratory where is principally operated the transformation of the albuminoidal alimentary matters into absorbable peptones. The state, still so imperfect, of comparative physiology does not permit us to draw here a complete picture of albuminoidal digestion in the whole of the animal kingdom. Nevertheless we know that in the inferior organisms, the invertebrates, for example, are for the most part carnivorous. It is manifest that, spite of the imperfection of their digestive system, they must succeed in effecting this transformation of albuminoids into peptones, which is achieved among the superior vertebrates only in a differentiated stomach provided with special secretory apparatus. In the transparent invertebrates we can demonstrate the transformation undergone by the aliments. Dugès has thus seen, through the tissues, in the planaries and the clepsines the blood swallowed losing by degrees its red colour in the stomach and changing into a greyish homogeneous matter.[1] While worms swallowed by the hydras underwent a similar transfigurement.

Moreover, Schweiger found in the digestive cavity of polypi, and in the canals connecting the individuals aggregated as a colony, a lactescent liquid, probably a nourishing liquid. It is this liquid which is absorbed by the animal, and which circulates through its organism, there where the stomach emits canalicules with cæcums, or even a sort of complete system of canals, as in the medusa.[2]

The isomeric metamorphosis of the albuminoids is therefore accomplished even in the rudimentary beings, but it demands for that purpose a time so much the longer as the living laboratory is the more imperfect. Thus after seven or eight days of abstinence food is still found in the digestive tube of caterpillars. Schweiger says that he found in the intestine of leeches blood sucked two-and-a-half years before. This extreme slowness of the digestive labour in the inferior animals helps us to understand how they are able to resist abstinence so long.

Absorption is slow in the digestive cavity of the inferior invertebrates, and, naturally, the movement of the alimentary mass is also very slow there. The contractility of the walls is aided in many of these animals by vibratile cilia, that is to say, epithelial cells, furnished with long mobile ciliary prolongations. These filiform appendices form on the stomachal and intestinal surface of numerous invertebrates a sort of living meadow where each blade is animated by a regular oscillatory movement, effected always in the same direction, and powerfully aiding the conveyance of the alimentary substances. This arrangement of the intestinal epithelium, frequent among the invertebrates, is also observed in the fœtal state among the selacians, the batrachians, and so on; but it is regularly lacking in birds and mammifers even at the embryonary period.[3]

[1] Dugès, *loc. cit.*, p. 331.

[2] *Handbuch der Naturgeschichte.*

[3] Leydig, *loc. cit.*, pp. 348, 375.

Nevertheless, at very different degrees of the zoological hierarchy the same mode of alimentation seems to bring with it the same structure. Thus the internal surface in certain carnivorous coleoptera, for instance the stag-beetle, has a reticulated aspect wholly analogous to that of the stomach of the mammifers. It is true that we have here to deal with a superior invertebrate in which the vitality is active, energetic, and the digestive system strongly differentiated.

Some other facts come to prove the feeble digestive power of the stomach in the inferior invertebrates. Trembley observed that the hydra neither altered nor digested one of its arms when it swallowed it along with a prey. The actinia devoured by a larger individual of its own species is often revomited safe and sound. It therefore resists digestion as the entozoa resist. The fact is not, however, general; and frequently shell mollusks, and whole crustaceans, swallowed by the actiniæ and the asteriæ are completely digested excepting the shells or the carapace, which are afterwards rejected.

The facts of resistance opposed to digestion by certain living animals are evidently due to the defect of energy in the digestive liquids secreted, that is to say, to the imperfection of the glandular system. Among the mammifers things go on differently, and thoroughly living tissues are promptly attacked and digested by the stomachal secretion unless they are protected by a thick epidermic or epithelial varnish. Claude Bernard having introduced into the stomach of a dog, through a gastric fistula, the posterior portion of a frog, saw that portion dissolved and digested, while the anterior portion continued to live.

The agent of this curious dissolution is a very active juice secreted by the stomachal glandules, and of which we have now to speak.

When the aliments are introduced into the stomachal cavity of a mammifer the mucous membrane is congested, grows red, and we see streaming out, when there is a gastric fistula, as in

Beaumont's famous Canadian, or in an animal prepared by vivisection, a juice abundant, liquid, limpid, acid, gathering into droplets and flowing along the wall: it is the *gastric juice.* This liquid attacks more or less energetically the aliments. It has not much action on the vegetal substances, and limits itself to dissolving the azotised substance of the cells. It is especially the albuminoidal aliments which more or less rapidly it dissevers and dissolves. It commences by saturating the animal tissues, which swell. The muscular fibres become more friable, grow softer, and their striæ disappear. The cellular tissue dissolves, and then the muscular fibrils are disaggregated, and end by being converted into a brownish pap.[1] The tendons, the aponeurosic membranes, are changed into a gelatiniform pap. The nervous glandular elements are also dissevered and softened. The bones themselves are attacked and pulverised; their organic mechanism is extracted and dissolved, and by degrees the terreous parts are likewise attacked and pulverised. A piece of beef-bone which an eagle was forced to swallow every day, and which he regularly vomited, disappeared in twenty-five days. The portion of the aliments merely softened swells, is saturated by the gastric juice holding in solution the substances already transformed into peptones, and the whole forms a species of soft mass called *chyme*, which gradually, in portions, after flowing for some time in the stomach itself, from the cardia or œsophagian orifice to the *pylorus* or intestinal orifice, ends by passing into the intestine.

The agent of this transformation, the gastric juice, owes its digestive properties principally to two substances, an acid and a ferment. It would be more correct to say, to acids—the lactic acid, and the chlorohydric acid; but the first, the lactic acid, comes in part from the evolvement of the starch, or rather from the sugar which proceeds from it, under the influence of the salivary ptyaline. The other acid is secreted by the glandules. Among animals with multilocular stomach, one of the parts

[1] Schiff, *Digestion, loc. cit.*, p. 145.

is specially entrusted with this acid secretion. Thus the stomachal juice is alkaline in the *paunch* and the *bonnet* of ruminants : it is acid, on the contrary, in the *caillette.*

In the mammifers, provided as man is, with a single stomachal pouch, the differentiation, which no longer exists in the general form of the stomach, continues in its apparatus of secretion. The innumerable glandules of the stomachal mucous membrane have not then the same function, and we can determine approximatively what region of the stomach secretes acid mucus, what other secretes fermented mucus, or *peptic mucus.*

In effect, if we subject to hot infusion in acidulated water the stomach of a mammifer, a part of the stomach disappears at the end of an hour or an hour-and-a-half : it is disaggregated first and liquefied afterwards. The other part, which is always the pyloric region, dissolves only with extreme tardiness. The part easily dissolved has probably the office of secreting the acid mucus.[1] The anatomical examination of the glandules in the one region and the other confirms this view. The glands of the region supposed to be peptic are almost always ramified into two or three small conduits interiorly lined with cells of cylindrical epithelium, but filled in the recesses by cells of globulous epithelium. These are the secreting cells, and we find their analogues in the caillette of the ruminants. The glands probably mucous are more rarely ramified, and do not contain cells of globular epithelium. We have seen that the mucous glands occupy especially the pyloric region of the stomach : as to the peptic glands they are especially abundant in the middle region.

The globular cells contained in the cul-de-sac of the peptic glands secrete during stomachal digestion a mucus, holding in solution a faint albuminoidal ferment, which has been called *pepsine.* According to M. Schiff pepsine is not secreted at the expense of the elements of the blood, except when this fluid is previously charged with substances which he calls *peptogens.*

[1] M. Schiff, *loc. cit.*, t. II., p. 238.

Dextrine, bread, the gelatine of bone, divers peptones, are peptogens.

The normal absorption of peptogenical substances into the small intestine has not apparently the result of charging with pepsine the stomachal glands, and yet this effect seems to be obtainable by the injection of these substances into the serous vessels, the subcutaneous cellular tissue, the stomach, and the rectum.

Pepsine is the special agent for the transformation of albuminoidal alimentary substances into peptones isomeric, but soluble, assimilable, and no longer coagulating through heat.

The two agents, acid and peptic, evaporate, moreover, in their transformation, and are both necessary. Without the acid the pepsine has no longer any action. Also when the alkaline bile accidentally reflows into the stomach and neutralises the gastric juice, the work of digestion is suddenly arrested. The acid disaggregates and softens the albuminoidal substances, and by this preparatory modification renders the peptic action possible.

By observing the phases and phenonmena of digestion in the stomach through the process of gastric fistulæ, and, better still, by recurring to artificial digestions, experimenters have succeeded in determining with sufficient precision the action of pepsine on the diverse categories of aliments.

In effect digestion is merely the result of simple, physical, and chemical phenomena, which are accomplished very well apart from the stomach. It is to Réaumur that the honour is due of being the first to make trial of artificial digestions.[1] But Spallanzani was the first who made successful trial thereof. Eberle prepared artificial gastric juice by softening a stomachal mucous membrane in water at 30° Réaumur, then in adding drop by drop chlorohydric acid, or acetic acid. Müller, Schwann, Tiedemann, and Gmelin, Burkinje, Pappenheim, Lauret and Lassaigne, made analogous experiments, which have since been often repeated.

It is generally of the caillette of a ruminant that we must

[1] *Histoire des l'Académie des Sciences*, 1752.

make use to obtain the necessary gastric ferment. M. Schi who has occupied himself a great deal with these artificial dige tions, employs often the stomach of animals which are n ruminants.[1] In that cases he detaches only the median pc tion, called *peptic;* he infuses it in from 500 to 600 grammes acidulated water, and lets it rest for five or six days. The liqu thus obtained has the property of transforming the albuminoid aliments into peptones isomerical, but soluble in water and ev diluted alcohol, moreover incoagulable by heat, and no long forming insoluble compounds with the metallic salts. In additio when these peptones are mingled with sugar, they, as we ha already maintained, mask the reaction of Trommer.

All the processes of absorption which we have just enumerat enable us surely to demonstrate that the only aliments trar formed by the gastric juice are the albuminoidal substances. Tl stomach absorbs besides a number of other substances contain in the aliments, but on condition that they are soluble in acid lated water, for the stomach does not modify them (salts, and on). The fat bodies, starch, are scarcely altered by the gastr juice; they pass with the chyme into the intestine, where, mor over, the transformation of the proteïc substances saturated wi gastric juice is gradually finished.[2]

According to M. Schiff,[3] and in opposition to an idea general received, the liquid albumine is more slowly digested than tl solid albumine; it demands more acid for its transformatio Moreover the stomachal acid is incapable of coagulating alb mine.

On the contrary, liquid caseine is promptly coagulated by tl gastric juice, and its previous coagulation is even a condition its transformation into peptones.

Vegetal legumine dissolves without difficulty in the acids, ar is afterwards transformed.

[1] M. Schiff, *loc. cit.*, t. I., p. 78.
[2] M. Schiff, *loc. cit.*, t. II., pp. 123, 124.
[3] *Ibid.* t. II., p. 150.

We have seen that the albuminoidal vegetal substances are generally contained in a membrane constituted by the cellulose; that is to say, that the vegetal aliments yield less easily than the animal aliments their proteïc principles. In effect, the digestive juices must first of all dissolve more or less the refractory envelopments.

If we are willing to depart from rigorous and literal strictness in definitions, there is a large amount of truth in the following general formula of digestion—Stomachal digestion is histogenetical, that is to say, that it prepares the quaternary azotised bodies destined to incorporate themselves with the anatomical elements. On the contrary, but in a more general manner still, intestinal digestion is thermogenetical; it renders absorbable principally the carburets of hydrogen, the ternary substances, destined in great part to undergo in the economy a complete oxydation, to be afterwards transformed into water and carbonic acid in producing heat.

Divers experiments of M. Schiff seem to prove that peptic secretion does not depend on the central nervous system. After the section of the pneumogastric nerves the pepsine does not the less continue to form itself in the glandular culs-de-sac, and the stomachal absorption itself is not diminished.[1] The case seems to be different in the *acid* secretion, and this explains how Wilson Phillip, Brachet, Breschet, Milne-Edwards were able to see digestion stop after the section of those same pneumogastric nerves. The first of these experimenters affirms even that he succeeded in restablishing the digestive activity by galvanising the peripheric tronçon of the cut nerves.

Besides, other experiments of M. Scheff show that the general functions of the stomach are far from being independent of the nervous centres. On dogs subjected to cerebral vivisections, to lamisections of the optic layers and of the cerebral peduncles, this physiologist demonstrated that the capillary vessels of the stomachal mucous membrane dilated in places, that at those

[1] M. Schiff, *loc. cit.*, p. 415.

points were produced first of all sanguineous stases, thereupon a softening of the mucous membrane, which then became incapable of resisting the action of the gastric juice, and was digested, whence ulcerations and even perforations of the stomach.

Certain general states influence, moreover, the peptic secretion; for example, according to the observations of M. Schiff, fever completely abolishes this secretion. The stomach is then absolutely incapable of digesting; it absorbs, but it requires aliments directly assimilable, such as dextrine glycose, the artificial peptones of Corvisart.[1]

C. *Intestinal Digestion.*—Intestinal digestion exists really, that is to say, with its distinctive characteristics, there only where exist also its true agents, that is to say, the hepatic and pancreatic glands. We have seen that the secretory elements of the bile seem to be at first in the echinoids, simple epithelial cells lining the internal surface of the intestine. The epithelial hepatic cells are, definitively, everywhere and always the creators of bile in the whole of the animal kingdom; but in proportion as we rise in the zoological series we see these cells becoming more and more numerous, and accumulating in special apparatus more and more complex. They are first of all simple cæcums, in the heart of which the cells have birth, fill themselves with bile, which they afterwards allow to escape, by dissolving, or, as in the insects, there are tubes in which urine and bile are generated side by side.

In the arthropods there exists a voluminous liver, especially in the crustaceans and the insects. The mollusks are also provided with a voluminous liver, dividing into lobes, fabricating at the same time sugar and bile like that of the superior vertebrates.

In these last the bile is secreted abundantly, sojourns in part at least in a special pouch called *biliary vesicle*, and is finally poured into the first part of the small intestine, the duodenum. The bile is, as every one knows, a yellowish or greenish liquid, bitter, holding in solution a colouring matter, the biliverdine,

[1] M. Schiff, *loc. cit.*, p. 269.

phosphates, chlorates, organic salts of soda, cholesterine, and so on. It is alkaline, but only during digestion. In addition, we do not find in bile any albuminoidal substance[1] and it does not coagulate through heat, the acids, the metallic salts.

We have seen that the liver of the superior mammifers is formed by the assemblage of two apparatus, the one glycogenical, the other biliary. The first acts at the expense of the venous blood, charged with nutriments brought to it by the vena portæ, that is to say, the common trunk of the intestinal veins. The secretory organ of the bile, on the contrary, elaborates its secretion at the expense of the general arterial blood, and we stop the biliary secretion by binding the hepatic artery. The same thing, moreover, takes place by absolute necessity in the inferior animals whose liver is not in contact with any vena portæ. Poured on the aliments the bile seems to act on the fat bodies which it emulsionises, and also on the albuminoidal matters impregnated with gastric juice. It neutralises them first of all, then stimulates their dissolution. But it seems to acquire all its properties only after blending with another liquid, the pancreatic juice.

Does the pancreas exist in the invertebrates? This is a question of comparative physiology which still waits for a reply. We have seen that we do not begin clearly to recognise the pancreas except in fishes, and then only in a rudimentary state. In the superior vertebrates, it is a large gland cluster-like, pouring an abundant liquid into the intestine through a conduit which sometimes anastomoses with the biliary or *choledochal* canal, and sometimes opens direct into the intestine, generally in immediate proximity to the biliary conduit, sometimes at a considerable distance from that conduit and below it. The pancreas secretes a colourless limpid, viscous and alkaline liquid. This liquid contains a special albuminoidal substance called *pancreatine*, and this substance is in such quantity there, that through heat or alcohol the pancreatic juice coagulates in mass

[1] Ch. Robin, *Leçons sur les Humeurs*, p. 501.

and entirely. After being thus coagulated and dried the pancreatine is susceptible of being redissolved into water, and we can by this means obtain an artificial pancreatic juice, and hereby the study of the properties of the pancreatic juice is much facilitated. As the pancreatic juice like the gastric juice is wholly formed in the cells of the pancreatic gland, we can also by simple infusion of the glandular tissue obtain an artificial pancreatic juice.[1]

First of all this liquid emulsionises the fat bodies surely and promptly. In addition it evolves in part some neutral fats (butyrine, oleïne, margarine, stearine) into glycerine and butyric acid.

Moreover the pancreatic juice transforms almost instantaneously the fecules into soluble glycose.

Lastly it accomplishes the liquefaction of the proteïc substances of the chyme.

We have seen that the bile by itself has a somewhat feeble fluidifying action on the chyme: but from its union with the pancreatic juice, which is besides more active, results a liquid endowed with very energetic properties.

It is easy for us now to form a sufficiently exact idea of intestinal digestion. The chyme impregnated with gastric juice, containing proteïc matters already attacked by the stomachal liquid, and in addition amyloidal and fat substances not modified, passes the pyloric orifice of the stomach and arrives in the first portion of the small intestine, the duodenum, where in most of the vertebrates it is saturated by the biliary and pancreatic juices. These two liquids being alkaline, the gastric acid is neutralised: the chyme itself becomes alkaline. Moreover in the intestine, as in the stomach, the presence of the alimentary mass causes the congestion of the mucous membrane, an abundant glandular secretion, and more vigorous movements of

[1] Ch. Robin, *Leçons sur les Humeurs*, pp. 532, 535.—Cl. Bernard, *Mémoire sur le Pancréas*, etc. (*Supplément aux Comptes Rendus de l'Académie des Sciences*, t. I.)—L. Corvisart, *Sur une Fonction peu Connue du Pancréas.*

the digestive walls. The afflux of the biliary and pancreatic liquids finishes and completes the chemical elaboration commenced in the superior portion of the digestive tube. The amyloidal and saccharine substances are metamorphosed into dextrine and glycose: the proteïc are transformed into peptones. The fat bodies are emulsionised.

All these important chemical modifications are effected, little by little, in the degree that by the action of its muscular layer the intestine contracts and thus makes the demi-liquid alimentary mass march on. By degrees the aliments become assimilable substances, veritable nutriments, Now these substances are in contact in the intestine with the villosities so richly vascular: they are therefore absorbed by endosmosis and thus pass into the circulation. It is possible, however, that the vascular absorption is not direct. There seems, in effect, to be incessantly on the intestinal surface a generation of epithelial cells, which on the one hand protect the mucous membrane, and on the other absorb the nutritive liquids, elaborate them perhaps, then surrender them by endosmosis to the capillary vessels of the intestine.[1]

The emulsionised fat bodies seem to be more specially seized by the fine lymphatic canalicules, then to be driven thence into the secondary lymphatic network. Therefore they give to this network the milky aspect, fill it with *chyle*, which, drawn into the grand lymphatic circulation, goes finally to pour itself into the venous blood.

The gastric biliary and pancreatic liquids are unquestionably the principal chemical agents of the transformation of the aliments: but they are not the only ones. Numerous small glands of diverse forms are disseminated in the substance of the intestinal mucous membrane. These glands secrete an important digestive liquid, the intestinal juice, and also simple mucus. It results from various experiments that the intestinal juice is the auxiliary of the other digestive liquids already

[1] Cl. Bernard, *Rapport sur les Progrès de la Physiologie*, etc. p. 199.

described. It transforms the fecula into dextrine and glycose: it emulsionises the fat matters; it metamorphoses isomerically the proteïc substances. But its digestive action diminishes and gradually ceases in the last portion of the intestine, the large intestine. Nevertheless the lymphatic vessels of this region are still charged with chyle exactly like those of the small intestine.

The origin of the large intestine is distinguished by a dilatation called *cæcum*, where the chymic mass sojourns for a longer or shorter period. This cæcal cavity acquires considerable dimensions in the herbivorous animals, and this is conformable to the general law according to which the digestive system must be so much the more complex the less assimilable are the aliments. Now nothing is less assimilable than the vegetal cellulose, which traverses in great part the digestive tube without being modified, not only in man but even in the herbivorous animals.

We have now indicated in large outline the diverse digestive processes, the principal metamorphoses which the alimentary subtances undergo to become absorbable and assimilable, to become nutriments after being simple aliments. We must next occupy ourselves with the second preparatory phase of nutrition, see by what series of anatomical and physiological processes the assimilable matters elaborated by digestion are brought into intimate contact with the tissues, with the histological elements, that is to say, give a general description of the circulation.

CHAPTER XIII.

CIRCULATION.

1.—*General Morphology of the Circulation.*

If we embrace with a complete and comprehensive glance the anatomy of the circulatory apparatus in the whole animal kingdom, we still see, as in other cases, specialisation gradually effected. Nature, as it was formerly the fashion to say, did not arrive at perfection at a bound. She made numberless attempts, long she groped, adding successively new pieces to the system, or complicating little by little those which existed already. In modern language, the circulatory apparatus, like all others, was perfected and differentiated more and more under the influence of the struggle to live, and of natural selection.

A first fact results from comparative anatomy, namely, that the circulatory system only shapes itself in living organisms, after the digestive system, of which at the outset it may be viewed as the appendix.

In the *protozoa* [1] there is not yet either digestion or circulation. The nutritive liquids absorbed pass direct into the nearly homogeneous parenchyma of the body. The liquids expulsed come forth in the same fashion by a sort of insensible perspiration.

In the *rhizopods* the intimate contact of the substances absorbed with the living substance is aided by the sarcodic contractions, by

[1] Gegenbaur, *Anatomie Comparée*, p. 103.

O

the emission and the retraction of the pseudopods. Liquid currents charged with granules are visibly produced.

In some infusoria we observe a rotatory movement, in a constant direction, of the intra-cellular liquid, that is to say, a phenomenon completely comparable with the intra-cellular gyration of the vegetal protoplasm (Paramæcia).

Immediately after the protozoa come the animals a little more complex, having a digestive cavity, but no other. These are the coral polypi or *anthozoïds*, the *hydromedusæ*, the *ctenophores.*

In a certain number of animals of this group we see the first rude beginning of the circulatory apparatus, consisting of ramifications more or less complex, emanating from the digestive cavity and communicating with it (medusæ, and so on,—see Figure 6). The digestion and the circulation being still confounded, we may characterise the whole of the system as *gastro-vascular.* The content of the apparatus is not yet specialised: it is chyme diluted with water.

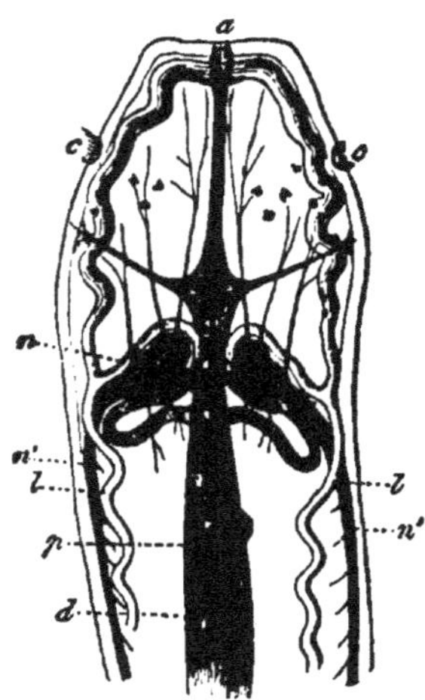

Fig. 15.

Anterior extremity of the body of a nœmertian (*Borlasia camilla*): *a* opening of the trompe; *p*, trompe; *c*, vibratile depressions; *n*, cervical ganglion; *n'*, lateral nervous system; *l*, sanguineous lateral trunks, which bend forward in order to unite: before this union they send round the brain a branch which joins that opposite in order to form the dorsal vessel *d*.

An organic progress is accomplished in the *nœmathelminths.* For the first time the nourishing liquid elaborated by the digestive apparatus is, before being assimilated, collected in a special cavity distinct from the digestive system, and surrounding it more or less completely. Figurate elements, a sort of sanguineous globules, float in this liquid in some nœmatods. In the *nœmertians* we already see appearing some vascular canals. (Figure 15.) In the *annelates* there is a tolerably regular network; there are also longitudinal vessels connected with each other, and of which one, very constant, is dorsal. This last is always contractile, and it drives the blood from behind forward. The liquid of the perienteric cavity is colourless, and we find in it figurate elements.

In the *hirudinates* the perienteric cavity exists only among the young. Habitually this cavity communicates with the exterior, and consequently the liquid which it contains is more or less mixed with water. In many of the *hirudinates* and of the *annelids* the blood is yellewish or more or less red. (Figure 16.)

FIG. 16.
Anterior portion of the vascular sanguineous system of a young *Sænuris variegata*; *d*, dorsal vessel; *v*, ventral vessel; *c*, transversal anastomosis enlarged; heart. The arrows indicate the direction of the blood.

In the *tunicates* there is constantly an impulsive organ, a heart, situated on the passage of the ventral trunk.

The circulatory system of the *echinoderms* is constituted by two vascular rings surrounding the orifice of the digestive tube. These rings are connected with each other, they emit radiating ramifications, and one of. them receives vessels coming from the intestine.

In the *arthropods* (crustaceans, arachnida, insects) there is still, as in all the preceding groups, a general cavity filled with blood. In all there exists an impulsive or cardiacal organ, whence proceed efferent vessels, called for that reason *arterial*. The blood returns to the heart by the lacunar spaces situated between the organs. These conduits without special walls debouch into a pericardiacal reservoir, and the blood penetrates afterwards into the heart by cardiacal clefts. In the *decapods* (Figure 17) the blood before returning to the heart is oxydised in passing through the branchiæ. In insects the vascular system with well determined walls is almost limited to the contractile dorsal vessel.

All the *mollusks* have a sanguiferous organ contractile or cardiactile, they have even sometimes several. (Figure 18.) There has been an attempt to show the distinct existence in the heart of the mollusks of a ventricle and of auricles; but the ventricle is only a dilatation of the dorsal trunk, and the auricles are only dilatations of the transversal vessel debouching into the pretended ventricle. In the *cephalopods* the dorsal vessel curves like a curl, the branchial veins debouch thereinto by dilating. The heart sends forth two arteries, the one cephalic, the other abdominal. There is a true network of fine capillaries between the last ramifications of the venous and arterial systems. Most of the cephalopods have cardiacal muscular dilatations on the branchial arteries. Spite of the relative perfection of this apparatus, the cavity of the body constitutes always a huge lacuna full of blood, which bathes the organs direct. This cavity communicates with the vessels; veins debouch direct thereunto.

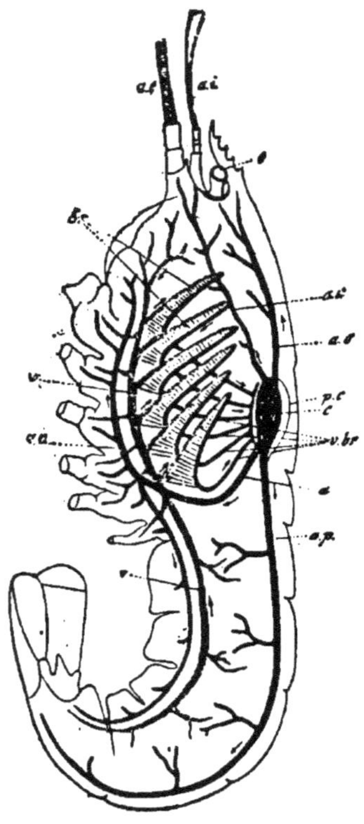

Fig. 17.

Schematic figure of the circulatory apparatus of the *lobster*: *o*, eyes; *a e*, exterior antennæ; *a i*, interior antennæ; *b r*, branchiæ; *c*, heart; *p c*, pericardium; *a o*, median and interior artery of the body; *a a* artery of the liver; *a p*, posterior artery of the body; *a*, trunk of ventral artery; *a v*, anterior ventral artery; *v*, ventral venous sinus; *v b r*, branchial veins. The arrows indicate the direction of the sanguineous currents.

The circulatory system attains its completion in the vertebrates, but it is still very imperfect in the first of them, the *amphioxus*. In effect in this animal, which Haeckel wishes to regard as a connecting point between the mollusks and the vertebrates, there is still no heart, and the impulsion is impressed on the sanguineous liquid by all the larger vessels, which are contractile.

In the other vertebrates there is a heart, but we see it perfecting

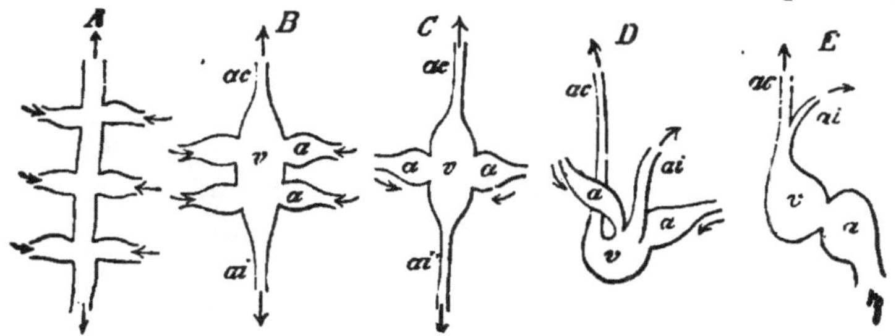

FIG. 18.

Schematic figures showing the comparison of the modifications of the circulatory centres in the *mollusks* : *A*, part of the dorsal trunk and of the transversal vessels of a *worm* ; *B*, heart and auricle of a *nautilus* ; *C*, heart and ventricle of a *lammelibranchian* or *loliginate* ; *D*, the same organ in an *octopus* ; *E*, heart and auricle of a *gasteropod* ; *v*, ventricle ; *a*, auricle ; *a c*, cephalic artery ; *a i*, abdominal artery. The arrows indicate the direction of the sanguineous current.

itself little by little. The simplest form of the heart in verte-

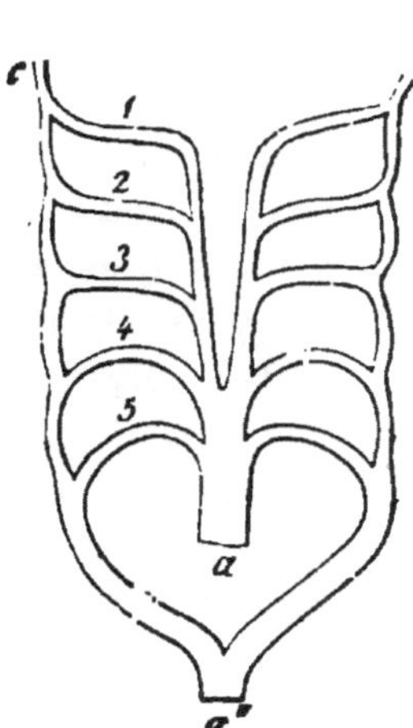

FIG. 19.

Schema of the framework of the great trunks, of the differentiated apparatus of the branchial vessels ; *a*, arterial bulb ; 1—5, arterial arcs ; *a a"*, aorta ; *c*, carotid artery.

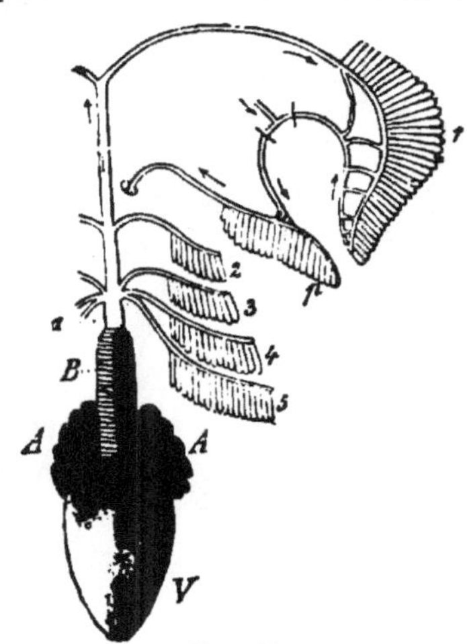

FIG. 20.

Heart, branchial arteries, and operculary branchia of the *Lepidosteus osseus* ; *v*, ventricle ; *A A*, auricle ; *B*, arterial muscular bulb ; *a*, trunk of the branchial arteries ; 1, accessory branchia (operculary) ; *p*, pseudobranchia (vent branchia) ; 2, 3, 4, 5, branchiæ of the arcs. The arrows indicate the direction of the blood.

brates exists in fishes. There, the cardiacal dilatation is merely divided into two parts, an auricle and a ventricle. The first part receives the venous blood coming from the rest of the body, and impels it into the second cavity, whence it is sent forth again by a system of vessels into the respiratory or branchial organs. In the branchial folds the blood traverses a system of fine capillary vessels; it is there charged with oxygen, and is afterwards poured into the general circulation to return anew to the heart. (Figures 19 and 20.)

When lungs take the place of branchiæ the heart commences to divide longitudinally into four cavities more or less incompletely separated. There are, first of all, in the auricles a reticulated tissue; in the ventricle protuberances, muscular projections (lepidosiren). In the reptiles there are three cardiacal cavities; two auricles, the one receiving the venous blood of the body, the other the blood returning from the respiratory organs. The separation of the ventricle into two cavities is indicated only by a reticulated tissue. Sometimes, however, these two reticulated halves do not contract simultaneously (tortoises). Here still, as in the inferior crustaceans and the arachnida, the totality of the blood is not constrained to traverse the respiratory surfaces before returning to the heart.

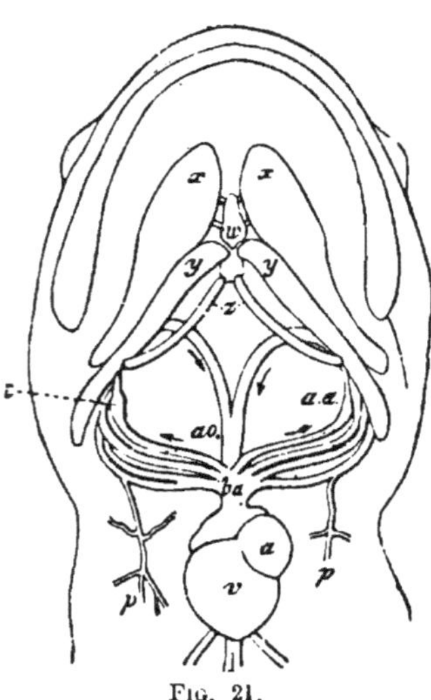

Fig. 21.

Heart and large vessels of *Salamandra maculosa*. The first arc of the aorta, *c*, in direct continuation with the carotid; *w*, *x*, *y*, *z*, apparatus of the hyoid bone; *c*, carotidian glands.

In birds the longitudinal separation of the heart is complete; there are two auricles, two ventricles. The blood vivified, oxygenised, returning from the lungs, the arterial blood, is no longer mingled in the heart with the blood which has served for nutrition, the venous blood, black and charged with carbonic acid. The case is naturally the same with the mammifers; in a cetacean, the *dugong*, the two hearts exist, but they are almost entirely separate. In proportion as the cardiacal cavities take distinct shape they subdivide each into two secondary cavities by means of membranous valves not contractile. Other valves of analogous structure are found at the orifice of the large vessels which go from the heart or debouch into it.

The circulatory system of the vertebrates forms an ensemble of closed canals, containing a special liquid. There are no longer any lacunæ, or at least they are merely exceptional and partial (Fig. 21). From the heart goes a system of efferent or arterial canals, while to the heart comes another afferent or venous system. Lastly, these two systems of canals, which in the superior vertebrates have a different structure, are connected with each other by very fine canalicules, called *capillary vessels*, with very thin walls. These capillary vessels can be classed in three principal networks, among the superior vertebrates. One of these networks extends on the surface of the respiratory organs: it is the one which gives issue outwardly to the carbonic acid: it is the one which gives at the same time access to the oxygen: it is the network of the *respiratory capillaries*. The second grand network of capillary canals exists in the liver: it springs from the subdivision of the intestinal venous system. The vasculary ramuscules which constitute it intertwine round the glycogenical cells of the liver, then unite anew to form veins (hepatic veins), which pour their contents into the general venous system. The group of the capillaries of the liver may be called the network of the *glycogenical capillaries*. The ensemble of all the other capillaries, belonging neither to respiration nor to glycogenesis, may be denominated the network of the *nutritive capillaries*. The capillaries of this last network plunge into the mechanism of all the organs, even into that of the respiratory organs and of the liver; it is through their walls that the nutritive exchange is effected, that the anatomical elements receive their assimilable substances, and reject their disassimi-

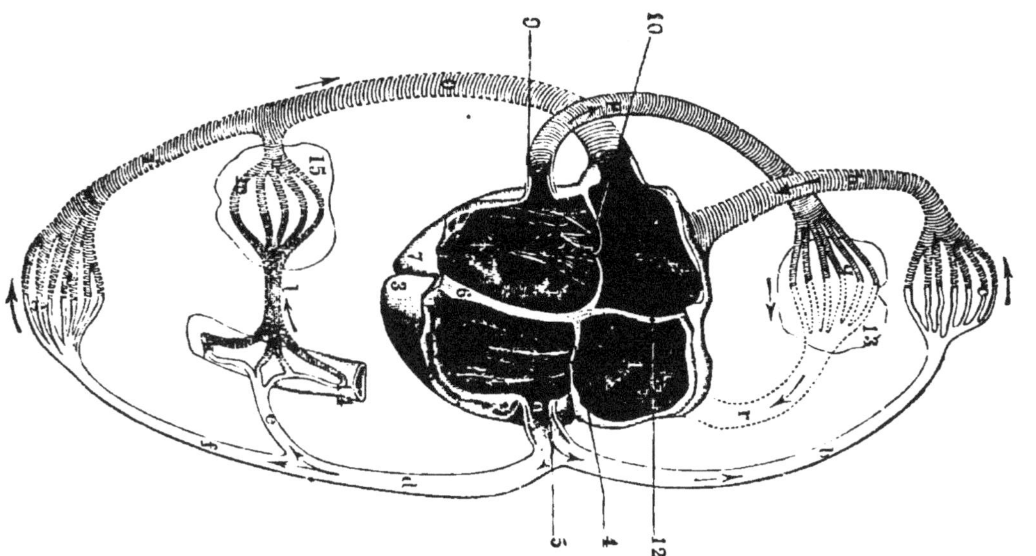

Fig. 22.

Schema of circulation in the vertebrates. The heart is sectionised longitudinally to show the cavities and valves which are shut. 1, left auricle; 2, left ventricle; 3, extremity of the heart; 8, right ventricle; 11, right auricle; 13, lungs; 14, intestines; 15, liver; *a*, sanguineous arterial current (aorta); *c*, capillary system of the higher half of the body; *d*, arterial current serving the lower parts; *e*, arterial current serving the digestive organs; *g*, capillary system of the lower part; *h* and *i*, venous currents superior and inferior; *k*, digestive capillaries; *l*, vena portæ; *m*, capillary system of the liver; *q*, capillary system of the lungs.

lated products. To be complete we must besides mention the numerous small capillary networks serving the secretions not glycogenical.

Assuredly in this summary, so rapid and so incomplete, we cannot have the pretension to expound in detail the physiology of the diverse organs and apparatus. Our duty is to keep to generalities, while merely borrowing from particular facts those of them which are indispensable to our exposition. Nevertheless, it will be opportune to describe briefly the mechanism of the circulation in the superior vertebrates. We shall thus have to speak of the heart, of the arteries, of the capillary vessels, and finally of the veins. But previously it will not be without utility to present in a few lines the idea of ensemble which springs from our rapid summary of comparative anatomy. The movement of the fluids is evidently a primordial law of nutrition. It is effected alike in the monocellular being and in man. First of all it is by direct absorption through the exterior wall of the cell. It has been observed that an active and visible exchange, when the protoplasm conveys granulations, is carried on between the nucleus and the protoplasm (hairs of the *Tradescantia virginia*). Here the circulation rigorously depends on the nutritive exchange, on the summons to action of the assimilable materials, on the rejection of the waste materials. In the polycellular organism, little or not at all differentiated, each cell acts nearly as if it were alone, and we have seen that, in plants, there is a rotatory movement of the protoplasm in each cell. In the inferior animal, in which already the nutritive labour begins to break into parts, there exists a gastro-vascular apparatus. A substance more or less assimilable passes into a system of fine canals in direct communication with the digestive pouch. Then these canals separate from the digestive apparatus. As in a system so complex, the nutritive appeal made by the anatomical elements and the capillarity would no longer suffice to impress on the nutritive liquid the requisite movement, the canals became more or less contractile. Then by a new specialisation the contractile power accumulates after a fashion at special points, where a muscular tissue more or less rich appears. Generally, then, there is a respiratory apparatus more or less complete, and as a result there are two species of blood more or less distinct, a blood oxygenised, and a blood which has need to be so. These two species of nutritive liquids finally separate, while there is progressive improvement, and though the second always comes from the first, it no longer blends with it. Then the propulsive agents are multiple. They are the nutritive call of the anatomical elements, the reflex impressed by them on the waste substances, the general contractility of the vessels, contractility more accentuated in the arteries, that is to say, in the canals which convey the vivified blood. Lastly, there exists a central propulsive organ, a hollow muscle, first with two, then with three, then with four cavities, communicating with each other two by two, but separated by valves which hinder the reflex. Each pair of these cardiacal cavities is, truth to speak, a special heart. One of these hearts is venous, that is to say, charged to impel rapidly the black blood returning from the tissues toward the respiratory apparatus, where it obtains a supply of oxygen. The other heart is arterial, and has for function to impel toward the tissues the oxygenised blood which is indispensable to them.

Naturally, the nutritive system is the better specialised, the better closed and the better differentiated the system is which contains it. First of all it is not distinguished from the aliment itself, then from the chyme; then it becomes an elaborated chyme; finally we meet with, therein, floating figurate elements. These elements, which at the outset are simple epithelial cells, become special globules, characterised by a particular form, a particular colour,—hæmatia,—as greedy for oxygen as they are prompt to relinquish it to the anatomical elements.

A. *Of the Heart.*—We have seen that the heart of the superior vertebrates is, in the animal series, bound to the contractile sanguiferous organs by a long chain of gradations. Moreover the same seriation is individually reproduced in the embryological development of every vertebrate. In these days it is merely to utter a commonplace to recall that the heart of every vertebrate appears first of all, in the embryon, under the form of a simple contractile vesicle, and represents tolerably well, in its ulterior development, first, the cardiacal type of the fish, then that of the reptile.

Further on we shall see that the anatomical muscular elements are distinguished into muscles of the life of relation and muscles of the nutritive life. The first constitute the muscular masses which obey the will; the others form the contractile elements of the viscera not subject to that influence. Now, by exception, the heart, the organ of the organic and independent life of the will, is constituted by muscular fibres of the animal life, by fibres called *striated*. This is the case among mammifers, birds, and reptiles. Nevertheless these cardiacal muscular fibres are not absolutely identical with those of the muscles of the animal life. They are thinner, have a granulous aspect; besides, they have a reciprocity of ramification and anastomosis. Striation, but of a very fine kind, is also seen on the cardiacal muscles of many invertebrates, especially insects, spiders, and crustaceans: in short, among the arthropods. There likewise the muscular fibres have a granulous texture and a more sombre hue.

The nervous apparatus of the heart is also altogether peculiar. By and by we shall find that the peripheric part of the nervous system is composed of afferent or motory fibres, of efferent or sensitive fibres, and lastly of a nervous network having some special anatomical features, and called the system of the sympathetic nerves. Now the heart, already singular by the nature of its muscular fibres, is also innervated in a particular fashion. It receives equally sympathetic fibres with ganglionary expansions and smooth fibres of the animal life. Finally, the distribution of all these nervous fibres is not made uniformly among the muscular fibres. If, employing a suitable sodic solution, we render transparent considerable portions of the auricles and the ventricles, we find therein few or no nerves. It is in proximity to the valves that these seem to be concentrated. Nevertheless sympathetic threads furnished with ganglionary expansions penetrate here and there into the muscular substance.

This structure of the heart is in perfect harmony with the functions of that organ. It is true that this organ belongs to the department of the nutritive life; but it has to contract with rapidity. Now the smooth muscular fibres of the animal life contract slowly. It was needful then that in animals with a circulation active in a degree, however small, the propulsive organ of the blood should be constituted by striated fibres. Furthermore it was indispensable for these fibres to have a great functional independence. In effect, the heart, which commences to beat the very first day of the embryonic life, does not definitively stop till death. Contrariwise to what takes place in all the organs of relation and in a considerable number of nutritive organs, the heart has no normal intermittence; it has no need of reparative repose. It is an indefatigable worker, and the labour it accomplishes during the life of a man is truly prodigious. From birth to death, it can without reposing contract three billion times, and impel into the tissues about 150 millions of kilogrammes of blood. If during sleep its movements are somewhat slower, it is because many organs at that time repose and so expend less. During the hibernal sleep when the nutritive movement falls to the minimum, and when the consumption of oxygen diminishes nineteen-twentieths (marmot), the number of beatings of the heart diminishes nine-tenths. If the heart were strictly enslaved to the nervous system, if it always obeyed it with docility, and waited always for a first excitation from it, then the heart would share the functional feebleness of the nervous system, which is the most intermittent of the organic systems: but this is far from being the case. The heart severed from the organism still beats for a longer or shorter period according to the species. Thus Scoresby saw the heart of a shark beating some hours after it had been torn from the body (Burdach, vi. 303). The heart commences to beat (*punctum saliens*) before there is in the embryon the smallest trace of nervous system, and at a later period, in the adult, when the motory nervous system is killed by the help of the poison curaré, the heart continues nevertheless to palpitate. It is even totally freed from dependence on the nervous centres, and we no longer succeed, as in the heart's normal state, in suspending the cardiacal beatings by irritating one of the principal sources of the heart, the pneumogastric nerve.[1]

In effect, in the state of integrity of the organism, the heart, spite of its independence, is bound by its nervous system to the rest of the organism, and specially to the brain. The rapid ingestion of a very cold drink into the stomach provokes a contraction of the vessels of the brain, thence anæmia of that organ, and sometimes stoppage of the beatings of the heart, whence death can result.[2] A sudden stoppage of the beatings of the heart can also be provoked by a shock on the epigastrium: it is produced when the heart and the viscera of a frog, being laid bare, we strike a violent blow on the abdominal viscera.[3] A sharp pain on the passage of a sensitive nerve produces the same effect.[4] If we electrise any nerve whatever the heart stops in its state of dilatation, in diastole. The result is the same in a frog, if we electrise the origin of the spinal marrow, the medulla oblongata.[5] But the stoppage of the heart in diastole is produced more surely still if we electrise the nerves which connect it direct with the brain, the pneumogastric nerves.

Moral impressions produce the same effect. Syncope frequently follows a strong emotion. This is a fact of common observation. Sometimes, on the contrary, emotions provoke a

[1] Saissy, Regnault.—See also Gavarret, *Des Phénomènes Physiques de la Vie*, p. 227.

[2] R. Ganz, *Ueber die Gefahr des kalten Trunkes bei erhitztem Körper* (*Pflüger's Archiv*, 1870).

[3] Brown-Séquard, *Archives Générales de Médicine*, 1856, t. VIII.

[4] Cl. Bernard, *Sur la Physiologie du Cœur*.

[5] Vulpian, *Leçons sur la Physiologie du Système Nerveux*, p. 853.

tumultuous acceleration of the cardiacal beatings, either primitively, or consecutively to a retardment. We can also directly excite precipitate beatings by touching the internal wall of a ventricle with a foreign body, for instance, a thermometer introduced into the heart. (Cl. Bernard.)

Apart from all these causes of perturbation, the heart exercises on the blood contained in its cavities a constant and very uniform pression. In the mammifers this pression is equivalent to scarcely more than a force capable of raising a column of mercury of 150 millimetres. Contrarily to all prevision, this pression is strikingly inferior to that which the blood undergoes in the arteries; and yet the blood is rapidly driven from the heart into the arterial canals. M. Schiff has given an ingenious explanation of this physiological paradox, by demonstrating experimentally, that the predominance of the cardiacal impulsion is due to the combined action of the shock, and of the elasticity of the walls of the heart.[1]

We can now form a sufficiently clear idea of the functional activity of the heart. The cardiacal cavities contract in couples, the auricles first of all, and together, then the ventricles.

In the simple heart of fishes the heart is never traversed except by venous blood, which flows first of all into the auricle, is driven by it into the ventricle, and thence by a ventricular contraction into the respiratory organs, the branchiæ. In the reptiles, in which there are two auricles and a single ventricle, the arterialised blood returning from the respiratory surfaces, and the venous blood charged with carbonic acid coming from the tissues, blend in this ventricle common to them both.

In the birds and the mammifers the venous blood penetrates into the right auricle, passes thence into the right ventricle, which drives it into the lungs. There it is oxygenised, becomes vermilion, arterial, returns to the heart; but this time, into the left heart. It penetrates into the auricle of that side, which sends it forth once more into the corresponding ventricle; whence

[1] M. Schiff (unpublished work).

it is impelled into the general arterial system, whither we have now to follow it.

B. *Arteries.*—The arteries, that is to say, all the vessels going by ramification from the heart to the capillaries, have all an elastic and contractile wall. Long there was a controversy regarding the nature of the contractile elements of the arteries till the moment, when in 1840, Henle demonstrated that the arterial contractility is due to anatomical elements identical with those of the muscles of organic life.[1] They are fusiform fibro-cells transversely placed relatively to the axis of the vessel, and furnished with a nucleus. They have a length of from five to seven hundredths of a millimetre, and have a breadth of from five to six thousandths of a millimetre on the level of the nucleus The mode of contraction is naturally in strict relation with the nature of these anatomical elements. The contractions of the fibro cells is effected slowly, progressively, after an excitation of a certain duration. But in requital, they persist during from ten to fifteen seconds at the least, and the artery returns slowly to its primitive state. This muscular layer is in general the thicker, the larger the calibre of the vessel is. It diminishes in the degree of the ramification of the arteries, and ceases altogether in the capillaries.

The arterial contractility is less independent of the nervous system than that of the heart. The experimental researches of modern physiology have, in effect, demonstrated the existence of motory nerves of the vessels, or vaso-motories. We shall have to speak of them at tolerable length. Their action determines the paralysis of the vessels, which dilate; for they then lose their tonicity, the state of demi-contraction which keeps always the arterial calibre in a certain degree of narrowness. If we cut in a rabbit's neck the cord of the great sympathetic nerve whence emanate the vaso-motor nerves of the ear, we see ceasing the rhythmical beatings of the auricular artery, which normally are effected five or six times a minute. The excitatory nervous

[1] Henle, *Wochenschrift für die gesammte Heilkunde*, 1840 (No. 21).

centre of these movements, as well as of almost the totality of the vaso-motory nervous network, seems to be situated in the upper part of the spinal marrow. According to experiments of M. Schiff the section of the cervical marrow abolishes in the rabbit the rhythmical movements of which we have spoken. The complete section abolishes them on the two sides; the hemisection abolishes them only on the sectionised side.[1]

If we can paralyse the vessels, especially the arteries, by sectionising the nerves which animate them, or the excitatory centre of those nerves, we can inversely excite arterial contractions by diverse means. The mere action of cold suffices to determine the contraction even of large arteries. Thus we can in a sheep stop the hæmorrhage produced by the section of the carotid artery by applying to this section a simple sponge saturated with cold water.[2] In an amputated limb, Kölliker det rmined the contraction of the poplitic artery, and of the tibial artery, by help of a magneto-electrical apparatus of induction.[3] This is an effect which we obtain in all the arteries by electrising them with interrupted currents. M. Vulpian has observed the contraction effected slowly at the outset between the two electrodes applied to the vessel, then propagated to the last peripheric ramifications, The continuous electric currents, whatever their direction might be, would act in an analogous manner.[4]

The arteries, like the heart, never rest during life. Their action appears even to survive that of the heart, which thus no longer deserves to be called *ultimum moriens*, a name long given to it. When the heart has ceased to beat the arteries still contract, and as their action is no longer counterbalanced by the cardiacal impulse, they tend to drive the blood into the capillaries and into the veins. Then when their muscular coat is dead in its turn, they

[1] Schiff, *Sur un Cœur Artificiel Accessoire dans les Lapins* (*Comptes Rendus de l'Acad. des Sc.*, t. XXXIX. 1834).

[2] Vulpian, *Leçons sur l'Appareil Vaso-moteur*, p. 63.

[3] Kölliker, *Zeitschrift für Wissensch. Zoologie*, 1849.

[4] Vulpian, *loc. cit.*, p. 56.

dilate; and that is the cause of their vacuity in the dead body.[1]

It seems idle at the present day to debate whether the arteries dilate when they are traversed by the sanguineous stream which each contraction or systole of the cardiacal ventricles drives into their canals. But the fact at one time was the subject of controversy among physiologists. Spallanzani demonstrated the dilatation of the great arterial trunks, and especially of the general trunk of the arterial tree, the aorta, by surrounding this vessel with a metallic ring.[2] He saw that during the cardiacal systole the aortical diameter augmented a third, in immediate proximity to the heart, and a twentieth only in the rest of its passage. Other experimentalists have demonstrated by analogous means similar facts in the large arterial trunks, for instance, the carotid; but the dilatation becomes less and less considerable in proportion as the artery is more distant from the heart and of a narrower calibre. In the small arterial branches, as Magendie has remarked, the dilatation is no longer perceptible.[3]

If the arterial dilatation is unquestionable in arteries of considerable calibre, and if it assuredly depends on the passage of the sanguineous stream, we may ask ourselves whether the pulsation, the arterial pulse has the same cause. The older physiologists, Haller, Spallanzani, and so on, have pretended, and wrongly, that the pulsation is simultaneous in the whole arterial tree, and consequently could not be attributed to the movement of translation of the sanguineous fluid, but rather to the shock, to the transmission from point to point in the liquid column, of the rapid agitation of the vibratory wave determined by the ventricular systole. We now know that the isochronism of the arterial beatings does not exist; but it is nevertheless probable that the vibratory transmission contributes its part to the production of the pulse.

[1] Vulpian, *loc. cit.*, p. 324.

[2] Spallanzani, *Expériences sur la Circulation.*

[3] Magendie, *Journal de Physiologie*, t. I., p. 113.

We can now form a sufficiently exact idea of the arterial circulation. At every ventricular systole a certain quantity of blood, on an average fifty grammes in man, is driven from the heart into the arteries. The valves situated at the orifice of the arterial trunks in the heart, the sigmoidal valves, rise to let the flow pass, and afterwards fall to hinder the reflux. The elasticity and the contractility of the arteries, vanquished by this sudden afflux, yield; the calibre of the vessels abruptly augments; but the muscular fibro-cells of the arterial walls, excited by their vaso-motory nerves, instantly re-act; the artery contracts, and impresses on the sanguineous wave a new impulsion. Naturally it is the large trunks, those which receive the shock the most directly, that dilate the most; but the sanguineous impulsion is expended in part for this dilatation; the arterial contraction which it has provoked comes to its aid, is a supplement thereto in a certain measure, and it acts with the more continuity the more distant the arterial vessel is from the heart, and is influenced less by the sudden effort of the cardiacal impulsion. Thus in the fine arterial ramifications the sanguineous current is perceptibly continuous and uniform; dilatation and pulsation scarcely exist in them. This is why we observe in these small vessels a considerable diversity of operation. Of two arterial ramuscules emanating from a common trunk, one is the seat of a rapid circulation, the other of a slow circulation. Other differences seem to depend on the organs, in the tissue whereof the arteries proceed to distribute themselves. If these organs are the seat of a great nutritive expenditure, of important material exchanges, the arterial contractions are the more energetic; for they must maintain a more active sanguineous current. Also the arteries of the brain, of the marrow, of the glands obey the artificial excitations better than the others; they contract more rapidly and more vigorously.[1]

The circulation varies also in a general manner in the different parts of the arterial tree; the course of the blood slackens in the

[1] Vulpian, *loc. cit.*, p. 64.

degree that it approaches the capillaries. It slackens, on the one hand, because the cardiacal impulsion is transmitted with a diminishing intensity; on the other hand, because the total capacity of the arterial ramifications is greater than that of the trunks whence they emanate.

At last the arterial blood, in a continuous and uniform current, reaches the fine capillary ramifications, penetrating into the web of the tissues, and connecting the arterial tree with the venous tree. But this fine capillary network is the seat of phenomena very important, and entirely bound up with the primordial acts of nutrition.

C. *Capillary Vessels.*—In many invertebrates, as we have seen, the circulatory system is composed solely of some principal vessels, beyond which the blood circulates in the interstices of the organs, in lacunæ, without peculiar membraneous partitions. On the contrary, in other invertebrates there are true capillary vessels. Thus it is, for example, in the annelates (*annulata*), in which the finest vascular ramifications have their autonomy. In certain insects there exist also fine canalicules, comparable with the capillaries. In the cephalopal mollusks, we see, at many points of the body, the arteries terminating in veritable capillaries formed of a homogeneous membrane, which is bestrewn with elongated nuclei.

Such is, verily, the texture of the true capillaries in the vertebrates. They are fine canalicules, having nearly the diameter of a sanguineous globule. The slimmest allow the globules only to pass in file. Their diameter varies between $0^{mm},007$ and $0^{mm},030$; they are constituted solely by a single coat, with a thickness varying from $0^{mm},001$ to $0^{mm},002$. This coat, everywhere homogeneous, is besprinkled with ovoid nuclei, the main diameter whereof is directed toward the axis of the vessel. If we follow the capillaries, either in the direction of the arteries or in the direction of the veins, we see their diameter enlarging little by little. It passes from $0^{mm},030$ to $0^{mm},070$, and a second membrane, or coat, clothes the first at the exterior.

This second membrane, closely joined to the first, contains fibre-cells, this time transversal, which already indicate the rise of the contractile tunic of the larger vessels. Finally, these capillaries, with their transversal fibro-cells, are continued along with vessels still larger, from 0^{mm},060, to 0^{mm},140, which, besides the two preceding tunics, have a third external tunic, a coating of laminous longitudinal fibres. These last vessels, already visible to the naked eye, have direct continuance in the arterioles and the veinules.

In diverse regions of the body, the arteries have, apart from the capillaries, transversal channels of communication, which connect them direct with the veins. These are small vessels furnished with a muscular layer, comparatively thick. If these vessels contract, the blood is compelled to pass through the muscular network; if, on the contrary, the last contractile rings of the arterioles shrink up and cause the occlusion of the capillaries, the blood passes through the collateral vessels, and, in that case, it is no longer serviceable to nutrition. This mechanism is observable in certain glands, and notably in the venous apparatus of the liver, where it serves to regulate the glycogenical secretion, to deprive of blood the secreting cells, or to allow the adequate ration thereof to reach them.[1]

We can now form a general idea of the province of the capillaries in nutrition. The sanguineous stream, driven first of all by the heart, then by the arteries, directs its course toward the capillary network, for a time in an abrupt and pulsatile fashion, then more and more regularly. Arriving at the capillaries, the arterial blood moves at a moderate and uniform rate, at least in the normal state. We have seen that in the finest canalicules the globules only pass one by one, and so to speak, by friction. Then they and the plasma which bathes them and carries them on are only separated from the anatomical elements constituting the tissues by a thin homogeneous membrane of which we have spoken. The osmotic conditions are therefore very favourable

[1] Cl. Bernard, *Leçons sur les Propriétés des Tissus vivants*, pp. 415, 416.

to the exchanges of fluids and of gases. Besides the blood is subjected to a tolerably strong pression. To these conditions already so favourablo to the osmotic exchanges are added electro motory actions. This exceedingly interesting point of the physiology of the capillaries has been elucidated by M. Becquerel in a series of important papers.[1] If by means of a voltaic pile we decompose water contained in a vessel divided into two compartments by a membrane, we see the level rise in the negative compartment. There is therefore material transport from the positive pole to the negative pole. Something analogous takes place in the capillaries. In consequence of the incessant chemical mutations which are effected in the tissues and the vessels, electrical currents are produced. The capillary vessel is electrised negatively on the exterior, and positively in the interior of its wall. The oxygen traverses the wall, and is deposited on the external surface. As to the globules, they enter into contact with the positive internal wall. The oxygen, once infiltrated into the tissues, oxydises them, gives rise to the production of carbonic acid, which the electrical current tends to impel into the interior of the capillary. In other terms, and leaving aside the old notion of the two electric fluids, there is between the contents of the capillary vessels and the histological elements of the tissues a circular current, an electrical circuit, which carries the oxygen to the outside of the vessels, and brings back the carbonic acid. We shall have to expound in detail what relates to the special office of the capillaries in the respiratory surfaces, and in the liver; we must content ourselves here with speaking of the capillaries distinctively nutritive, of those instrumental in the general exchanges between the blood and the tissues. But on this point our work is already done, and it is sufficient to refer the reader to the preceding chapters, in which nutrition has been discoursed of in a general manner. We have indicated there what substances come forth from the capillaries to be assimilated by the anatomical elements, what others are yielded to the blood

[1] *Mémoires et Comptes Rendus de l'Académie des Sciences*, from 1867 to 1870.

by those anatomical elements; and when, speaking of diffusion and osmosis, we have passed in review the principal physical conditions of this nutritive barter. Cl. Bernard very justly compares the anatomical elements to animals fixed at the bottom of the sea, and the oxygenised sanguineous wave to the flood which comes incessantly to bathe these animals, bringing to them aliments and respirable air.

The diverse manifestations of life, and life itself, are intimately dependent on the passage of this nourishing stream. If it is for a moment interrupted in an organ, the functions of that organ are at once suspended, and if the interruption is of some duration, the death of the anatomical elements is inevitable. A ligature applied to the general trunk of the arterial tree, the aorta, in an animal, paralyses the animal instantaneously. Inversely, we can make to revive members amputated and already in a state of cadaveric rigidity; we resuscitate an animal's head severed from the trunk by injecting into the arteries, and consequently into the capillaries, either complete blood, or even defibrinised but oxygenised blood.[1]

But the capillary networks, exquisitely fine, are not rigid and inextensible. If they are not contractile, they are assuredly elastic. They can dilate or narrow, according as the sanguineous torrent which traverses them is more or less abundant. Now we have seen that the origins of the capillary vessels themselves, and especially the arterioles whence they spring, are furnished with a contractile tunic. Consequently the calibre of these vessels, and therefore the amount of the sanguineous current in the capillaries, must be very variable; and in truth this is what happens. But the history of the principal conditions of these variations forms one of the most interesting chapters of general physiology, and also one of the most important, forasmuch as it is, so to speak, the mechanism of nutrition which is here concerned.

Numerous agents can produce either the contraction or the

[1] Brown-Séquard, *Recherches sur le Rétablissement de l'Irritabilité Musculaire* (*Comptes Rendus et Mémoires de la Société de Biologie*, 1851).

dilatation of the capillaries. Physical, chemical, mechanical irritants can redden the skin, that is to say, cause the dilatation of the capillaries. The application of a refrigerant mixture provokes first of all the dilatation of the capillaries, then their contraction. The skin becomes pale and exsanguine. On the contrary, heat dilates the capillaries and congests them.

Bright light acts as heat acts. An electric light produced by 120 Bunsen elements has been seen to cause a cutaneous erythema. From diverse experiments made by M. Bouchard with prisms and lenses, it results that it is the chemical rays of the luminous spectrum which produce redness and vesication soonest.

The dilatation of the capillaries can be provoked by direct irritation without the intervention of the nervous system; for in the embryon of a chicken still destitute of nerves we need only deposit a drop of nicotine on the *area vasculosa* to produce a very beautiful congestion.[1]

This fact seems to demonstrate that the capillaries, though lacking muscular fibro cells, have a contraction peculiarly their own. At all events, the capillaries with simple homogeneous wall dep nd on larger capillaries, furnished with fibro-cells, and especially on the arterioles and arteries, which are most evidently contractile. It is therefore very natural that in acting on the nervous system we can cause either the contraction or the dilatation of the capillaries. A sharp irritation of the skin, or a violent pain, brings on the contraction of nearly all the capillaries.[2] If we excite in a dog the central end of the sciatic nerve sectionised, we see contracting first the vessels of the other member, then those of the whole body.

On applying ice to the cubital nerve Waller saw the little finger and the fourth finger grow red, and their temperature rise. The sanguineous vessels depend, then, in a large measure on the nervous system. Modern physiology has put beyond doubt this interesting fact, and it has demonstrated the existence

[1] Vulpian, *Leçons sur l'Appareil Vaso-moteur*, t. I., p. 171.

[2] *Ibid.* p. 217.

of special nerves, of motory nerves of the vessels, or vaso-motories. The function of the vaso-motory nerves is of such importance that it is indispensable briefly to describe it.

The vague notion of nerves specially charged to regulate the distribution of blood in the vessels is very old in science. Already even in 1727, Pourfour du Petit had noted the fact which in our own days has led M. Schiff in Germany, and M. Cl. Bernard in France, to the discovery of the vaso-motory nerves. Whosoever has opened a treatise on anatomy knows that the peripheric parts of the nervous system, situated apart from the nervous column—the nerves, properly so called—are distinguished into two great categories. One of these categories comprehends cords, smooth, cylindrical, and appertains to motility and to sensibility. The other is composed of cords along whose passage enlargements or ganglions are found; this is the ganglionary nervous system, which has for a long time been called the *sympathetic nervous system* or the system of *organic life*, because its principal branches are especially distributed to the viscera. We propose to give further on a somewhat less succinct description of the ganglionary nervous system. For the moment we must satisfy ourselves with saying that as a whole it is composed of two chains of ganglions placed on each side of the vertebral column.

These ganglions, connected with each other by vertical nervous cords, are, on the one hand, united by other nervous branches to the spinal marrow, and send forth numerous branches, distributing themselves to the viscera, penetrating even to all the tissues and organs of the body, and clinging more or less straightly to the arteries. In the cervical region in man, we count three cervical sympathetic ganglions, of which the higher one sends forth fibres which adhere to the intracranian and extracranian arterial branches. Now, as early as 1727, Pourfour du Petit had remarked that the section of the great sympathetic nerve in the neck provoked congestion of the eye and the redness of the ocular conjunctiva. To think that the cutting of the nerve had paralysed the vessels, and that consequently vaso-motory nerves existed, there was only a step, apparently easy, to take; and yet that step was not taken till our own day. The priority of precise and well-interpreted observations and experiments seems to belong to M. Schiff (1845).[1] In France, M. Cl. Bernard, who appears to have been ignorant at that time of the labours of his predecessor, was the first to give a theory of the vaso-motory nerves complete, demonstrative, and based on well-made experiments.[2] We have now to indicate, without restricting ourselves to follow the chronological order, the typical facts and their signification.

The section of the sympathetic nerve in the neck, or rather the outplucking of the higher cervical ganglion, causes the dilatation of the capillaries of the cerebral pia mater, of the conjunctiva, of the ear of the corresponding side, and so on. This dilatation is accompanied by a very notable elevation of local temperature (5, 10, 15 degrees), demonstrated for the pia mater and the brain; in short, by all the phenomena of simple congestion, without inflammation. A wound in the ear of the side thus injured bleeds more rapidly and more abundantly. The blood in the veins of the side affected contains more oxygen and less carbonic acid than the ordinary venous blood. The sensibility is at the same time intensified. Analogous phenomena are observed in all the organs whose sympathetic nerves are cut, the liver, the veins, the lungs, the intestines, &c.[3]

This congestion is not inflammatory. It persists for days and for weeks without modification, and we might without temerity affirm that it results from a paralysis, if we merely relied on the facts which we have just enumerated; but the counter-proof has been made, and leaves no room for doubt. If, in a dog whose

[1] *De Vi Motoria Baseos Encephali*, Bockenhemii, 1845.

[2] Cl. Bernard, *Influence du Grand Sympathique sur la Sensibilité et la Calorification* (*Comptes Rendus de la Societé de Biologie*, 1851).

[3] Vulpian, *Leçons sur les Nerfs Vaso-moteurs*, t. I., pp. 94, 95, 106, 112, 531, 561.

great cervical sympathetic nerve we have sectionised, we provoke the redness of the ocular conjunctiva by instilling into the eye a drop of ammonia, we can dissipate this redness by galvanising the higher end of the cut nerve. In the same way the electrisation of the great cervical sympathetic nerve, or rather of the higher cervical ganglion, determines the contraction of the pia mater of the same side.[1] In dogs M. Vulpian has seen the section of the renal nerves followed by red colouration of the kidney, and by polyuria, with the passage of the albumine of the blood in the urine. Inversely the electrisation of the nerves made the renal gland grow pale and abolished the secretion. In short, the electrisation of the sympathetic nerves and ganglions produces results diametrically opposed to those of their section and of their destruction, that is to say, paleness, the lowering of the temperature, the abolition of the secretions, the diminution of the sensibility.

These phenomena are invariable. They are observed everywhere and always in all the vertebrates. There are consequently vaso-motory nerves; and M. Schiff thinks they should be divided into vaso-constrictor nerves and vaso-dilatator nerves.

Anatomically the remote vaso-motory nervous filaments have been investigated as far as the capillaries of the second degree, already furnished with fibro-cells. Evidently the nerves cannot have direct action on the finest capillaries altogether destitute of contractile elements. It is in acting on the arteries, the arterioles, and the capillaries of the second and third degree that they obstruct or facilitate more or less completely the passage of the sanguineous current, modifying thus more or less the alimentary ration of the organic elements.

We have seen above that the ganglionary or grand sympathetic nervous system draws from the spinal marrow numerous and

[1] Nothnagel, *Des Nerfs Vaso-moteurs des Vaisseaux du Cerveau* (analysed in *Gazette Hebdomadaire*, 1867), and Callenfels, *Ueber den Einfluss der vasomotorischen Nervens auf den Kreislauf und die Temperatur* (*Zeitschrift für ration. Med.*, 1855).

important roots. It is therefore very natural that lesions of the nervous centres should react on the vaso-motory nerves, and that diverse excitations should be able to transmit themselves indirectly to those nerves, by reflecting themselves on the nervous centres, that is, by reflex action.

In truth, a lesion of the cerebral peduncles and of the optical layers determines a dilatation of the abdominal capillaries, especially of those of the liver and of the stomach. The electrisation of the marrow on the level of the first dorsal vertebræ acts exactly as the direct electrisation of the cervical and sympathetic cord acts. It provokes the contraction of the vessels of the head, the lowering of the temperature, the dilatation of the ocular pupil, and so on. The electrisation of the spinal marrow at its origin, the rachidian bulb, determines the constriction of all the vessels of the body (Bezold-Ludwig, etc.).[1] On the contrary, puncturing, chemical agents cause a general dilatation of the arterioles. It is for this reason that Cl. Bernard, by puncturing the roof of the fourth cerebral ventricle, was able to occasion saccharine diabetes, and so on. By practising from below to above hemisections and total sections of the spinal marrow, M. Schiff was led to consider the rachidian bulb as the general centre of all the vaso-motory nerves of the body.

As the vaso-motory nerves have their excitatory centre in the bulb and the marrow, or at least have very intimate bonds with those regions, it is very natural that they should be very susceptible of undergoing reflex actions; and in effect all excitations of the central end of a sectionised nerve reacts on the vaso-motory nerves, generally by determining vascular contractions. Analogous effects are moreover produced incessantly, normally, and without the slightest vivisection, in every vertebrated organism. All sensitive excitation a little strong, having its point of departure either from the cutaneous surface or from a mucous surface, or from one of the organs of the senses, can react on the capillary circulation. All the world knows that

[1] Vulpian, *loc. cit.*, pp. 210, 218, 220, 221.

strong moral impressions, that emotions produce the same effects; the change of the coloration of the cheeks is a familiar sign of strong emotions, especially in the young and in women; that is to say, when the impressionability is the more intense and the reflex actions the more easy. The changes of coloration are far from coinciding always with the abnormal cardiacal pulsations; their physiological reason is a reflex vaso-motory action.

In physiology and in pathology the office of the vaso-motory nerves and of the capillaries is enormous. The variations in the calibre of the small vessels regulate the distribution of the blood even in the innermost part of the tissues, and thus act direct on nutrition, and consequently on the operation of the organs. As a general rule any moderate dilatation of the capillaries of an organ has for effect the superactivity of that organ, whatever it may be, gland, muscle, brain, and so on. On the contrary, every contraction too great brings with it a nutritive retardment and a functional diminution, forasmuch as it lessens the quantity of arterial blood passing in a given time within range of the anatomical elements, and consequently the proportion of liquid and gaseous exchanges of which the capillaries are the principal seat.

After having, by circulating in the capillaries, supplied the anatomical elements with assimilable materials, and retaken the waste substances; after having yielded at the same time the oxygen needful to the nutritive chemical transformations, and received in return the carbonic acid, product of the vital combustions, the arterial blood, formerly vermilion, becomes venous blood, black blood, and passes from the capillaries into vessels more and more voluminous. These vessels are at first similar to the ultimate arterioles which pass into the capillaries; then little by little their muscular and elastic tunics grow thinner; some longitudinal contractile fibres cross the circular fibres; the vascular walls grow less rigid; the vessels have become venous. They go on thus anastomosing each other, augmenting always in calibre, to reduce themselves in the mammifers to two great trunks called *venæ cavæ*. Of these two trunks the higher one receives the venous blood of the higher part of the body; the other is the general trunk of the veins of the lower half. Both throw themselves into the right auricle of the heart, whence the venous sanguineous wave passes into the corresponding ventricle, which drives it in its turn into the lungs, where it is oxygenised, arterialised, to return into the right auricle, to pass into the ventricle of the same side, and recommence the cycle.

The impulsion of the heart, not very perceptible in the capillaries, where the blood moves in an unintermitting flow, seems to be almost null in the veins. There the causes of the circulation are especially the reflux of the liquids exhaled by the anatomical elements, and principally the elasticity and the contractility of the venous canals, in most of which, besides, all retrogradation of the sanguineous current is rendered impossible by valves arranged in such a manner that they leave the path free in the direction of the heart, but can shut it in the direction of the capillaries.

The grand system of sanguineous irrigation which we have just described functionates consequently without repose, without pause, as long as life endures. Unceasingly a stream of nourishing blood traverses it. This stream wastes its strength in the tissues, regains its strength at the expense of the substances elaborated by the digestive apparatus, renews its vivifying gas in the respiratory capillaries, exhausts itself in the capillaries of the glands. It is literally a living river. But by the side of the grand sanguineous system exists another network with which we have now to occupy ourselves.

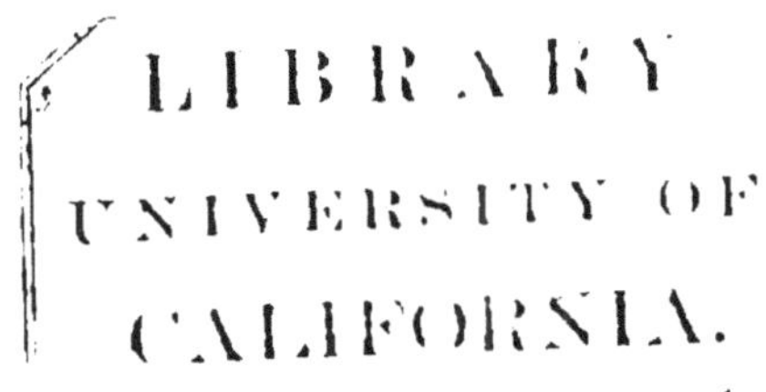

CHAPTER XIV.

LYMPHATIC CIRCULATION.

THE lymphatic circulation does not seem to exist in the invertebrates, unless we consider as lymph the colourless nutritive liquid which travels along the rudimentary vessels in many of them. Nevertheless certain persons have professed to recognise a lymphatic circulation in the hirudinates, which are moreover provided with a tolerably complete circulatory system.[1]

The lymphatic apparatus is assuredly an apparatus of perfectionment; for it is entirely lacking in the amphioxus, and does not appear in the vertebrated embryon till the late period when the venous and arterial networks are already differentiated.[2] The manner in which the lymphatic system complicates itself by degrees in the veins of the vertebrates is interesting; for it suggests conjectures respecting the physiological province of this system.

In the inferior vertebrates, the lymphatic vessels are especially represented by a kind of sheaths, surrounding the sanguineous vessels, especially the arteries (Fig. 23). In fishes and the batrachians we find this arrangement extending even to the aorta. In all the vertebrates, furthermore, the fine lymphatic networks, the radical networks, a kind of lymphatic capillaries, are joined to the sanguineous capillaries, so that these last form a part of their walls. In all the vertebrates,

[1] Leydig, *Traité d'Histologie de l'Homme et des Animaux.*

[2] Gegenbaur, *loc. cit.*, p. 9.

too, the lymphatic vessels sheathe the sanguineous vessels of the nervous centres which they isolate thus from the nervous substance. It is this intimate union of the sanguineous and lymphatic networks which made many anatomists believe that the brain had no lymphatic vessels.[1]

But the fine lymphatic canals do not seem to have always walls of their own; in that case they must be a kind of lacunæ of the conjunctive tissue.

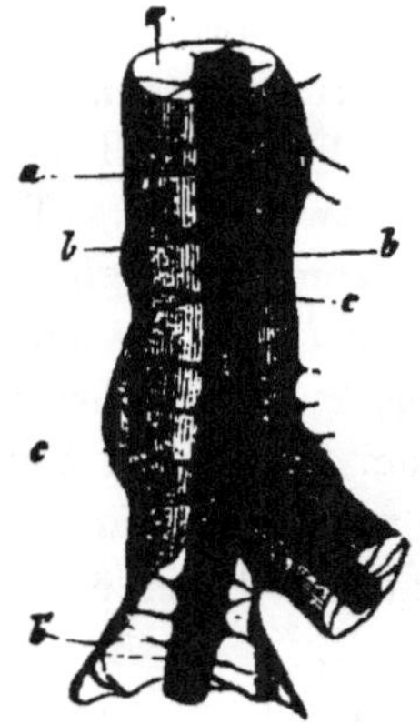

FIG. 23

Fragment of the aorta of a tortoise (chelydra) surrounded by a large lymphatic cavity: *a*, aorta; *b*, external wall of the lymphatic cavity removed at *b'*; trabecules fastening the sanguineous vessel to the walls of the lymphatic cavity.

FIG. 24.

a a, caudal sinus; *b*, transversal anastomosis; *c*, lateral vessels; *d* origin of the caudal vein in the *silurus glanis*.

In the inferior vertebrates the lymphatic vessels communicate with the veins by numerous branches. In the reptiles the principal lymphatic vessels have just got so far as to have some rare and insufficient valves.

Numerous reptiles have lymphatic hearts, which we find, moreover, in certain fishes, in amphibians, and even birds (Fig. 24). These hearts are partial dilatations of the vessels,

[1] Ch. Robin, *loc. cit.*, pp. 214, 315.

and are animated by regular pulsations. Sixty beatings a minute have been counted in a frog. In these special points the lymphatic wall, which elsewhere has nearly the same anatomical texture as the veins, is clothed with a muscular layer thick and striated.

It is well known that in the mammifers and in man the lymphatic system is constituted by very fine capillary networks, situated in the depth of the organs, in the various membranes, and in the skin. From these networks proceed vessels which throw themselves into each other, forming trunks larger and larger, and less and less numerous. Finally all the system is connected with the venous system by two canals. The smallest of these canals receives the lymph from the right half of the trunk and of the head; it throws itself into the subclavial vein on the same side. The other, larger, known under the name of *thoracic canal,* receives the lymph from the rest of the body, and has its conflux in the left subclavial vein. In man the lymphatic network does not seem to have any other direct communications with the sanguineous system.

After every repast, the lymphatic vessels of the intestine, the chyliferous vessels, swell and become lactescent. If we bind the lymphatic canal we see it swelling below the ligature; and if we prick it we see the lymph spurt out. This alone suffices to prove that the lymph pours itself into the blood; and the arrangement of the lymphatic valves also confirms this fact. But on its passage the lymph encounters special organs, distentions, or closed glands, from which it seems to obtain a supply of white globules.

These important organs are also subject to the law of gradual differentiation. In fishes we find at distant intervals, on the passage of some lymphatic vessels, simple enlargements, where globules form in meshes of cellular tissue.

In the reptiles there are already more prefect formative organs; they are the closed follicles or *Peyer's glands*, disseminated in the depth of the intestinal mucous membrane. These follicles are constituted by closed vesicles filled with nucleated cells;

they are more numerous in the mammifers, and largely aided in their cellular production by organs of analogous structure, varying in size from that of a lentil to that of a nut, and disseminated along the passage of the lymphatic vessels; they are lymphatic ganglions, which are met with already in birds, in the region of the neck. The lymphatic vessels are subdivided into fine afferent canalicules before approaching a ganglion, and on the other side of the gland the vessel only forms again at the expense of analogous efferent canalicules. As to the ganglions they are like the Peyer's glands, essentially constituted by vesicles, at least the tenth of a millimètre large, and filled with nucleated cells, with a diameter of about five-thousandths of a millimètre. By the rupture of the vesicles the cells become free and are borne along by the lymphatic plasma.

As we have seen, the lymph is a living liquid, playing in nutrition an important part; thus it is that the ligature of the thoracic canal causes the rapid emaciation of the mammifers, and their death at the end of a small number of days.

Trusting to the preceding data, we can form conjectures sufficiently probable respecting the physiological province of the lymphatic system. The lymph has in the vertebrated organism a double origin. A notable fraction of the substances which constitute it is drawn from the intestinal mucous membrane. It is principally in this way that the fat, emulsionised bodies penetrate into the circulation. But the networks, lymphatic in origin, of the rest of the body, appear to gather a part of the plasma completely elaborated, which pierces by osmosis through the wall of the vessels, and especially of the fine arteries and capillaries. To this sanguineous plasma comes probably to join itself a part of the intercellular blastema, that is to say, of the assimilable liquid which bathes the anatomical elements apart from the vessels. The whole forms, first of all, a sort of blood destitute of globules, a limpid plasma; but at the expense of this plasma white globules form in the ganglions. Then the lymph is complete, and it has the greatest analogy of chemical and morphological composition with the blood, in which it ends by throwing itself. Red globules are lacking thereto; but it is probable that these are formed in the sanguineous vascular system at the expense of the white globules of the lymph, and of those which are generated in diverse closed glands, and of a general texture analogous to that of the ganglions. We allude to the glands called *sanguineous*, to the spleen, to the thyroid gland, and so on.

The metamorphosis of the white lymphatic globules into red sanguineous globules is almost directly established by the following fact:—

Burdach, having placed on a watch-glass some drops of the blood of a dog, plunged the whole in a vessel filled with oxygen, which he closed hermetically. He says that at the end of twenty-four hours there was coagulation, and a *red* clot floated in the serum. Burdach further says that through the help of the microscope he saw in this clot globules, yellowish, ovalar, biconvex, smaller than the sanguineous globules.[1]

On the whole, the lymphatic system is an auxiliary of the venous system, with which, moreover, it largely communicates in the inferior vertebrates. But its function is not simply to convey to the heart and to the respiratory organs blood vitiated by the activity of nutrition; it furthermore undertakes the duty of making up for the imperfections of the sanguineous circulatory system by gathering a part of the plasma not utilised, and finally, at the expense of this plasma, and of the assimilable liquids obtained in the intestine, it fabricates white globules.

The movement of the lymph in the lymphatic vessels is so curious as to deserve remark. Of the causes of propulsion which we have indicated in the sanguineous circulation, one, the cardiacal contraction, is here absent. Nevertheless the lymph marches, and even with a notable swiftness, since it bursts out when we prick the thoracic canal below a ligature. Now its

[1] Burdach, *Traité de Physiologie*, t. II., p. 541.

movement of translation has only two possible causes, the pression exercised by the absorption itself in the fine canalicules, the *vis à tergo*, and, as an auxiliary force, the contraction of the walls; that is to say, the principal agents of circulation in the veins. We may therefore, by analogy, conclude that in the sanguineous circulatory system the heart is simply a reinforcing organ, and that a very great part of the propulsive labour is accomplished by the vessels themselves.

CHAPTER XV.

OF THE RESPIRATORY ORGANS IN THE ANIMAL KINGDOM.

THE fundamental phenomenon of respiration is very simple: it is merely the exchange of the carbonic acid and the water contained in an organism for the aërian oxygen. It is a physical phenomenon governed in its ensemble by the laws of osmosis. In respiration, the water which escapes from the organised body is in the state of vapour. But the carbonic acid expulsed results from the oxydation of the organised substance; consequently respiration is very intimately related to nutrition. It has for object to ventilate the organism, to renew the air of the interior medium. As there is no nutrition without absorption of oxygen, respiration is a primordial function; it is effected as an essential act of every living substance. It has been objected to the generality of this law that certain infusoria, for instance, certain vibrions, and so on, can live in a medium charged with carbonic acid and destitute of oxygen. The exception cannot be admitted without solid proofs which are still lacking. Of these pretended animalcules some are provided with chlorophyll, and are rather to be regarded as plants. If the statement were true for the others we should simply have to conclude that they have the faculty of taking their oxygen from the carbonic acid itself, by decomposing it. The whole organised world absorbs oxygen and needs to absorb it. Not without reason the ancients made synonyms of the verbs *to breathe* and *to live.* Every organised being, vegetal or animal, dies more or less rapidly when it is robbed of air; and aërian abstinence is always supported for an infinitely shorter period than alimentary abstinence; always it is the more rapidly mortal, the richer the nutrition.

In the protozoa the exchange of gases is made by the whole body. The oxygen impregnates the small protoplasmic mass, transforms itself there into carbonic acid, and comes out as it went in by the simple action of osmosis. Moreover, the protoplasmic movement, especially where there are emission and retraction of pseudopods, aids much the accomplishment of the phenomenon, by multiplying the contacts. This process, infinitely simple, does not essentially differ from the processes with complex differenciation so much as at first sight might appear. Definitively, in superior organisms, every anatomical element holds the same relation to the air which is observable in the amorphous or monocellular protozoa. What constitutes the difference is the apparatus thanks to which the anatomical element is brought into contact with the exterior atmosphere.

Every organised substance not clothed with an impermeable glaze, natural or artificial, respires more or less.

Placing in closed vessels reptiles from which he had torn the lungs, Spallanzani discovered that in these animals the skin acted as a respiratory organ more vigorously than the lungs themselves.

W. Edwards saw some frogs, whose lungs had been plucked out, and whose neck was tightened by a ligature, live from twenty to forty days when placed on humid sand in a chamber whose temperature was + 12 degrees. If, on the contrary, the frogs were plunged in water when their teguments were no longer in contact with the dissolved air, they perished in three days.

Humboldt and Provençal kept in vessels thoroughly separated the body and the head of a tench, and they found that the body absorbed oxygen,—in a word, respired.

Even in mammifers in which exists a respiratory apparatus voluminous and with a vast surface, the skin plays an important respiratory part. A mammifer grows cold and dies when its skin is covered with an impermeable varnish.[1] If in a mammifer we keep by means of a receiver air in contact with the skin, this air is modified as absolutely as in the lungs.

But, as we have previously seen, the fine capillary vessels are of all the parts of the body of complex animals the best organised for the exchanges, whatever they may be, between the interior medium and the exterior medium. Consequently an organic surface in contact with the atmosphere breathes the more freely and energetically the better provided it is with capillary vessels and the more immediate the contact of these vessels is with the exterior air. (Fig. 34). Rich capillary vascularisation, easy contact with the aërian gases; such are the two principal conditions of a good respiratory surface. And these conditions can be met with apart from the respiratory organs properly so called. It is thus that the pond leech (*Cobitis possilis*) breathes in part through the intestine. In this animal, the intestinal mucous membrane being excessively vascular, almost entirely constituted by sanguineous capillaries connected by a small quantity of conjunctive substance, the animal makes use of this substance as of a respiratory organ; he swallows the air, which, in contact with the intestinal capillaries, is exchanged for the carbonic acid which the intestine expulses.[2]

Nevertheless, every animal in any degree complex has points of his exterior or interior surface specially charged with the respiratory gaseous exchange. It follows, as a matter of course, that for the respiratory system, as for all others, specialisation is the more perfect, the higher the animal is exalted in the hierarchy. Before passing in review, in this respect, the diverse groups of the animal kingdom, it is useful to define some terms often recurring in the description.

The name of *branchiæ* is usually given to all the appendices,

[1] Fourcault, *Influence des Enduits Imperméables*, etc. (*Comptes Rendus de l'Academie des Sciences*, t. XVI.)

[2] Leydig, *Traité d'Histologie*, etc., p. 417.

all the vascular surfaces serving aquatic respiration, all the apparatus specially charged to absorb the oxygen dissolved in the water, and to relinquish in exchange carbonic acid. The organs of aërian respiration can be divided in a general manner into two categories, the tracheæ and the lungs. The tracheæ are canals more or less ramified, always open, plunging into the interior of the body, and communicating with the exterior air only by narrow openings. The lungs are internal sacs, sometimes simple, sometimes subdivided into compartments more or less numerous; they communicate also with the exterior air by orifices or conduits more or less narrow.

Like all organic classifications, that of the respiratory ap-

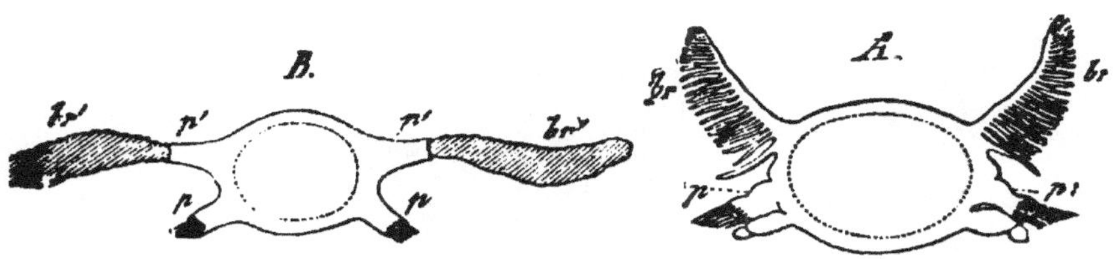

FIG. 25.

Transversal sections of annulated worms, to show the homology existing between the branchiæ and the cirrhi.—*A*, section of *Eunice*; *B*, section of *Myrianide*; *p*, abdominal parapod; *p'*, dorsal parapod; *br*, branchiæ; *br'*, cirrhi.

paratus suffers a number of exceptions. There are mixed forms, transitory or confused types. Thus the *tracheæ* are aërian respiratory organs according to the definition. But we can bring into affinity with them the vascular ramified systems, in which the water circulates in many of the invertebrates. The lungs of spiders can be considered as modified tracheæ; and so on.

In many of the inferior invertebrates (turbellariates, næmertians, annelidæ) there are no special organs appertaining to the respiration. The gaseous exchange is effected by the teguments, furnished often with vibratile cilia. This last anatomical arrangement is found, as we shall see, in the mucous membrane of the respiratory passages of many superior animals. When water penetrates into the cavity of the body the surface of this cavity plays also a respiratory part, a part the more important the more rapid the renewal and the circulation of the water are.

The branchiæ appear in the annelates. They are, as in most of the invertebrates, appendices projecting from the external tegument. Different organs, particularly the dorsal cirrhi, grow vascular and become branchial in tufts more or less ramified (Fig. 25).

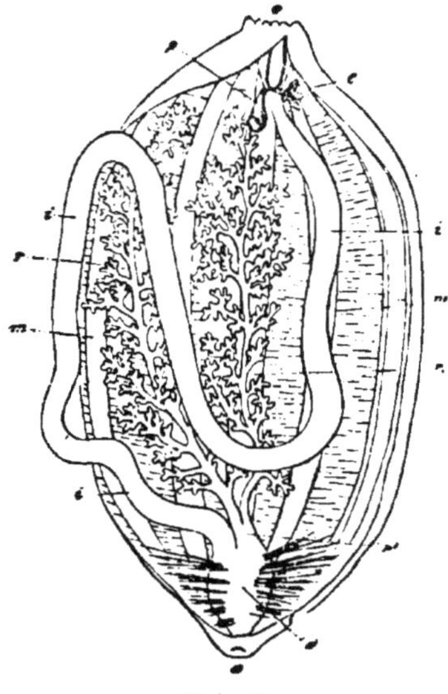

Fig. 26.

Intestinal canal and ramified organs of a *holothurion*: *o*, mouth; *t*, intestinal tube; *d*, cloaca; *a*, anus; *c*, ramified stony canal; *p*, Poli's vesicle; *rr*, arborescent organs; *r'*, their union at the point of intersection above the cloaca; *m*, longitudinal muscles of the body.

In the *tunicians* there is a sort of respiratory pouch; it is the anterior part of the digestive tube. This portion dilates into a sac, at the bottom of which the true digestive tube commences.

In the *holothuria* exists a system of aquiferous canals, probably respiratory; it is a double tree, the two orifices whereof open into the cloaca. This cloaca receives the water, and projects it vigorously outwards, on an average three times a minute

(Fig. 26). Analogous systems exist in many other echinoderms, and often they are bedecked with vibratile cilia.[1]

The *crustaceans* breathe through the branchiæ, which often are external appendices. Sometimes they are abdominal appendices, sometimes they are claws, or portions of claws modified. In the decapods we find branchiæ inclosed in cavities furnished with eddying organs, making the water circulate.

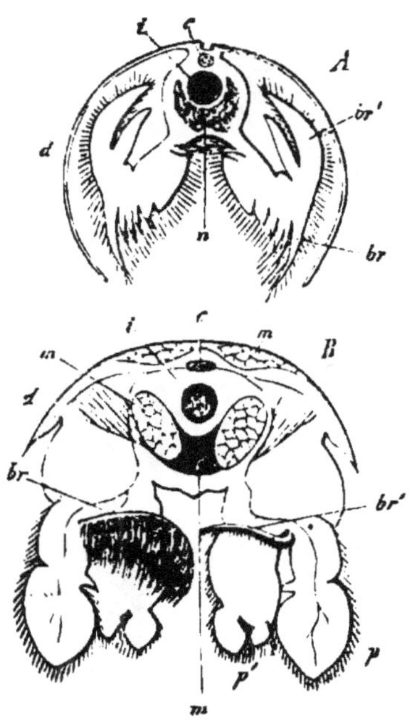

FIG. 27.

A, transversal section of a phyllopod (Limnetis), The section passes through the part which bears the first pair of feet.—*i*, intestinal canal; *c*, heart; *n*, ventral marrow; *d*, folding of the teguments forming a shell concealing the members; *br*, natatory feet. —*B*, transversal section of *squalla* (through the abdomen), *i*, *c*, *n*, as in *A*; *m*, muscles; *d*, tegumentary fold; *p*, external lamellary parts; *p'*, internal lamellary parts; *br*, *branchiæ*.

In a certain number of arachnida, in the myriapods, and especially in the insects, respiration is accomplished by means of the tracheæ, which the nutritive liquid of the animal bathes exteriorly (Fig. 28). The tracheæ form then a system of canals, whose ultimate ramifications constitute a fine network, analogous, as to arrangement, to the capillary networks.[2] The principal trunks are kept open by a chitineous bandelette rolled spirally and following the skeleton of the tracheæ. These tracheæ communicate outwards by means of determinate orifices called *stigmata*. The introduction and the expulsion of the air of the tracheæ seem to be helped by regular movements of the abdominal walls. These rhythmical movements are frequent; we can count on an average twenty-five in the stag-beetle, and from

[1] Dugès, *loc. cit.*, t. II. p. 355.
[2] Leydig, *loc. cit.*, p. 440.

fifty to fifty-five in the green locust. Spite of the comparative perfection of their respiratory apparatus and the activity of their life, insects resist asphyxia a long time. Thus Lyonnet saw caterpillars revive which had remained under water for eighteen days.

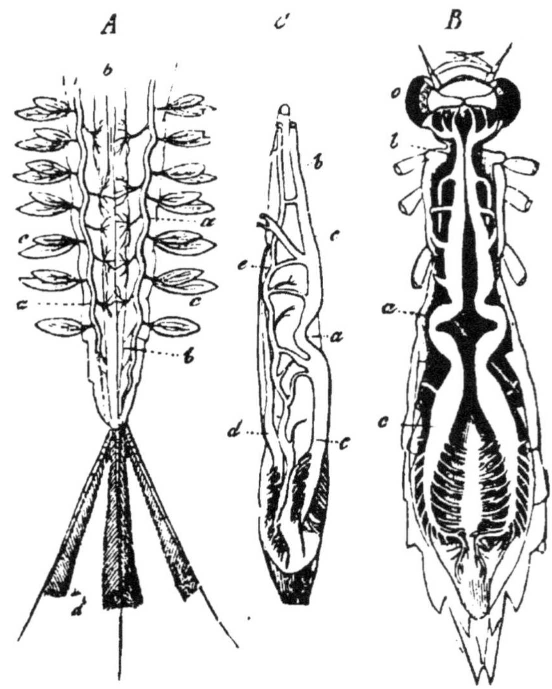

Fig. 28.

A, posterior part of the larva of *Ephemera vulgata*:—*a*, longitudinal trachean trunk; *b*, intestinal canal; *c*, trachean branchiæ; *d*, plumous appendices of the tail.—*B*. larva of *Æschna grandis* (the dorsal part of the teguments is removed):—*a*, superior longitudinal trachean trunks; *b*, thin anterior extremity; *c*, their superior part ramifying above the rectum; *o*, eyes.—The figure *C* in the middle represents the intestine of the same larva, seen sidewise:—*d*, inferior lateral trachean trunk; *e*, communications with the superior trunk; *a*, *b*, *c*, as in figure *B*.

Must we, with Blainville, consider as aërian branchiæ the wings of insects, which are often the seat of an active circulation?

The *mollusks*, being for the most part aquatic, breathe through the branchiæ, which, in general, are visibly prolongations, cutaneous appendices. These appendices are sometimes simple folds, sometimes foliated or pectiniform prolongations. They are usually traversed by vessels whose blood afterwards returns to the heart (Figs. 29 and 30). When the vessels are lacking there are at least lacunar cavities. But though these branchiæ are visibly respiratory organs, they are far from being the only

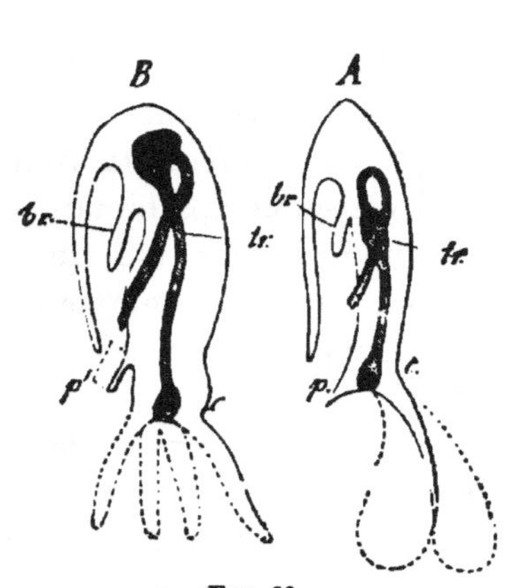

FIG. 29.

Schematic vertical sections of the types *pteropod* (*A*) and *cephalopod* (*B*). The heart is directed downward. *c*, cephalic part with indication in *A* of the fins, which form part thereof, in *B* of the arms; *tr*, intestinal canal; *br*, branchiæ, *p*, foot.

FIG. 30.

Polycera cristata, seen on the dorsal side:—*a*, anal orifice; *br*, branchiæ; *t*, tentacles.

organs performing the work of respiration. The skin comes to their aid and respires also. Sometimes there are no branchiæ; the respiration is wholly effected by the cutaneous surface and by the intestinal canal. This last surface is, moreover, clothed with a vibratile epithelium, like that of the branchiæ, and the oscillatory movements of these cilia are continued from the anus to the stomach. The anus opens and shuts by a rhythmical movement, and an uninterrupted current penetrates to the vicinity of the stomach (Gegenbaur).

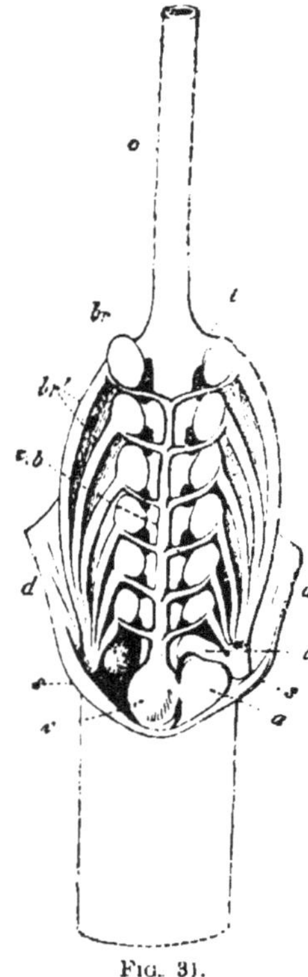

Fig. 31.

Respiratory organs of the *Myxine glutinosa* seen on the ventral side; *o*, œsophagus; *i*, internal branchial canals; *br*, branchial sacs; *br'*, external branchial canals uniting in a general branchial conduit in *s*; *e*, œsophago-cutaneous canal; *a*, auricle of the heart; *v*, ventricle; *ab*, branchial artery sending a branch to each branchia; *d*, wall of the body thrown outward and backward.

In a small group of mollusks the pulmonary organ appears. In these *pulmonate* mollusks the lung is only a simple sac, a cutaneous depression communicating with the exterior by an orifice. Some, as for instance the *ampullaria*, have at the same time a branchia and a pulmonary sac with contractile orifice. Of these pulmonate mollusks some are terrestrial, the others aquatic; and these last often come to the surface of the water to breathe. They need the free atmosphere, and in a closed vessel, as Spallanzani saw, they, without immediately suffering asphyxia, transform into carbonic acid the whole of the oxygen contained in the confined medium.

The two chief respiratory modes, the branchial mode and the pulmonary mode, are observable in the vertebrates, but in a higher degree of perfectionment. Here the external tegument still respires but accessorily, and the chief part of the function is accomplished by special organs. Nevertheless the respiratory apparatus is always more or less connected with the intestinal canal; it seems to be a diverticulum thereof. In fishes, in which the branchial apparatus attains its maximum of development, it is supported by arcs appertaining to the visceral skeleton (Figs. 31 and 32). The water penetrates by the mouth of the animal, and is not expulsed till it has traversed the branchial clefts, and bathed the surface of the respiratory mucous membrane. The amphibians have external branchiæ under the form of foliations or of ramified filaments, supported by the branchial arcs.

The lung, in its most elementary form, is a simple vascular sac. It appears first of all among the amphibians, where it co-exists with the branchiæ, of which we have spoken above, and which are sometimes temporary, sometimes permanent (perennibranchials). The lungs of the *proteus* tribe are not more than elongated sacs. Many reptiles have also simple lungs. In the anourians the lung is already subdivided into secondary cavities or alveoli, which augment considerably the respiratory surface.

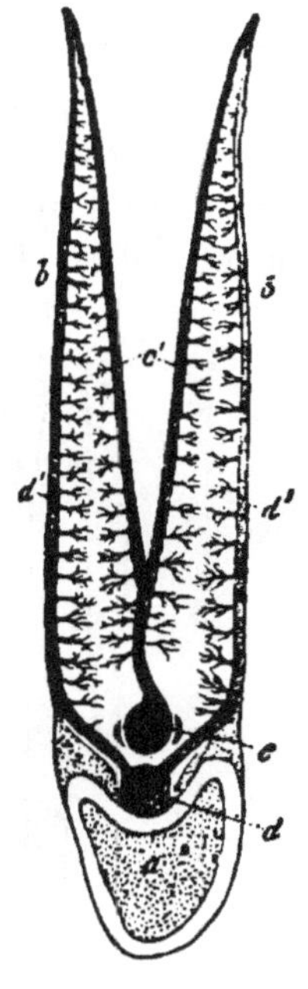

FIG. 32.
Vascular distribution in the branchial laminæ: *a*, section of osseous branchial arc; *bb*, the branchial lamellæ; *c*, branchial arteries; *c'*, ramuscules of the arteries in the lamellæ; *d*, branchial veins; *d'*, *d'*, venous ramuscules in the lamellæ.

There has been a disposition to consider as a rudimentary respiratory organ the velatory bladder of fishes (Fig. 33). It is certain that we find in this cavity carbonic acid, azote, and hydrogen; we also meet with enormous proportions of oxygen. According to Biot and Laroche, this proportion can rise to 70 in 100 in the fishes caught at more than 50 mètres deep, and only to from 26 to 29 in 100 in the others, though the deep layers of water are not more oxygenised than the layer on the surface.

In reality, and making abstraction of the general form of the apparatus, there is no fundamental difference between the branchiæ and the lungs. The branchiæ can even act in the open air on the single condition of being kept in a state of sufficient humidity, and on the other hand, the surface of the pulmonary mucous membranes is constantly lubrified by a liquid secretion. We can feed carps in the open air, in moss saturated with water;

we can even thus make them travel considerable distances without killing them. Branchial surfaces and pulmonary surfaces absorb oxygen, exhale carbonic acid, and transform more or less perfectly arterial blood into venous blood. This double current can be seen, so to speak, when we examine, by means of the microscope, the branchiæ plunged in water. There is established, from their contact, in the liquid medium a sort of circular movement; the corpuscules floating in the water seem to be attracted on one side and refilled on the other.[1]

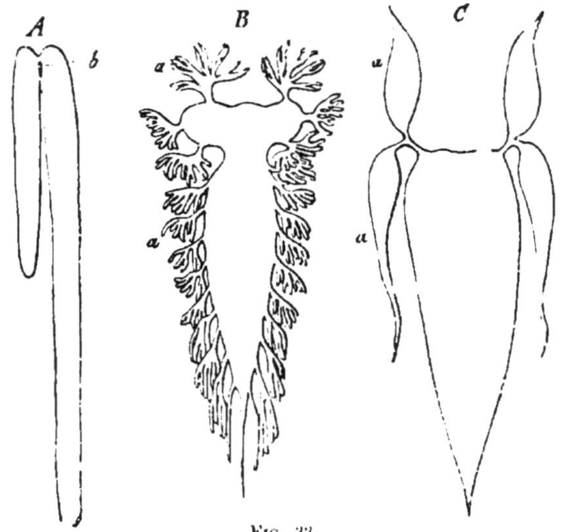

Fig. 33.
Diverse forms of natatory bladders :—*A*, *Polypterus bichir*; *B*, *Johnius lobatus*; *C*, *Corvina trispinosa*; *a*, annexes of the bladder; *b*, its orifice.

We have not here to describe in detail the pulmonary apparatus of the superior vertebrates. Let us merely say that the more the animal is perfect the more the pulmonary pouch is divided and subdivided into lobes, lobules, terminal *culs-de-sac* or cells, in ever increasing number, whence in the superior mammifers an enormous respiratory surface (Fig. 34). At the same time that the pulmonary sac grows more perfect it separates

[1] Dugès, *loc. cit.*, t. II., p. 522.

more and more from the digestive tube ; it acquires aërian canals more and more long and ramified, kept constantly open by a cartilaginous skeleton, and supporting at one or more points of their passage the organs of the voice. Over all its surface, the aërian tree, trachea, and bronchi, is clad with vibratile epithelium ; and this is the case for all surfaces in the whole animal kingdom. The vibratile vestment is lacking only on the surface of the cells or pulmonary alveoli.

Muscles, more or less numerous, impress on the costal frame

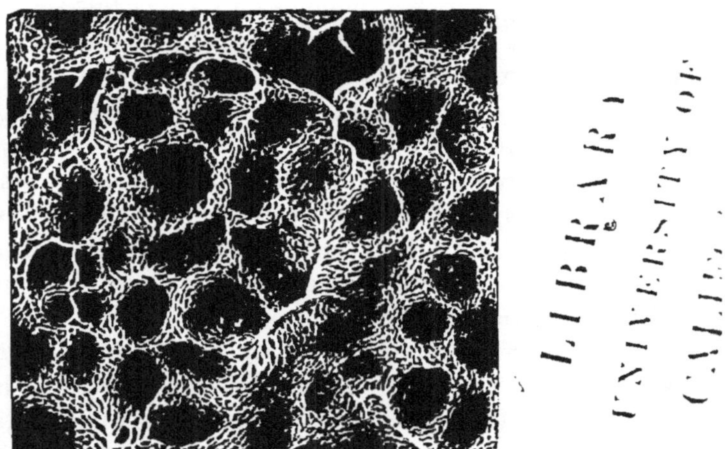

FIG. 34.
Capillary network of the pulmonary vesicles.

which covers the lungs and keeps them dilated alternative movements of inspiration and expiration; but the diaphragmatic partition which so powerfully aids inspiration in man and the mammifers is completely lacking in reptiles, and is only in a very rudimentary state in birds.

We are compelled, that we may not indulge in a parergon, to renounce all detailed anatomical description. Nevertheless we must dwell at rather more length on the physiological province of respiration, on the indispensable aid which it brings to the work of nutrition in the complex animals.

CHAPTER XVI.

OF THE PHYSIOLOGICAL OFFICE OF RESPIRATION.

It is necessary to distinguish between the fundamental biological fact of the absorption of oxygen, which is one of the first conditions of nutrition, and the physiological function properly so called, which is simply the physiological process employed to render this absorption possible and easy. The absorption of oxygen is a general fact; every organised being absorbs oxygen under penalty of death; this is true as regards the most humble vegetal and animal organisms, as well as the highest. It is true of every living being in all the stages of its existence. William Edwards and Colin[1] have shown that seeds do not germinate in a vacuum, and it was sufficient for Réaumur to cover an egg with an impenetrable varnish to prevent the embryon from developing.[2] But respiration, properly so called, only and truly exists in those cases where a special branchial, trachean, or pulmonary apparatus is the means of furnishing with respirable air the entire organism, and of exhaling those gases which are hurtful to or useless for the maintenance of life.

Whatever may be the diversity of the organological processes, the object of all respiratory apparatus is to bring the capillary vessels containing blood more or less vitiated by nutrition into as easy contact as possible with the atmosphere. If we were to employ the language of the ancient natural philoso-

[1] *Comptes Rendus de l'Academie des Sciences*, 1838.
[2] *Ibid.*, year 1735.

phers, who attributed intentions to nature, we should say that this was the idea of respiration.

The most easy conditions of the exchange of gases being best realised amongst the higher mammifers, we can take as a type respiration in man, the only one indeed which has yet been thoroughly studied. Moreover it will be easy for us to indicate, in the course of this exposition, the analogies or differences observable in the remainder of the animal kingdom.

The capillary vessel is preeminently the seat and agent of material exchanges in the animal organism. Wherever it may be, it adapts itself readily to the osmotic phenomena. In the web of the tissues, when the capillary partition separates liquids, it is these liquids which mingle through the thin intermediate membrane. In the lungs of man, the wall of the capillaries separates the sanguineous liquid, holding gases in solution, from the exterior atmosphere, and the exchange is made between the gases.

In the capillaries of the lungs (Fig. 34), as in those of the tissues, the electro-motor currents play an important mechanical part. But whilst in the tissues they are the agents of the expulsion of the oxygen across the wall of the capillaries, in the lungs they are precisely the reverse. In effect, in the lungs, the oxygen is situated outside the circulation; consequently the direction of the electric current is inverse. The electro-motory capillary-current tends therefore to cause the expulsion of the carbonic acid.[1]

In spite of the action of breathing in the lungs of the vertebrated animals, and especially of the mammifers, though there are constantly and regularly an alternate entrance and exit of gas in the pulmonary cavity, respiration, gaseous exchange through the capillary wall, is nevertheless a continuous uninterrupted phenomenon. In effect, at each expiration the lungs expulse only a very small portion of their gaseous contents. According

[1] Becquerel, *loc. cit.*, and *Des Forces Physico-chimiques*, 8vo. Paris, 1875. Onimus, *Des Phénomènes Electro-capillaires* (*Revue Scientifique*, 1870, nº 42).

to the calculations of Menzies, Goodwin, Davy, &c., this portion may be valued at a fifth, a seventh, even an eighth of the pulmonary capacity. The absolute volume of gaseous contents varies remarkably, like the pulmonary capacity, according to individuals and ages. Hutchinson has given the following averages of this capacity :—[1]

	litres
From 15 to 25 years	3·590
,, 25 to 30 ,, ...	3·623
,, 35 to 40 ,, ...	3·720
,, 40 to 45 ,, ...	3·459
,, 50 to 55 ,, ...	3·215
,, 55 to 60 ,, ...	2·970

In a gigantic and athletic American the pulmonary capacity rose to 7·082 litres.

The pulmonary capacity, then, bears a very obvious and constant relation to youth and strength, or more generally to the degree of vitality. Of this Hutchinson's American is a striking example. After two years of idle and dissolute life, his enormous pulmonary capacity fell first to 6·364 litres, then to 5·222 litres, and finally, a little later, he succumbed to a tubercular malady.

The mean volume of gas inhaled at each inspiration may be estimated at about one-third of a litre, or from nine to ten cubic metres in twenty-four hours. As to the volume expired, it naturally is much the same as the volume inspired, with a slight diminution of a fiftieth to a seventieth ; for there are other outlets.

Evidently the pulmonary gases, expelled at each inspiration, are especially those nearest to the outside, those which fill the trachea and the large bronchiæ. As to the air inclosed in the pulmonary cells and in the fine bronchial ramifications, it renews itself by diffusion, by the gaseous mingling from point to point; so that at last the respiring surface, the vascular wall of the

[1] Hutchinson, *On the Spirometer*, 1846 (analysed in *Arch. Gen. de Med.*, 1847.)

pulmonary cells, is in contact with a gaseous mixture of which the composition is obviously always the same. The exchange of gases can, then, be effected regularly without interruption, and even without any sudden variation.

The lungs receive the exterior air, that is to say, a gaseous mixture of about 21 of oxygen and 79 of azote, besides a small quantity of vapour of water and about four ten-thousandths of carbonic acid. They restore to the atmosphere air remarkably impoverished of oxygen, since it scarcely contains more than 18 in 100 (Dugès), but impregnated instead with carbonic acid and the vapour of water.

From numerous observations, among which we must first cite those of Lavoisier, next those of Regnault and Reiset,[1] it is proved that the quantity of oxygen contained in the carbonic acid exhaled by the lungs is sensibly inferior to the quantity of atmospheric oxygen absorbed. Lavoisier had already noted this interesting fact. MM. Regnault and Reiset saw that this proportion varied very little in the same animal species, living under normal conditions. The proportion between the oxygen of the carbonic acid exhaled and the oxygen absorbed has been, according to their observations, from 0·743 to 0·750 in the dog, 0·920 in the rabbit, &c. Atmospheric oxygen indeed ought to be considered as a true aliment; it combines with the plasmas, with the matter of the anatomical elements, in a word, with the living substance; it enters into numerous compounds, and is eliminated through diverse channels. The organism only takes from the atmosphere that quantity of oxygen which is necessary for it: it rations itself. In an atmosphere containing 40 and even 60 in 100 of oxygen, the absorption of this gas is not sensibly more considerable than in the common air, unless there is variation in the regimen. In this last case, on the contrary, the quantity of oxygen absorbed varies; for there must be more or less of it, according to the variation in the proportion and the nature of the substances to be assimilated in the sanguineous

[1] *Annales de Chimie et de Physique*, 2e serie, t. XXVI.

liquid. Thus, the deficit in the quantity of oxygen given back to the air by the exhalation of carbonic acid is greater in the carnivora than in the herbivora. If animals are fed on grain, the proportion is sometimes inverse; there is then an excess in the exhaled oxygen.

The exchange of gases through the pulmonary membrane is much more a physical phenomenon than a physiological one; the azote must, then, also penetrate the blood through the wall of the capillaries; this is, in effect, what happens. But here, reversing what takes place in the case of oxygen, the pulmonary exhalation restores to the air more azote than it has borrowed from it, at least in most cases.

Regnault and Reiset found that all animals subjected to their habitual regimen always generated an excess of azote. Despretz demonstrated the same thing by more than two hundred experiments.[1] Boussingault, proceeding by the indirect method, that is to say, by comparing the quantity of azote introduced by the aliments with the quantity expelled in the solid and liquid matters, observed that the first was always greater than the second.[2] We can thence infer that a certain portion of the azote of the aliments is exhaled by the lungs. In effect, if an animal is made to respire in an atmosphere composed of seventy-nine parts of hydrogen, and twenty-one parts of oxygen, there are seen to be a simultaneous absorption of hydrogen, and an exhalation of azote. In this case the production of azote must evidently be ascribed to the organism. Besides, the absorption of hydrogen changes into certainty the great probability of the absorption of azote in the normal air. We must, then, admit, with Edwards, that in the ordinary atmosphere the disengagement and absorption of azote by the pulmonary surface always take place simultaneously, and that the only variation is caused by the augmentation or diminution in the quantity of azote exhaled.

In effect, it is the quantity of azote exhaled which varies;

[1] *Annales de Chimie et de Physique*, 2ᵉ série, XXVI.

[2] *Economie Rurale*, t. II.

sometimes there is an excess, sometimes a deficit. Thus, there is a retention of azote in the organism during inanition, and even several days after its cessation. It is the same when the animal under observation is suffering, perhaps because then it submits itself spontaneously to a relative inanition.

The pulmonary membrane, offering to contact with the air a large surface at a tolerably high temperature, is naturally the seat of transpiration, of an abundant aqueous evaporation. Lavoisier and Seguin, whom we must always quote when respiration is in question, have estimated at 7·90 grammes in an hour, or about 569 grammes, the quantity of water exhaled in twenty-four hours by the pulmonary surface of man.[1] The average of the numerous experiments of Valentin[2] is 540 grammes; consequently, very nearly the number indicated by Lavoisier and Seguin. This quantity is besides very variable. It may be augmented at will in animals, for example by injecting water into their veins.

Complex reactions take place in the blood, still as little known as those which cause the elaborated sap in plants. The grand fact is the fixation of oxygen; but we have seen that there were also the absorption and exhalation of azote. Lavoisier and Seguin had already deduced from their observations that a portion of the water exhaled by the pulmonary surface was directly formed in the blood by the oxydation of the hydrogen. Modern observations have confirmed this view, which may, with great likelihood, be extended to aqueous excretion in all its forms; for the lung is only one of the numerous channels through which water goes forth from animal organisms.

The most complete researches have been made, touching the physical phenomena of respiration, especially with regard to animals having lungs, birds and mammifers; but these phenomena, viewed in a general manner, differ little in branchial respiration. On the

[1] *Premier Mémoire sur la Transpiration* (*Mémoires de l'Académie des Sciences de Paris* (1790), and *Annales de Chimie*, t. XC.).

[2] *Lehrbuch der Physiologie*, t. I.

surface of the branchiæ, as on the surface of the lungs, there is a continual exchange of gas between the sanguineous liquid and the aquatic medium. The quantity of air dissolved in water is not considerable; but this air is respired by cold-blooded animals, and it is besides much richer in oxygen than the atmospheric air. In effect, the air dissolved in the water is composed of 29·8 of oxygen, 66·2 of azote, and 4 of carbonic acid. Now Humboldt and Provençal observed this air to contain only 2·3 of oxygen, 63·9 of azote, and, in compensation, 33·8 of carbonic acid, after having been respired a certain time by some fishes.[1]

If observation has been able to scrutinize and measure the respiratory phenomena, properly so-called, that is to say, the exchange of gases through the capillary membrane, it has not succeeded so well as regards the nutritive part of the respiration. Nevertheless, interesting facts have been collected, and they permit us already to form a general idea of the office which absorbed oxygen fulfils, and of the succession of chemical transformations with which they co-operate.

The first and most striking phenomenon is the sudden change of colour which the sanguineous liquid undergoes immediately after absorbing the atmospheric oxygen. The blood, which arrived in the capillaries in the state of black or venous blood, becomes rutilant and vermilion, as soon as it has exchanged its carbonic acid for oxygen; it is then called red or arterial blood. The change of colour is instantaneous. For example, the blood in the carotid artery of an animal blackens and reddens alternately and suddenly, according as we prevent or permit the access of air to the respiratory apertures.

Of the oxygen absorbed by the blood, a part remains in a state of solution in the sanguineous plasma, and probably co-operates with the oxydations in the isomeric transformations of the nutriments; but the larger portion of this oxygen is fixed by the hæmatia. These corpuscules absorb many volumes of it; they are truly the vehicles of the oxygen. They impregnate

[1] *Mémoires de la Société d'Arcueil*, t. II.

themselves with it, and convey it to the capillary vessels of fine calibre, which penetrate the very web of the tissues. But this part of the system of circulation is the special field of nutritive exchanges. There, in effect, the blood is no longer separated from the anatomical elements and the liquids which bathe them, except by the thin homogeneous partition of the capillaries. The blood then yields to the tissues assimilable materials and takes back disassimilated materials. In the first category is comprehended oxygen; in the second, carbonic acid. Without the incessant access of the first, without the incessant expulsion of the second, the succession of chemical reactions and transformations, which, as a whole constitute nutrition, would cease to take place; the whole vital movement would be retarded and stopped.

The oxygen is not simply dissolved in the substance of the sanguineous globule; it seems to be combined with this substance. In effect, if we inject into the veins of a dog an eager absorbent of oxygen, for example, pyro-gallic acid, the animal does not appear sensible of it; its globules do not part with their provision of oxygen.[1] On the contrary, certain substances, which are capable of yielding oxygen, are deprived of it in the blood; the globules carry it away from them. Thus, per-oxyde of iron injected into the circulation, is found again in the urine as protoxyde. The globules may be considered as condensers of oxygen; they are pre-eminently the exciting agents of life; also, in the operation of transfusion it is sufficient to inject into the veins defibrinated blood; probably the globules alone would suffice. Certain substances have the property of killing the globules, of rendering them for ever unfit for the absorption of oxygen. Such is particularly oxyde of carbon; but even in cases of poisoning, by this substance, persons have often succeeded in reviving the poisoned animals, in resuscitating them by having recourse to transfusion, that is to say, by replacing the dead globules with living ones.

[1] Cl. Bernard, *Leçons sur les Liquides de l'Organisme.*

The nutrimentary matters, prepared by the labour of digestion and diffused into the circulation, are exchanged through the wall of the capillaries with the products of denutrition. These last are the result of an oxydation, the principal seat of which is probably in the very substance of the anatomical elements, and not in the capillaries themselves, as is still affirmed in a number of special treatises. The capillaries are organs marvellously adapted to the exchanges between the blood and the anatomical elements; but these last are the laboratory itself, where the principal transformations are accomplished. The comparison of the two kinds of blood permits us to fix, in some degree, the balance of gain and loss. After having served as nutrition, the ternary matters not azotized are brought back to the state of carbonic acid and water; their combustion is, then, complete. They have attained their maximum of oxydation. On the contrary, the quaternary or albuminoidal substances only undergo an incomplete combustion, of which the product is a certain quantity of carbonic acid and water, besides a residue of crystallizable quaternary elements, of which we have already spoken, and which are eliminated by the various emunctories of the economy.

From the comparison of the two kinds of blood springs another interesting fact, namely, that the arterial blood has a composition sensibly identical in the various regions of the organism, whilst the composition of the venous blood varies much according to different organic localities. In effect, the arterial blood is the general nutritive reservoir; from it each anatomical element satisfies its needs; but these needs vary according to the chemical constitution and function of the elements. Each anatomical element, then, takes, according to its affinities, one substance in preference to another. Besides, and above all, it makes the matters which it absorbs undergo a special elaboration, and restores to the circulation also a special nutritive residue, whence results necessarily the local diversity of the venous blood.

On an average, respiration, or at least the primordial fact of respiration consists, according to Regnault and Reiset, of an absorption of oxygen, varying from 10 grammes to 0·09 in an hour, and in a kilogramme of living matter. Besides, this oxygen transforms itself into carbonic acid and water in the proportion of 0·800 of a gramme combined with the carbon, and of 0·200 of a gramme combined with the hydrogen. Finally, there is an exhalation of azote representing about six-thousandths of the weight of the oxygen consumed. These rough figures must be accepted with great caution. Respiratory combustion does not simply produce water and carbonic acid. These two last bodies only result from the combustion of ternary substances; but quaternary substances oxydize themselves also, and their oxydation gives birth to crystallizable, azotized elements, eliminated by other outlets besides the pulmonary surface.

However this may be, an important general fact results from the observations of Regnault and Reiset, corroborated besides by those of many others; namely, that the energy of the vital combustion has a close connection with the nutritive and functional movement.

Thus, the quantities of oxygen absorbed, and of carbonic acid and azote exhaled, are about seven times greater amongst birds than amongst mammifers. In compensation, respiratory intensity, estimated after the same manner, is nearly ten times stronger in the mammifers than in reptiles (Regnault and Reiset).

The respiration of insects, when these animals are in full activity, has plainly the same energy as that of the mammifers, whilst the earthworms do not respire more than reptiles; but in these comparisons the size of the animal must be taken into account. In effect, the smaller the animal the larger the proportion which the surface bears to its bulk; consequently oxydation, which, as we shall see, is the principal source of animal heat, must be more active since the coldness caused by peripheric radiation is greater.

Andral and Gavarret have made some very excellent observations upon respiration in man.[1] These observations bear upon man from sixteen to thirty years of age, the experiment lasting from eight to thirteen minutes, between one and two o'clock in the day, during the same interval between repasts, and under the same conditions of muscular exertion. According to them the consumption of carbon in an hour undergoes, in proportion to age, the variations indicated in the following table :—

From 8 to 15 years it is	70·42
,, 15 to 20 ,, ,, ...	10·76
,, 20 to 30 ,, ,, ...	12
,, 30 to 40 ,, ,, ...	11
,, 40 to 50 ,, ,, ...	9·17
,, 50 to 60 ,, ,, ...	11·07
,, 60 to 70 ,, ,, ...	10·25
In an old man of 76 years, it is	6
,, ,, 92 ,, ...	8·5
,, ,, 102 ,, ...	5·9

Andral and Gavarret not having noted the weight of the persons under observation, the figures above quoted lose much of their value; nevertheless, they furnish the confirmation of the general law which we have enunciated above, namely, that the respiratory activity varies in proportion to the intensity of the nutritive movement, that is to say, of life.

The observations of Scharling complete those of Andral and Gavarret, for they furnish an indication of the weight of the persons. Now these observations prove that oxydation is more energetic in the child than in the adult. A child of nine years consumed by kilogramme and by hour 0·25 of a gramme of carbon; adults of from twenty-five to twenty-eight yearsonly consumed on an average 0·12 of a gramme.

Also, according to Scharling, the difference of sex influences the energy of vital combustion, even before puberty; in the human species girls consume less carbon than boys. But Andral and Gavarret have moreover shown that the difference is notable

[1] *Recherches sur la Quantité d'Acide Carbonique Exhalé par le Poumon* (*Ann. de Chim. et de Phys.*, 2ᵉ série, 1843).

at the middle period of life. During the whole period of ovulation, from the first appearance of the menses to their cessation, the exhalation of carbonic acid is less in woman than in man; but the equality is sensibly re established during pregnancy and after the critical age.

Many other causes vary the exhalation of carbonic acid. Thus, in animals having a fixed temperature or hot blood, it augments in proportion as the exterior temperature decreases. On the contrary, in cold-blooded animals, respiratory combustion diminishes in proportion as the temperature is lowered. This is because the first resist refrigeration by a more active combustion, whilst the second, being incapable of producing a sufficient internal warmth, grow gradually colder till the moment when they fall into a state of hibernal sleep or die.

Every functional activity, whatever it may be, has as corollary, or rather cause, a more active combustion within the organ or functional apparatus.

After having eaten copiously insects respire more energetically, and they also die more promptly in confined air, or in an irrespirable gaseous medium.

It has been observed that there is a proportion between the digestive activity in frogs and the quantity of oxygen contained in the ambient medium.

According to Vierordt, Valentin, Scharling, and Horn, the exhalation of carbonic acid augments rapidly in man after a repast. Analogous facts have been stated by Spallanzani concerning snails when fasting, by Story of various insects, and by Boussingault of turtle-doves.

In inanition, there is a lesser absorption of oxygen, and likewise a lesser exhalation of carbonic acid. The animal derives from the atmosphere more oxygen than it restores to it. This is simply because then it nourishes itself at the expense of its own tissues; it eats itself, and consequently respires as a carnivorous animal, but one badly nourished. In effect, Regnault and Reiset have seen that animals fed upon butcher's meat and aliments rich in fatty matters, exhale less carbonic acid, and, consequently, burn more hydrogen.

The fewer the functions exercised by an inanitiated animal, the less carbon it expends. This is the reason why Ridder and Schmidt have observed that when animals were deprived of sight the degree of their diurnal carbonic exhalation tended to equal that of their nocturnal exhalation.

Moleschott has shown that the action of light upon the skin notably augments the intensity of the respiratory phenomena.[1]

During the day, under the influence of various kinds of activity, the production of carbonic acid is, according to Scharling, less by a quarter than during the nocturnal slumber. But it is especially during the hibernal sleep that the absorption of oxygen is reduced to its minimum. Regnault and Reiset have seen it fall to a twentieth part of what it was in a waking state.

Contraction and muscular effort demand a large absorption of oxygen, as Lavoisier and Seguin had already observed. This is because, in a given weight of living matter, respiratory combustion increases in direct proportion to the muscular activity.

Behind every biological activity there is an oxydation of the anatomical elements. No organ escapes this law, and the nervous centres are as much in subjection to it as the other organic apparatus. Every thought, every volition, every sensation, corresponds to an oxydation of the living substance, as well as every secretion, every movement, &c.

In speaking of innervation, we shall have to consider, in connection with it, the cerebral functionment. Lavoisier, after having pointed out the relation which exists between muscular activity and the exhalation of carbonic acid, wrote thus: "This kind of observations leads us to compare the displays of force, between which no connection would seem to exist. We may know, for example, to how many pounds in weight the efforts of a man reciting a discourse, of a musician playing an instrument would correspond. We can even estimate how much there is of

[1] *Wiener Medicinische Wochenschrift*, 1855.

the mechanical in the labour of the philosopher who reflects, of the literary man who writes, of the musician who composes. These effects, considered as purely moral, yet possess some physical and material attribute, which permits us to compare them with the efforts of the labouring man. It is therefore not without justice that the French language has included under the general denomination of *labour*, the efforts of the mind as well as those of the body, the labour of the cabinet and the labour of the artisan."[1]

Truly it is singular and regrettable that, even in our days, and in spite of the brilliant triumph of the doctrine of the correlation of the physical forces, no physiologist has brought this beautiful idea of Lavoisier into the domain of facts, rigorously and minutely observed. The last champions of vitalism and animism always repeat that thought escapes the law of the correlation of physical forces, that its mechanical equivalent cannot be determined. Now it is incontestable that every cerebral activity answers to an absorption of oxygen, which manifests itself by a larger exhalation of carbonic acid, and by the renal excretion of certain products of oxydation. It would surely be possible to estimate approximately in calories, and consequently in kilogrammètres, this process of oxydation. Thus the mechanical equivalent of thought, and even of the different modes of thought, would be determined.

In vertebrated animals with lungs, the movement of the ventilating apparatus of the lungs, of the thoracic frame, are generally executed automatically; but they are much more subject than the movements of the heart, to the will, which can accelerate, retard, or suspend them; it is, then, natural that certain lesions of the nervous centres immediately affect, not directly indeed the exchange of gases on the pulmonary surface, but the alternate movements of the thoracic frame. In effect, certain wounds of the spinal marrow, in rabbits and guinea-pigs, cause a slackening of the respiration, and a gradual coldness; in a word, a state very

[1] Lavoisier, *Mémoires de l'Académie des Sciences*, 1789.

analogous to hibernation. In the mammifers it is a very limited region, situated at the origin of the spinal marrow, in the rachidian bulb on a level with the V of grey substance, comprehended in the posterior angle of the fourth ventricle. Now the integrity of this region is so necessary to the accomplishment of the respiratory movements that a section performed at this point at once destroys both respiration and life. This celebrated point in the spinal marrow is known as the *nodus vitalis*. Galen had already pointed it out; but only in our days, through the minute researches of Flourens, has its precise position been determined.

In reptiles, the cutting of the nodus vitalis has a much less striking effect than in the mammifers; it effectually destroys the respiratory movements, but does not immediately kill the animal, a result very easy to comprehend, and which is attributable to the feebler centralization, the smaller energy of the respiratory apparatus. Here the skin comes to the aid of the lungs in a great measure; whence batrachians are of all reptiles those which survive the longest the cutting of the nodus vitalis. This section besides, not only destroys the respiratory movements, but also the whole of the voluntary movements of the trunk, the members, and the head, since it interrupts most of the conscious and voluntary communications between the brain and nearly the whole of the other organs.

Our object being, not to write a special treatise on physiology, but rather to give a general idea of the process by which nutrition is accomplished in the animal kingdom, we shall here terminate this brief but adequate description of respiration. We have now seen how the animal organism elaborates and absorbs the materials which it borrows from the outer world; how these materials, having become assimilable, circulate throughout the whole of the animal mechanism, and how atmospheric oxygen, indispensable to the accomplishment of the primary phenomena of nutrition, is absorbed. It remains for us to examine how the organisms free themselves from the non-gaseous products of dis-assimilation; how also they fabricate, at the expense of the plasmas, certain substances destined afterwards to play more or less important parts in the different functions. Thus it is that we have to speak of excretion and secretion.

NOTE.

A *litre* is about a quart English. A *kilogramme* is a thousand grammes. According to the new system of weights and measures introduced into France at the Revolution, *deci* as a prefix means tenth; *centi*, hundredth; *milli*, thousandth, when the principal unit is progressively fractionised :—*deca* as a prefix means tenth; *hecto*, hundredth; *kilo*, thousandth, when the principal unit is progressively augmented or multiplied. Thus *millimètre* is the thousandth part of a *mètre*; *hectogramme* a hundred grammes, and so on. In Professor Roscoe's *Lessons in Elementary Chemistry* a comparison of the French metrical with the common measures is given.—*Translator.*

CHAPTER XVII.

OF SECRETION AND EXCRETION IN GENERAL.

Every organised and living element assimilates and dissimilates incessantly. But in proportion as the organism is complicated and differentiated, the conditions of assimilation become more difficult; and this function needs for its accomplishment the auxiliary apparatus, digestion, circulatory and respiratory, which we have passed in review. Besides, assimilation would be impossible if the waste of the immediate principles were not expulsed in the degree of its production. In short, dissimilation is the indispensable corollary of assimilation, and like this last it needs special apparatus wherever the structure of the organism is too complex. These special apparatus are the glands of which we have by and by to give a general description. Nevertheless along with the special organs of expulsion, the elementary property of simple exhalation still continues. The most marked example of this primitive mode of elimination is certainly the gaseous and aqueous exhalation, which is so extensively effected by means of the lung.

But besides the glands serving for simple expulsion, there are others whose distinctive office consists in forming, at the expense of the nutritive liquids, special substances, destined to play a part in the accomplishment of one of the grand functions of the living organism. The glands of the first type are called glands of *excretion*, the others, glands of *secretion*. We may give as an example of the first the glands which excrete urine and sweat, that is to say the kidneys and the sudoriparous glands. The salivary glands and the liver represent well the glands of the second type.

The higher the perfection of an organised being the more its organs of excretion and secretion are numerous and specialised. From this general proposition, applicable moreover to all organic apparatus, we may infer that the organs of elimination are very rudimentary in the vegetal kingdom. Lacking there completely are the complex glandular organs, furnished with canals of discharge, such as exist in animals. We might bring the vegetal glands into relation with the glands closed or without the excretory conduits of animals. In both cases, in effect, the secretory organ is represented by cells elaborating at the expense of the nutritive liquids special substances, which come forth from the cells, either by exosmosis or by rupture of the cells which have formed them, but are not then received into special conduits; for the network, called network of *secretion*, into which is poured the essential oil of the *helianthus*, the resinous gum of certain umbelliferæ, the limpid balm of the conifers and the terebinthaceæ, is merely composed of meatus more or less ramified, due to simple écartements of cells.

The secreting cells of vegetals are sometimes isolated and sometimes grouped. As example of the first we may cite the spherical cell, filled with a viscous or odorous substance, which terminates certain vegetal hairs. Other glandular cells, isolated, are dispersed over the parenchyma of the tissues, for instance the camphoric cells of the leaves of the *Camphora officinarum*. Often the glandular cells are grouped in small masses, for instance, that which contains the essential odoriferous oil in the rind of the citron. Sometimes the walls of these cells open or are reabsorbed, and the place they occupied is transformed into a reservoir. But these are special secretions. As to the form under which are eliminated the products of disassimilation of the plant, it is still very little known. Must we regard as products of excretion the matters resinous, cereous, or slimy, and so on, which are spread over the surface of certain plants? But we do not find them in all plants? yet all plants must excrete. Must we admit, with certain botanists, that the products of disassimilation are borne along with the descending sap, and expulsed by the roots themselves? But all this is very improbable. In reality vegetals disembarrass themselves ill and imperfectly of their disassimilated products. Water indeed is exhaled on the surface of the leaves and of the stalk. Certain semifluid substances are also expulsed by simple exosmosis. But considerable residua, notably a great quantity of mineral substances, remain in the vegetal tissues, encumber them, and determine the death of the anatomical elements which they contain. This is probably the cause of the death and the fall of the leaves, of the decay and disappearance from within to without of the ligneous elements in the centre of the dicotyledonous trees. Also the blackish cadaveric detritus which often bathes the internal wall of hollow trees contains products of decomposition, ulmine and its derivatives.

It is in the animal kingdom alone that we find organs of secretion numerous and perfectionated. Here the life is more intense, and special organic agents are necessary to regulate the expulsion of the disassimilated products, and form humours or substances indispensable to the accomplishment of physiological acts of the first order.

As we have already remarked, a distinction is made between the secretion which fabricates new principles, and the excretion, which merely furnishes passage to the materials of disintegration. But the secretion itself is effected according to two principal modes. It is operated either by glands without excretory conduits, or by glands with excretory conduits.

The difference between these two glandular types is however purely morphological. In the glands, with excretory conduits, the product of the secretion is poured by a special canal on the surface of the body, or of a mucous membrane (Fig. 35). On the contrary, the closed glands act on the composition of the plasmas; they fabricate, at the expense of the blood, special products, which pass afterwards into the circulation by osmosis. But essentially closed glands and glands with canals are very analogous. In both the secreting elements are cells, anatomically comparable with the epithelial cells covering the mucous membranes and the skin. The mode of action of these cells has, moreover, nothing exceptional. Every anatomical element is a living laboratory, borrowing by election from the nutritive liquids materials which it fashions and transforms. This is what the glandular epithelial cells do (Fig. 35); but as in them the vegetative properties are very energetic, they transform with a greater activity. The special substance, as soon as elaborated, is expulsed from the secreting cell either by simple osmosis or set at liberty

FIG. 35.
Human gastric gland, with acid secretion, as example of simple gland.

by rupture or resorption of the cellular wall. In the closed glands the product of the secretion traverses the wall of the capillaries, which are in contact with the secreting cells, and is afterwards conveyed into the circulation. This is what takes place, for instance, with the glycogenical cells of the liver, which we have already mentioned. The cells elaborate the materials which the blood of the intestine brings them, and restore to the fine capillaries in contact with them those same materials metamorphosed into sugar, which serves for the general nutrition. But if the glycogenical function of the liver is now well elucidated, this is far from being the case with many other closed glands. It would be needful to make for each of those glands the precise and methodical researches, thanks to which Professor Maurice Schiff has thrown so much light on the function of the spleen, or at least on one of its functions.

That the spleen and the pancreas could co-operate for the same general function might be deduced from comparative anatomy, as in certain vertebrates these two glands form only one. But the physiological bond uniting them was most difficult to discover.

Excellent experiments of Corvisart had already established that the pancreatic juice and even the infusion of the pancreas had the faculty of transforming the albuminoidal substances into peptones incoagulable by heat. In addition Meissner had shown that the formation of the pancreatine in the pancreas is intermittent, and that when a person is fasting the pancreas has no longer any digestive power. The case is moreover the same with the stomachal pepsin, which is formed intermittently at the expense of the peptogens introduced into the blood. But the formation of the pancreatine is subjected to physiological conditions far more complex; forasmuch as it is absolutely dependent on the integrity of the spleen. After the extirpation of this last gland, in effect:—

1. The neutral or acidulated pancreatic infusion no longer digests the smallest trace of albumine, whether the pancreas is taken in a fasting animal or an animal in full digestion;

2. In these conditions, albumine introduced by a fistula into the duodenum, when this is bound at its two extremities, no longer transforms itself into peptone, either in the fasting animal or the animal in full digestion;

3. Finally, always after the extirpation of the spleen the albumine introduced into the bound duodenum is well digested slowly by the duodenal juice, but is no longer transformed rapidly into peptone after the first hour of stomachal digestion, as happens always in animals still possessing a spleen.

To make amends, the peptic glands finding in the blood a richer provision of peptogens, forasmuch as the pancreas no longer absorbs any, secrete with much greater activity, and thus compensate in a large measure for the inaction of the pancreatic gland.[1]

These valuable researches, which put beyond doubt the fact of the intimate solidarity between the spleen and the pancreas, show how very complex the phenomenon of secretion can be. The cells of the spleen do not fabricate pancreatic ferment, and there is no trace of this ferment in the blood; nevertheless the function of the pancreas is abolished when the spleen is extirpated. In order that the pancreatic glands may elaborate pancreatine at the expense of the albuminoidal substances, or of certain of these substances, it is needful, previously, for the materials, absorbed by election into the pancreatic gland, to undergo, in the cells of the spleen, a first modification, a first process, the nature of which is still unknown.

But who can say how many of these physiological correlations are still unknown, and how fertile the study of the other closed glands may be, undertaken from that point of view?

The glands with excretory conduit must be considered as folds, as diverticules of the cutaneous or mucous teguments. The labour of specialisation bears especially on the epithelium or the epidermis of the tegumentary membrane; the mode of nutrition of this epithelium is modified, and its cells acquire the faculty of elaborating at the expense of the blood, here saliva, there gastric or pancreatic juice; at one point bile, elsewhere spermatic liquid, and so on.

The humours *secreted* have all two general characteristics:—

1. None of the humours is living; none is endowed with the property of continuous renovation which we have signalised in the plasmas. Their water is in great part free; it is not a constituent water, and it is charged with salts directly dissolved.

2. All these humours contain one or more quaternary principles,

[1] A. Herzen, *Sulla Digestione dell' Albumina Effettuata dal Succo Pancreatico e sulla Funzione della Milza.*

which we do not find in the blood (peptine, pancreatine, and so on).[1]

In reality, the secreting glands are at the same time organs of excretion. All of them take from the blood water and salts, substances to which they offer a passage without in any respect changing them. But besides they form, at the expense of the sanguineous materials, a special azotised product. Very habitually the agent of these chemical transformations is the epithelial cell. Sometimes, however, the metamorphosis can be accomplished in the wall itself of the gland, as is the case, for instance, with the mammary gland.[2] But the secreting element by excellence is the epithelial cell, whose numerous varieties have each their special affinities.

When neither the wall of the glandular organ, nor the epithelial cells which are contained therein, exercise any modifying action on the materials of the blood, but fulfil simply the office of a filter, offering a passage to certain substances and refusing it to others, there is merely excretion. Excretion is a biological act more simple than secretion, and comparable with the exhalation which goes on, for instance, on the pulmonary surface.

The excretory glands are never closed. Always they pour the humour which they filtrate on some point of the tegumentary, cutaneous, or mucous surface, and this humour is destined to co-operate with no ulterior physiological function. It is a dead product, whose expulsion is necessary, and whose retention in the organism would be fatal. It is the residuum of nutrition.

The excrementitial humours, of which the sweat and the urine are the types, are solely constituted by the water, holding in solution saline principles, and also crystallisable azotised substances, which, formed in the anatomical elements themselves by disassimilation, pass first of all into the blood, whence they are extracted and excreted by the glands.

[1] Ch. Robin, *Leçons sur les Humeurs Normales et Morbides*, pp. 29, 30, 32, and Introduction, p. xxvi., xxvii.

[2] Ch. Robin, *loc. cit.*, pp. 17, 18.

Neither the secreted humours nor the excreted humours are living, but the first have a composition perceptibly stable, reactions nearly fixed. On the contrary, the composition of the excreted liquids is variable, according to the greater or lesser activity of the gland, the proportion and the nature of the substances absorbed by digestion, the state of activity or of inaction of such or such organic apparatus, and so on.

We can range in graduated series the diverse modes of elimination, of excretion, and of secretion realised in the organisms.

The mode the most simple, the only one existing in most vegetals and inferior animals, is the direct passage of the expulsed substances across the tissues, without the intervention of any special organ; it is simple transudation or exhalation, such as continues to be in the superior animals on the pulmonary surface.

Then comes the excretory mode, that is to say, a sort of transudation through particular glands; such are the urinary and sudorific excretions.

At the degree, immediately superior, is found the secretion which is called *direct.* In effect, the gland drains, direct from the general mass of the materials of the blood, substances which it transforms and which have a physiological use. We may cite, as example, the salivary and biliary secretions.

The mode the least simple is that in which the open gland, furnished with a canal, needs the co-operation of a closed gland which prepares materials for it. The only example well demonstrated of this mode of secretion so complex, is that offered us by the pancreas and the spleen. But it is very possible that other correlations of the same kind exist in the superior organisms, that the salivary secretion, for example, is connected with the thyroidal body, and so on; and that the most of the closed and open glands are thus coupled together.

CHAPTER XVIII.

OF THE SECRETIONS AND EXCRETIONS IN PARTICULAR.

No exposition of a subject as a whole can attain a sufficient degree of clearness, unless it rests on the description of some particular facts. It is therefore not unsuitable after the general outline, contained in the preceding chapter, to consecrate some pages to the most important and most typical particular secretions and excretions.

Evidently we must place in the first line the hepatic secretion, of all the most complex. In effect, in the liver, the division of labour is far enough from being distinct, and this gland is simultaneously a closed gland, a secretory gland with excretory conduit, and a gland of excretion. Comparative anatomy, moreover, reveals to us a still higher degree of confusion. In truth, while in diverse vertebrates, for instance, the lizard, the spleen is merely adjoined to the pancreas, in the *Chimæra monstruosa* liver, spleen, and pancreas are united into one mass.

In the mollusks the same epithelial cells seem to be charged to secrete at the same time both sugar and bile; for this last humour always contains sugar. Besides there is no vena portæ, and the saccharo-biliary secretion is accomplished at the expense of the still badly arterialised blood.

In the superior mammifers the glycogenical cells of the liver are sometimes rounded, sometimes polyhedrical; their diameter is about 0.02 of a millimètre (Fig. 36). They contain granulations and one or two nuclei furnished with nucleoles. These cells are in contact with the finest capillaries of the vena portæ, from which they borrow the materials necessary for the fabrication of their amylaceous product (Fig. 37). This product, denominated *glycogen*, is analogous to the cellulose of vegetals; it is uncrystallisable, like starch, and, like starch, susceptible of transforming itself isomerically into glycose, fermentiscible and crystallisable. Once formed by the hepatic cell, the glycogen is yielded by it to the blood of the super-hepatic capillaries, and transforms itself into glycose, which normally is either destroyed

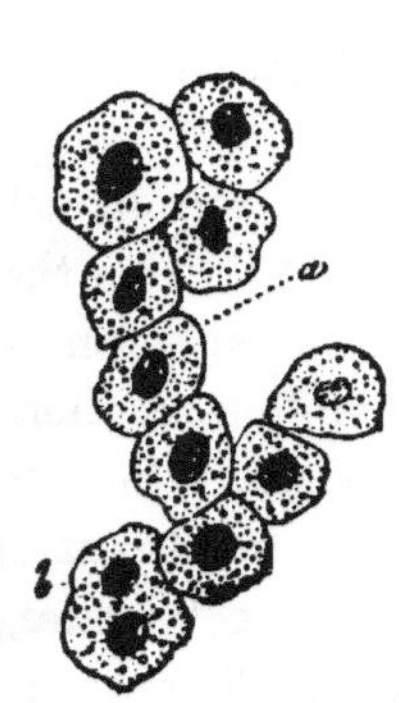

FIG 36.
Isolated cells of the liver; *a*, with simple nucleus, *b*, with double nucleus.

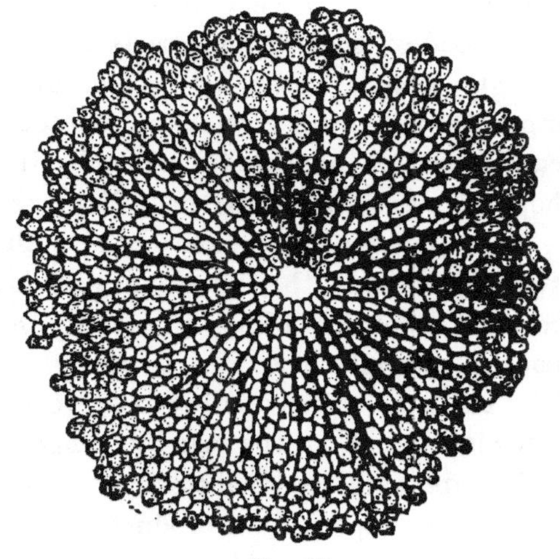

FIG. 37.
Arrangement of the cells of the liver in a lobule cut transversally, with the section of the hepatic vein in the centre.

by oxydation in the blood or utilised for the nutrition of the anatomical elements.

It is by no means certain that the glycogenical cells do not also contribute their share to the secretion of the bile. If they are in contact with the sanguineous capillaries, they are likewise in contact with the fine ramuscules of the biliary canals, a sort of excretory capillaries. Finally, we sometimes find in the normal state, often in the pathological state, yellowish granules and fine droplets of the same tint, which seem to be the colouring matter of the bile.

But this contribution, if it exists, is assuredly accessory, for the bile is specially formed in the fine culs-de-sac or *acini* whence go forth the biliary canals. These culs-de-sac receive thin arterial ramuscules, from which their walls and the epithelial cells, which clothe them, borrow the materials which they elaborate and transform into bile.

Between the glycogenical cells and the secretory acini of the bile there is not a very intimate physiological bond. In effect the blood of the vena portæ is indispensable to the activity of the glycogenical cells. But a ligature of this vein by no means stops the biliary secretion, alimented almost exclusively by the capillaries of the hepatic artery.[1] Inversely the ligature of the hepatic artery dries up the biliary secretion.

The glycogenical formation and the biliary secretion do not suffice to exhaust the functional activity of the liver. From the researches of M. Flint, it results that one of the substances poured out with the bile is so by an act of simple excretion. We mean a ternary substance, very rich in carbon, which is likewise found in the vegetal kingdom among the mushrooms. This substance is cholesterine, which, among the vertebrates, seems to be especially a product of disassimilation of the nervous system. The venous blood returning from the brain, that of the jugular vein, for instance, contains 0,801 of cholesterine, while the arterial blood darted from the heart towards the encephalon, that of the primitive carotid, contains only 0,774. Now this excrementitial product, which can be met with, moreover, in diverse secretions, is especially taken from the blood by the liver, which excretes it into the biliary liquid. There is here an act of simple election. The liver lets the cholesterine filter through, in the same way that to certain salts, for instance the iodide of potassium, it offers that free passage which it refuses to certain others, notably calomel.

[1] Oré, *Journal de l'Anatomie et de la Physiologie*, Paris, 1864.

The function of the salivary glands is much more simple than that of the liver. The culs-de-sac or *acini* of the salivary glands in cluster are lined with epithelial cells which have a single nucleus. These cells, flattened and pavimentous, in the state of repose of the gland, swell and soften during the period of secretion. It is then that they take from the blood the materials necessary for the fabrication of ptyaline; also they become loaded with granulations, which give them a slighly opaque look.

Though identical in appearance, the diverse salivary glands of man and of the superior mammifers secrete, that is to say fabricate, coagulable substances specially appertaining to each gland. There are here imperceptible particularities, belonging to the molecular acts themselves of nutrition. The fact becomes more striking still if we range the poison glands of reptiles with the salivary glands; a classification justified moreover by anatomy.

The salivary cells swell assuredly during the period of glandular activity, but it is especially during repose that they seem to detach themselves from the partition and to accumulate in the culs-de-sac. Then, when the gland commences to operate, a liquid flow, borrowed from the blood, traverses the wall of the acini, and goes to dissolve and to carry along the salivary ferment previously formed. Analogous phenomena are produced in the lactiferous glands, and perhaps in most of the glands. In the lacteal glands the secretory cells load themselves with droplets of fat, which their rupture sets at liberty.

The fundamental structure of the excretory glands does not perceptibly differ from that of the secretory glands. The mechanism of their functions is also the same. In excretion as in secretion there are always cells, called *epithelial*, which borrow from the blood of the capillaries certain substances; but the secretory cells content themselves with giving a passage to the substances substracted without perceptibly modifying them. The most important of all the excretions is certainly the urinary or renal excretion.

Organs analogous by function to the kidneys of the superior vertebrates exist in many invertebrates, notably in insects and the arachnida. In these animals the secretory cells are even sometimes very voluminous. Those of the *Coccus Hesperidum*, for instance, are so large that they can only place themselves in single file in the vessel of Malpighi, to which they give a knotted aspect. These uriniferous cells dissolve and set at liberty the numerous granulations of uric acid and of urate which they contain.

The schema of the kidney of the vertebrates is found in the myxinoids.[1] The apparatus is composed of a long canal comparable with the ureter, that is to say, with the long and narrow conduit, which, in the superior vertebrates, unites the kidney to the bladder. In the myxinoids this canal shows from distance to distance culs-de-sacs narrowed at the neck. At the bottom of every cul-de-sac is found one of those small clusters, consisting of intricated capillaries, which are called *glomerules of Malpighi*, in the kidney of the superior vertebrates (Fig. 38). This arrangement of the renal capillaries has for result to slacken the course of the blood in the glands, to multiply the contacts of the vessels with the excretory cells, and consequently it is very suitable for facilitating the urinary excretion. In effect this excretion is of supreme importance, forasmuch as it represents a grand current of expulsion, thanks to which the principal mass of the mineral principles and of the quaternary azotised principles, disassimilated, useless or hurtful, is driven from the animal economy.

The urinary liquid represents the residuum the most general and most abundant of disassimilation, that is to say, that its composition is alike very complex and very variable. With the exception of a small quantity of colouring matter or some products of vesical secretion in the animals that have a bladder, we find in it no immediate principles of the third class. It is merely a solution much charged with mineral or mineralised

[1] Leydig, *loc. cit.*

substances, whose proportion incessantly varies. We find in it a quantity of mineral salts, first of all chlorure of sodium, then chlorure of potassium, chlorohydrate of ammonia, sulphates of soda and potash, phosphates of lime, of soda, of potash, of magnesia, ammoniaco-magnesian phosphate, carbonates of lime, of potash. The organic salts and the quaternary products resulting from the disassimilation are lactates of soda, of potash, of lime, urates of lime, of magnesia, of soda, of potash, of ammonia, a

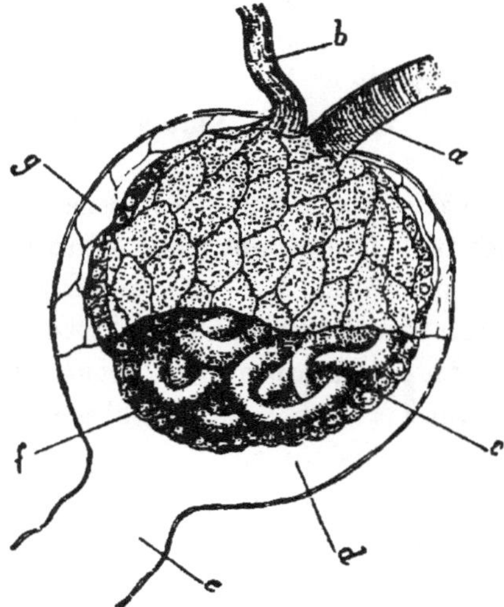

FIG. 38.

Schematic representation of a Malpighi glomerule of the bladder :—*a*, sanguineous vessel entering ; *b*, sanguineous vessel leaving ; *c*, clusters of the capillary vessels in the interior of the glomerule ; *d*, lower part of the capsule drawn without its epithelium ; *e*, commencement of the urinary canal ; *f*, internal epithelium of the mass of the capillaries ; *g*, internal epithelium of the capsule.

little uric acid, and besides, especially in the herbivorous animals, hippuric acid and hippurates. Finally, we must add to this long, though incomplete list, the most important products of excretion, urea, creatine, creatinine, immediate principles of the second class derived direct from the oxydation of the albuminoidal substances in the concatenation of the tissues.

The composition of urine varies under the influence of all the causes capable of acting on the nutrition of the animal organism.

It is denser, more charged, more coloured during digestion, more colourless, on the contrary, and holding in solution fewer products of excretion during the first infancy, when assimilation predominates over disassimilation, and when the needs of growth fix in the economy, in great quantity, the nutriments which penetrate thereinto.

After violent exercise, the urea, the phosphates, and the sulphates notably augment in the urine; for there is a greater consumption of albuminoidal substances.

Abstinence brings the same results; for then the animal is nourished at the expense of his own tissues; but in this last case the augmentation of the phosphates and sulphates is simply relative.

Excellent observations and analyses, especially those of Dr. Byasson, have put it beyond doubt, that all intellectual activity, how little soever energetic, has as corollary a corresponding augmentation of the urinary phosphates. Every organic act has, in truth, for basis an oxydation of the tissue, of the anatomical element which operates. The brain is subject to this law, like every other organ; but, as we shall see by and by, it is very rich in phosphates. It is therefore perfectly natural that its disassimilation should give birth to phosphorised products of excretion.

As we have already mentioned in a preceding chapter, the urine of the herbivora is generally alkaline, and richer in hippuric acid and in hippurates. That of the carnivora is habitually acid, and the urates and the uric acid are in larger proportion therein. But, as we have indicated, there is nothing fixed and immutable. Every herbivore, subjected to an animal alimentation, has the urine of a carnivore, and simple abstinence itself suffices to produce this result.

An abundant animal alimentation has for result greater production and excretion of uric acid and urates. On the contrary, in abstinence the uric acid disappears and the urea is excreted in greater quantity. The urea, product of an oxydation more profound of the organic substances, is formed by disassimilation in the anatomical elements themselves. Possibly, however, the uric acid, the result of an incomplete oxydation, is produced for the most part in the blood itself, when the fluid is surcharged with peptones which the tissues are not able to assimilate.

But how can we explain the fact that uric acid is found in such enormous proportion precisely in those of the vertebrates which are endowed with an extreme respiratory activity, namely, in birds, which, however, as regards uric secretion, resemble testaceous reptiles? Verily in both the uric acid is found in so great a quantity, that it is concreted and crystallised even in the interior of the urinary canals, and gives to the urine the appearance of a whitish pulp. Among the invertebrates, concretions and crystals of like nature are not met with even in the cells of the urinary glands.

We can view as having affinity to the urinary excretion the sudoral excretion, operated by millions of cutaneous glandules, each of which may be considered as a small kidney excreting an aqueous solution, charged with substances analogous to those which are found in the urine. But the sudoral excretion is much less rich in salts and in organic matters than the urine. We usually find in it urea, but no uric acid or urates.

Secretion, we have said, can be traced back to phenomena of nutrition, that is to say, to molecular acts, effected in the midst of glandular cells, which means that it can be accomplished without the intervention of the nervous system. Such is evidently the case with vegetal secretions. But in the superior animals, having a complete nervous system and suitable capillary vessels, the secretion evidently depends on the degree of repletion of these vessels, and on the rapidity with which the sanguineous current traverses them. It is therefore indispensable that the secretion should be influenced by the vaso-motory nerves; and this is in effect what experiment demonstrates. Normally, secretion is always accompanied by a dilatation of the capillaries of the gland, by a congestion, and every congestion artificially provoked by the secretion of the vaso-motory nerve or by the excitation of the vaso-dilatatory nerves, which seem t exist at least in certain glands, has for result a greater secretor activity.

Incitations arising or provoked in the nervous centres, eithe directly or by reflex action, can also transmit themselves to th glandular vaso-motory nerves, and react on the secretion. Thu we determine an abundant secretion in a dog, chained an famished, by placing a piece of roast meat before him. All th world knows likewise with what facility certain strong emotion act on the biliary secretion. These are examples of reflex physiological acts. The direct excitations of the nervous centre can, in their turn, modify or trouble the secretions. As early a 1845 M. Schiff had demonstrated that lesions of the cerebra peduncles render the urine acid and albuminous. Punctures o the roof of the fourth cerebral ventricle provoke saccharine diabetes (Cl. Bernard). Lesions of the isthmus and of the lowe part of the cervical marrow can abolish the urinary excretion, can produce anuria.

Brief and incomplete though it may be, the exposition which precedes suffices to make the mechanism and the importance of secretion understood. We have therefore now passed in review everything relating to nutrition, to its modes, to the diverse biological contrivances which render it possible in the essential being of complex organisms. Consequently we can forthwith enter on the exposition of the chief properties of organised matter. We know how organised beings are nourished; let us now see how they grow, and how they are reproduced.

BOOK III.

OF GROWTH.

CHAPTER I.

OF THE PROCESSES OF GROWTH.

The organised individual, vegetal or animal, is perpetually mutable and always perishable. It is born, it grows, it dies, after having more or less painfully maintained its organic equilibrium in the midst of the exterior medium. We have ere long to formulate the general laws of the birth, of the generation, of organised beings. Let us now see how these beings grow. Every living creature being definitively constituted by anatomical elements, it can only increase in volume by the growth or the multiplication of those elements.

As to the development in volume of anatomical elements already existing, the phenomenon is relatively simple. There is scarcely anything more before us than a particular case of nutrition with a certain predominance of movement of assimilation over that of disassimilation. The anatomical element or elements acquire more than they lose, and their mass more or less augments. If a complex or organised being is concerned the growths of the elements, isolately considered, totalise themselves, and the whole individual grows and greatens.

But evidently this mode of growth is insufficient to account for the phases which every organised being traverses from birth

T

to death; for putting aside for the present the cases of spontaneous generation with which we have not to occupy ourselves here, every complex individual of the two organic kingdoms springs from a simple cell. Growth can therefore be effected only by an enormous multiplication of the anatomical elements. This, in effect, is what happens. But as to the mode of generation of these histological elements, we find ourselves in the presence of two great rival theories which we have already signalised. The one more especially defended in Germany is the theory of *cellular generation.* The other, maintained chiefly in France by M. Ch. Robin and his school, is that of *spontaneous genesis.*

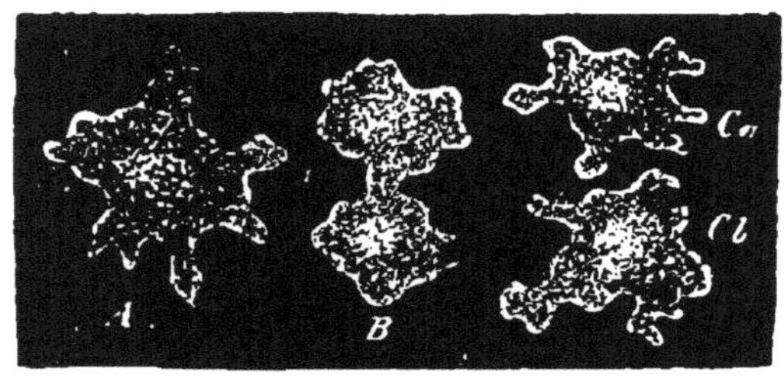

Fig. 39.

Reproduction by segmentation of an elementary organism, a moneron: *A*, entire moneron (protamœba); *B*, the same moneron divided into two halves by a median fissure; *C*, the two halves have separated, and constitute now independent individuals.

According to the cellular theory, maintained in all its vigour by M. Virchow, and still admitted by the majority of German naturalists and physiologists, every cell comes directly and strictly from a pre-existing mother cell. *Omnis cellula e cellula;* such is the formula which sums up the doctrine. Whatever the biological process may be, simple division (Fig. 39), or the budding from the mother cell, &c., there is always *proliferation* alike in the animal kingdom and in the vegetal kingdom.

The doctrine of spontaneous genesis is less exclusive. Without denying the fact of cellular proliferation, which is very generally observed in the vegetal kingdom, at the outset of the embryological evolution, as well as for certain species of histological elements, the partisans of spontaneous genesis, supported moreover by numerous and precise observations, affirm that for the most part in the animal kingdom, and here and there in the vegetal kingdom, new histological elements appear spontaneously in the intercellular blastemas.

Even if we admit that the defenders of spontaneous genesis have generalised a little too much the application of their doctrine, it is certainly on their side that there is the largest amount of truth. We must, in any case, admit two grand processes of histological generation: proliferation, or multiplication by extension and division of substance and genesis, or multiplication by a sort of living precipitation in the heart of the blastemas. The two processes are observable in the two organic kingdoms, but in very different degrees, each of these two great modes of histological multiplication being dominant in the one kingdom and exceptional in the other. In short, proliferation and genesis are the rule, the first in the vegetal kingdom, the second in the animal kingdom.

CHAPTER II.

OF GROWTH IN THE VEGETAL KINGDOM.

CELLULAR proliferation is far from being accomplished in a uniform manner. It results from various processes, which sometimes even are more or less combined, but of which the principal may be arranged under the following heads—1, simple *division* or *segmentation, scission*; 2, *budding, germination* or *surculation*; 3, *copulation* or *conjugation*; 4, *endogenous generation.*

In cellular segmentation it is first of all the azotized utricule, the inner envelope of the cellular protoplasm, which depresses itself, narrows itself into a circular furrow. Afterwards, the external tunic in cellulose places itself in its turn in the depression, which, becoming deeper and deeper, at last forms a partition at first incomplete and pierced with a circular hole, afterwards complete. While this division is taking place, a second nucleus appears in that one of the two cavities which was primitively destitute of it. The result of this very simple evolution is the formation of a new cell, which divides itself in its turn. M. Mohl was the first to observe closely this cellular unfolding in the terminal cell of the confervæ. It has since been affirmed regarding most of the vegetal tissues. It is thus that the ligneous cell of the phanerogams form their layers and their fibrous bundles; it is thus that the cells multiply themselves in the sporangia, and the spores of the algæ, &c., &c. Sometimes, as in the cells of the pith of the dicotyledons, the division of the nucleus precedes that of the cell.

In germination, the cell emits from one point of its surface a prolongation, a kind of protoplasmic rupture, which acquires a nucleus, separates itself by partitionment from the rest of the parent cell, and lives afterwards an independent life. It is by this process that certain unicellular plants, for example the *Vaucherias*, reproduce themselves.

In the monocellular algæ the spores and sexual cells, containing antherozoïds, analogous to the spermatic animalcules, form themselves by the terminal partitionment of one of the ramifications emitted by the cell.

It is also in the algæ, the conjugated diatomous algæ, that the phenomenon of conjugation is especially remarkable. Two neighbouring cells each emit a prolongation; these projections meet; their partitions are reabsorbed at the point of contact; the protoplasms of the two cells intermingle; soon the cells are completely amalgamated. They then form only one cell, which is a reproductive one, a spore or zygospore (*spirogyra longata*).

The multiplication of the cells, by endogenesis, is characterised by the formation of a greater or lesser number of daughter cells, which grow, burst the tunic of the parent cell, and live in their turn an independent life. This mode of reproduction exists in some unicellular vegetals, for example, in the protococcaceæ.[1]

To sum up, all these diverse modes of proliferation do not essentially differ from eath other. In every case the essence is the same; it is a particle of living matter, individualised, which assimilates beyond what is necessary for its own maintenance, and becomes a true organic centre. Thus is produced an exuberance of force and matter; whence the tendency to the formation of fresh centres.

The growth of a vegetal, by the simple increase of volume of the histological elements, may be considered as the first step, the prelude to that development by cellular multiplications which we have just described. Here also there is a predominance of the movement of absorption, of assimilation. The nutritive liquids

[1] Naegeli, *Gattungen einzellingen Algen*, Zürich, 1847.

diffuse themselves into the substance of the anatomical elements; living molecules of new formation intercalate themselves by intussusception with the molecules primitively existing; the cellular membrane extends; the protoplasm grows; then the form of the histological elements often modifies itself. It is in this manner, for example, that the ligneous fibres form themselves. The cells, at first spherical, elongate, and become cylindrical; their extremities adhere to each other, either squarely or obliquely. Concentric layers deposit themselves successively on the inner surface of the partition, which they thicken.[1]

In other tissues the spherical or polyhedrical cell grows simply by increasing in age without losing its first form. Sometimes, even, we can determine the relative ages of the cells of a tissue by the variations in volume. In the pith of many plants the cells go on regularly and gradually decreasing in volume from the central region, where they are the most aged, to the circumference, where they are youngest. The same fact may be observed in comparing the medullary cells of the same stem at different heights. Dutrochet has observed that in certain plants the gradual decrease of the diameter of the stem from the base upwards is owing solely to the decreasing volume of the medullary cells, beginning at the neck of the root.[2] In effect, slices of the pith of the elder-tree, cut at different heights, contained manifestly the same number of cells.

In the inferior vegetals, especially in certain families of algæ (conjugated, diatomous, siphonated), there is neither division of labour nor specialisation of tissue. The cell then lends itself to various functions, metamorphoses itself, adapts itself to multifarious uses. Besides, whatever may be the degree of complication of its structure, it generally preserves the faculty of segmenting itself, of multiplying itself. It is quite otherwise

[1] Ch. Robin, *Des Eléments Anatomiques* (*Bibliothèque des Sciences Naturelles*).

[2] Dutrochet, *Mémoires pour Servir à l'Histoire Physiologique et Anatomique des Animaux et des Végétaux*, t. II., p. 139.

with the complex phanerogamous plants. There the faculty of self-multiplication by division is in some degree the special appanage of incompletely developed cells, always young. These are the cells which issue from the mother cells, which diversify themselves and form the different histological vegetal types, of which we have before given a succinct account. From them issue the tubes, the ligneous cells, the chlorophyllian cells; but all these derived and special elements have, on the other hand, habitually lost the faculty of division and of multiplication.

Growth is only an exaggeration, an outcome of nutrition; that is to say, it is dependent upon the grand conditions of the medium which regulate this primordial property of organized matter.

Nevertheless, it appears only to depend indirectly upon light. In effect, it is often in the night that seem to go on most strongly the growth and segmentation of the vegetal cells. The necessary conditions are simply the previous existence, in the tissues of the plant, of a store of materials, elaborated, assimilable, and of a sufficient quantity of aqueous vehicle to dilute and convey these materials.

The action of light on the chlorophyll is necessary to the formation of the immediate assimilable principles in the plant; but assimilation, properly so called, is carried on very well in darkness. In effect, the subterranean parts of plants live and develop themselves; the phenomena of germination are perfectly accomplished in darkness; truffles and all tuberaceous plants effect subterraneously all the phenomena of their development; mushrooms, deprived of chlorophyll, assimilate even when the organic materials are prepared by other organized beings.

Light seems even to retard the development of the plant. It is in the morning, towards sunrise, that the internode in its process of growth offers its maximum of horary augmentation. If a stem receives upon different sides of its surface luminous rays varying in intensity, it curves itself towards the side on which there is most light, because on that side the anatomical elements increase and multiply least actively. It is only the very refrangible rays, blue, violet, and ultra-violet, which exercise this retarding action. Nevertheless, the mean elongation is ordinarily greater during the twelve hours of the day than during the twelve hours of the night; but this result is due to the mean elevation of the diurnal temperature, and it is always in the morning at sunrise that growth reaches its maximum.[1]

A. de Candolle, De Vries, Köppen, Sachs, &c., have occupied themselves with the influence of the temperature upon the germination and elongation of different parts of plants. Below 0 and above 50 degrees, vegetal life is generally impossible, and it is ordinarily at about 30 degrees that it attains its maximum of activity. This proposition has, however, only a general application, as, for each species, there are special maximal and minimal temperatures, and also a certain temperature particularly favourable.

At this suitable temperature, development is as rapid as possible; nevertheless it still takes place above and below it, within sufficiently wide limits, but much more slowly.

From observations made by M. Boussingault upon grains of barley, it has been concluded that to develop itself, a vegetal requires a given and nearly uniform degree of heat.[2] This given quantity of calorific vibrations transforms itself into nutritive and evolutive equivalents. A lower temperature produces the same effect as the more favourable temperature, but naturally in a longer time. According to this hypothesis, very high temperatures must be considered as agents of perturbation, as forces too great, hurtful by their very excess, and of a nature to shackle the phenomena of nutritive chemistry.

In treating of nutrition and respiration, we have stated what

[1] Sachs, *Traité de Botanique*, pp. 886-890.

[2] Boussingault, *Economie Rurale, Considérée dans ses Rapports avec la Chimie, la Physique, et la Météorologie*, t. II.

an indispensable part is filled by the atmosphere in the life of organised beings. It is evident that in an atmosphere unsuitable for the maintenance of life, or in a vacuum, growth is arrested, the floral and foliaceous buds no longer develop themselves, as Saussure and many other experimentalists have moreover shown.

The principal facts and general conditions of vegetal growth being determined, we can now describe or point out a certain number of particularities connected therewith.

The epoch of blossoming often coincides with a more or less rapid acceleration of growth; and, at that time, we see, in the greenhouses, agaves lengthen by more than two décimètres in twenty four hours. The *Corypha umbraculifera* has been seen to grow forty-five times more during the four months preceding its blooming than it had done in the same lapse of time in thirty-five years.[1] Mushrooms grow with extreme rapidity. In three or four days the *Lycoperdon giganteum* develops into a sphere three décimètres in diameter. Certain algæ, for example the confervæ, formed of a single row of cells juxtaposed, end to end, lengthen almost visibly by cellular division.

In the most complex vegetals, growth has been closely studied. It is produced by the descending and elaborated sap. In the arborescent dicotyledons, the *duramen*, the ligneous portion, is no longer traversed by the descending sap; moreover, the fibres which compose this ligneous portion are half mineralized; they have ceased to multiply themselves, and even to grow. It is between the wood and the bark that the flow of descending sap finds an easy passage, and it is there, in fact, that every year a layer of fresh tissue is formed. Most botanists admit that the creation of the new histological elements takes place by simple division of the old, especially of those which constitute the inner-most layer of the bark, since, according to the experiments of Duhamel, a shred of bark, only connected with the rest by its upper part, or else separated from the aubier by a plate of pewter, still effused cambium on its inner surface. According to M.

[1] Treviranus, *Biologie*.

Trécul, there is probably also an advent of new elements by spontaneous genesis.[1] In the opinion of this observer, when the buds are produced, there is also an effusion of gelatiniform blastema between the bark and the aubier. This effusion forms a mamelon; then in this mamelon appear little cells. M. Ch. Robin goes farther. According to him differentiated histological elements spring forth also by genesis in the centre of the cellulous mamelon. We should observe the appearance of ovoïd cells arranged in a single cluster, and having from the beginning a reticulated aspect; then afterwards cells with a spiral thread spring forth. When once the evolution of the bud is more advanced, when once the leaves are formed, the primitive, single cluster ramifies to send little clusters into each leaf.[2]

The pith grows energetically when it is not imprisoned in a rigid ligneous tissue, which isolates it from the cortical system. In this last case, on the contrary, it often dies, is destroyed, and the stem becomes fistulous. But when the pith is bound by transverse radii to the bark and to the cambium, it most frequently grows by cellular division. Notwithstanding, Dutrochet has affirmed that he saw new cells spring forth, by spontaneous genesis, in the cellular interstices of the pith. The medullary cells are generally soft, and very osmotic; they readily imbibe water or sap, become turgescent, and fit to produce generative blastemas.

The *duramen* is a kind of senile state. Trees which are destitute of it, as the poplar and the maple, grow generally with greater rapidity than others; for growth then takes place throughout the whole thickness of the trunk, and not only between the bark and the aubier. In fact, there is growth wherever the elaborated sap penetrates. Dutrochet cites, as an example, the radiciform stem of the beetroot, composed of layers of loose cellular tissue, between which the cambium circulates.[3]

[1] Trécul, *Annales des Sciences Naturelles*, 1846.
[2] Ch. Robin, *loc. cit.*, p. 39.
[3] *Loc. cit.*, t. II. p. 162.

The development of the plant in length is accomplished in two ways; on the one hand by terminal elongation, due to the evolution of the bud, which produces internodes or *merithalli* of new formation; on the other, an elongation is effected of the merithalli, already formed, by the lengthening of their vascular or cellular organs.

The roots scarcely lengthen except from the point, since two ligatures placed across their course near their extremity never separate from each other (Duhamel).

Most vegetals grow during the whole of their life, either specially in length, as the monocotyledons, or in both dimensions, as the dicotyledons. Growth only slackens more and more in proportion as the vegetal tissues become mineralised, and its complete cessation generally coincides with the death of the vegetal. Death, besides, may be partial. In effect, there is no centralisation in the plant. Every vegetal may be compared to a polypier. Each bud has its individuality, its own existence; it attracts and elaborates the nutritive fluids; also when a terminal bud dies the branch which bears it dies also.

The fall of the decayed leaves is determined much less by the autumnal cold than by the termination of the foliaceous growth, due, doubtless, to an excessive degree of mineralization. Many decayed leaves fall before the appearance of the cold. As to the leaves termed *persistent*, they scarcely differ in this respect from the so-called *caducous* leaves. Instead of falling in a mass, they fall one by one, when their growth has reached its bound.

In the ligneous stems of the dicotyledons, to speak correctly, death incessantly accompanies life. Every year a certain number of anatomical elements lose the faculty of growth, of multiplication, and at last of receiving nutriment. They become first what are called definitive tissues, then little by little are destroyed; but, on the surface of the aubier, where the flow of cambium, of elaborated sap, passes, a new living stratum is formed. In proportion as death invades the centre of the stem, vitality takes refuge and renews itself on the surface, and these two regio are bound together by a graduated succession of layers, mo vitalized in proportion as they approach the bark, more dead proportion as they approach the centre.

NOTE.

The French word *aubier* expresses better than any English word the whi fresh tender wood between the body of the tree and the bark.—*Translator.*

CHAPTER III.

OF GROWTH IN THE ANIMAL KINGDOM.

WHEN molecularly the male fecundating substance has blended with that of the female ovulum (Fig. 40), has impregnated it, the work of evolution, that is to say, of generation and multiplication of the anatomical elements, has commenced in that ovulum. Growth commences in the ovulum by the inferior process of *segmentation* or *fractionment* (Fig. 41). Cells, all like each other, juxtapose themselves to form a membrane, in a point of which appears the rudiment of the embryon. The primitive membrane is the *blastoderm;* the first indication of the embryon is called the *embryonary spot.* In this point the blastodermic membrane is double; it is composed of an *external* vestment, *serous* or *animal;* and of an *internal* vestment, *mucous* or *vegetative.* From the first vestment specially proceed, in the vertebrates, the teguments and the organs of the life of relation; the second vestment gives, above all, birth to the apparatus of vegetative life. These first phenomena of development are perceptibly the same in the whole animal kingdom, from the lowest point to the highest.

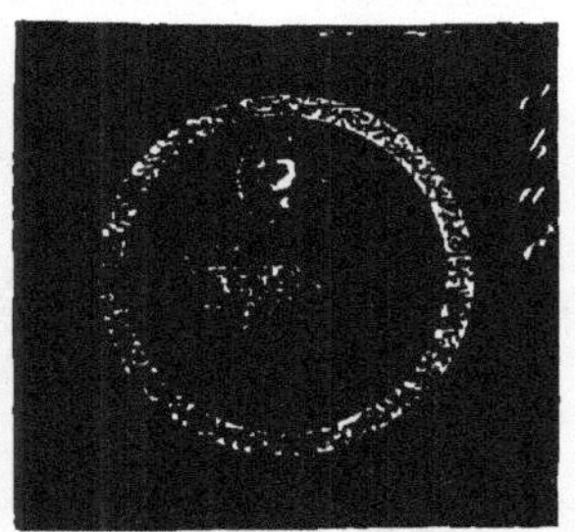

FIG. 40.

An ovum of mammifer (a simple cell): *a*, nucleole (*nucleolus*) or germinative point of the ovum; *b*, *nucleus*, or germinative vesicle of the ovum; *c*, cellular substance or protoplasm, yolk of ovum; *d*, enveloping membrane of the yolk in the mammifers; it is called *Membrana pellucida*, on account of its transparency.

This preparatory labour once accomplished in the ovulum, the process of segmentation makes way before the process of genesis. The cells of the blastoderm gradually disappear in proportion as spring up among them by genesis the anatomical elements destined to constitute definitively the new individual. It is between the two blastodermic vestments that arise these first elements. Thenceforth the difference is marked between the invertebrated animals and the vertebrated animals. In these

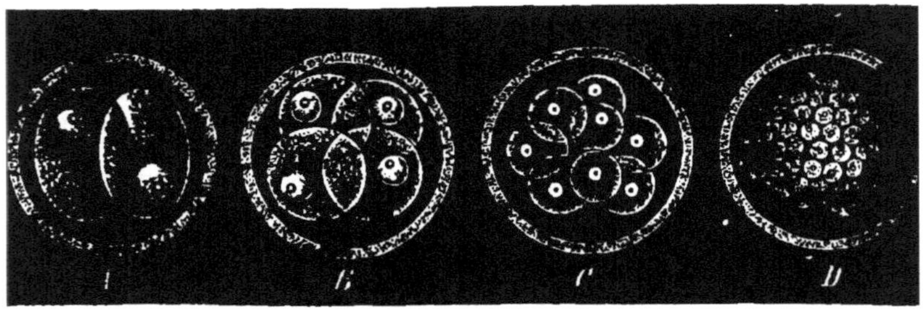

Fig. 41.

First stadium of the evolution of a mammifer, "segmentation of the ovum," multiplication of the cells by reiterated scissions: *A*, the ovum divides by a first fissure into two cells; *B*, the two cells divide into four cells; *C*, these last divide into eight cells; *D*, the segmentation, indefinitely reiterated, has produced a spherical mass of numerous cells.

last, in effect, we soon see traced the rudiment of the vertebral column, the cells of the *notocord* (Fig. 42). This notocord ensheaths itself. Then we behold the formation of the first cartilaginous bodies of the vertebrates, and of the first elements of the central nervous axis; and so on. All the tissues and all the organs thus appear little by little. At the moment of their birth the anatomical elements are already for the most part well characterised. Nevertheless they undergo another evolution; they develop themselves; azotised granulations, nucleoles, are formed little by little in their substance; their volume augments. But, once arrived at their complete development, they are stable, or at least no longer modify themselves, except regressively, to disappear and perish. Never is an animal histological species metamorphosed into another; never, for example, does a muscular fibre become a nervous fibre, &c.[1]

The nucleus is usually the centre of genesis and of evolution of the anatomical element. It is the nucleus which habitually appears the first; it is after the nucleus that the nucleoles show themselves. Most of the elements destined to a prompt destruction, such as the hæmatia, the leucocytes, have no nucleus. If

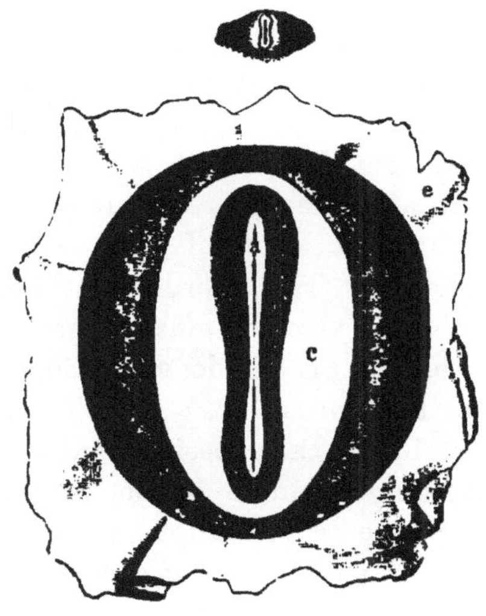

FIG. 42.

Ovum of dog. The embryon in form of a shoe sole is in rudimentary shape: *a*, dorsal fissure; *b*, dorsal plates; *c*, clear (or bright) area; *d*, opaque germinative area; *e*, membrane of the germinative vesicle.
The small superior figure is of natural size. The inferior figure is magnified.

the primitive hæmatia, those which show themselves first of all in the embryon, are furnished with a nucleus, it is probably because they have those functions of which, at a later period, they are deprived. In effect, in the embryon, at the moment of the genesis of the first hæmatia, there are not yet any of those sanguineous glands called *closed glands*, in which new hæmatia are subsequently to be fabricated. Also the first hæmatia are completer elements, endowed with an intenser vitality, and

[1] Ch. Robin, *Des Eléments Anatomiques*, pp. 47—54.

capable of engendering by segmentation other hæmatia similar to themselves.

In the animal, moreover, the histological multiplication by scission, by segmentation, is scarcely observable, except in elements not much differentiated; for example, in the hæmatia, the leucocytes, the epithelial cells. The superior histological elements, the nervous cell, the muscular fibro-cell, are never segmentised.[1]

When segmentation is observable, it is produced only in elements fully developed, adult in some fashion. The small nuclei, the small cells are not segmentised. In truth, in every anatomical element growth has two phases: the first, during which the element augments merely in volume; the second, during which it multiplies. Both of them have equally as cause an excess of assimilation. But in order that the second may be produced, the anatomical element must naturally have attained its limit of growth, and can no longer assimilate into its substance molecules of new formation.

In the adult vertebrate, the genesis of the white globules, the leucocytes, is especially produced in the lymphatic ganglions; and in the sanguineous closed ganglions, the spleen, for example. Recent observations have shown that the venous blood coming forth from the spleen is much richer in globules than the arterial blood entering it.

There is great probability that the new red globules proceed simply from the modification, the transformation of the white globules. Thus, the venous blood of the spleen contains numerous globules intermediate between the leucocytes and the hæmatia. These mixed forms are met with also in great number in the blood after a hæmorrhage, or frequent bleedings, when the sanguineous liquid is in process of regeneration.

We have previously cited a curious fact observed by Burdach, the coloration into red of a lymphatic clot. Analogous facts have been noted by Virchow and Friedreich. These physiologists have likewise seen the lymph redden in the air.

[1] Ch. Robin, *Ibid.*

It would seem at the first glance as if the question of knowing whether the anatomical elements arise by scission or by genesis belonged purely to the domain of science, and were dependent solely on observation and experiment. This however is by no means the case. This is one of the subjects which have the privilege of bringing into play passions by no means scientific. In effect, the spontaneous genesis of the anatomical elements in the blastemas seems to have a close connection with the theory of spontaneous generation, which by some is so much execrated and mocked. But if we contemplate the matter with composure all this fury seems to have very little justification. Two facts are very certain: the first, that anatomical elements multiply; the second, that they multiply by diverse processes, but in general conditions which are identical. These principal facts dominate all others. Little it matters whether a cell arises by scission, by germination, by endogenesis, or by genesis. Definitively the contents of an anatomical element do not differ essentially from the environing blastema; in both cases there are organised substances, in the midst of which is effected the incessant molecular movement which constitutes the fundamental phenomenon of life. If assimilation prevails over disassimilation there must necessarily result the formation of new living centres, of new anatomical elements, and from the philosophical point of view it matters little whether this new formation is effected without or within a cell.

Observation in effect shows that in the living tissues the processes of scission and of genesis are sometimes associated, and sometimes succeed each other. Nuclei, cells even, arising from genesis, for instance leucocytes, can afterwards multiply by segmentation and germination.[1]

According to M. Robin the advent of epithelial layers on the surface of the cutaneous dermis and the mucous membranes

[1] Ch. Robin, *loc. cit.*, pp. 47, 48.

begins by the genesis of the nuclei. Between the nuclei it produced a layer of amorphous matter, which is segmentised afterwards and individualised into cells. As to the aristocratic elements, such as the nervous cells, the fibre cells, they are more the seat of this proliferation by scission.

CHAPTER IV.

OF THE GENERAL CONDITIONS OF GROWTH.

Growth is essentially the result of the predominance of assimilation over disassimilation, that is to say, it cannot be effected without the concurrence, the immediate presence of a sufficient provision of assimilable materials. But the living anatomical elements seem sometimes capable of profoundly modifying and metamorphosing these materials in a great degree. According to M. Cl. Bernard, the larvæ of flies form their organised tissues from substances soluble in alcohol, and consequently deprived of albuminoidal substances, properly so called. Notwithstanding, as a general rule, a certain analogy of chemical construction is necessary between the different kinds of histological elements and the blastemas, at the expense of which these elements live and develop themselves. If, for example, we transplant, by animal grafting, the anatomical elements of one animal to another, the operation will be likely to succeed in proportion as the animal species are analogous.

As to the general conditions of heat, light, oxygenation, and alimentation, we must refer to the chapters treating of nutrition, and limit ourselves to pointing out the particular facts relating directly to growth itself.

It has been pretended that certain animals, certain tissues, can live and develop themselves without oxygen. These paradoxical facts must only be received with extreme reserve. They have, without doubt, been insufficiently observed, badly elucidated, and are included probably in the general law, according to which oxygen is indispensable to the nutrition, and consequently to the life of organised beings and their anatomical elements. Truly, we may say of oxygen, as the ancients did of the air in general, that it is the *pabulum vitæ.*

Like the vegetal anatomical elements, the animal elements can only develop and multiply within certain limits of temperature. There are, for animal and vegetal development, thermic cardinal points which seem very near each other. The temperature of from 30 to 35 degrees seems to be one of the most favourable, since it is this which animals, with warm blood, maintain, in spite of the thermic variations of the exterior medium. In his experiments in embryogeny, M. Dareste has observed that elevated temperatures constantly determine, first an acceleration of the evolutive phenomena, then their premature stoppage; whence *nanism.* On the contrary, lower temperatures much retard the progress of development; they even stop the evolution, and do not permit the embryon to advance beyond a certain period.[1]

The influence of the seasons is attended by analogous results. In man, and most animals, the maximum of grow this in summer, and the minimum in winter.

If, by means of an artificial hatching apparatus, we, as M. Dareste has done, limit the influence of the source of heat to a fixed point of the embryon, we can, by varying the position of the ovum, obtain all the types of simple monstrosities described in the treatises on teratology.

The rapidity of nutritive exchanges, and consequently of growth, is closely connected with the abundance of liquids within the living tissues. In ligneous vegetals, the fibres have less vitality in proportion as they harden. Plants, with loose tissues, loaded with juices, grow more quickly than vegetals with dense tissues.

In the same way, amongst animals, the histological elements are less impregnated with humidity in proportion as the animal

[1] *Exposé des Titres et des Travaux Scientifiques* de M. Camille Dareste. Paris, 1868.

is aged. The muscular fibre of a young animal contains 26 in 100 of water; that of an adult of the same species only 23,5. Haller, comparing the different degrees of cohesion of human hair according to age, found that at eight years of age this cohesion was represented by 10, at twenty-two by 17, and at fifty-five by 25.[1]

The degree of fluidity of the anatomical elements is great in proportion to the youth of the animal, and the rapidity of renovation and of growth is correlative with it. According to the tables of Quételet, the growth of man is two-fifths the first year, a seventh the second, an eleventh the third, a fourteenth the fourth, a fifteenth the fifth, and eighteenth the sixth and seventh. It is only a sixty-eighth at eighteen years, and a two hundredth at nineteen.

Upon the subject of the advantageous influence of suitable humidification upon growth, Burdach points out that, in many aquatic animals, fishes, amphibious, and cetaceous animals, growth lasts as long as life, and is only gradually and more and more retarded.

The nutritive characteristic of youth is the rapidity and facility of nutritive exchange. This is why the phenomena of colouration and discolouration of the bones by madder are effected with much greater rapidity when the animal is young.

With the advance of age, all the secretions are impoverished, especially the cutaneous perspiration ; at the same time the general oxydation of the tissues lessens, the production of heat diminishes; all the functions become less energetic; some are extinguished by degrees, especially those of the most differentiated histological elements, those most noble, least vegetative, those of the nervous and muscular elements. Strength decreases, and intelligence is blunted.

This is because the ruling property of the living substance gradually declines. Nutrition becomes less and less active, and all the other properties of which it is the support decline with

[1] Burdach, *Traité de Physiologie*, t. V. p. 491.

it. In the higher vertebrates this decline of nutrition is immediately attended by slackening of the pulse and of the respiration. The animal offers less resistance to low temperatures, often succumbs to them.

During youth, assimilation prevails over disassimilation. In old age the case is reversed. The tables published by Quételet show that the weight of the body diminishes from the fiftieth year in man, and the sixtieth in woman. At ninety years this weight is reduced in the former from 136 to 123 lbs., in the latter from 120 to 103 and a half.

The average duration of life varies, in the two organic kingdoms, with the species.

In general, the lower organisms live a shorter time than the higher organisms. Less richly endowed with organs and apparatus, less differentiated, they harmonise less readily with the exterior medium and its variations. At the longest, mushrooms only live a few days. The infusoria sometimes complete the cycle of their life in a few hours, and most of the invertebrated animals have only a short existence.

In general, life is short in proportion to the rapidity of growth and of embryonary evolution. Aptitude for generation being in some degree a sign of the full development of the organised being, it is natural that its tardy appearance should be connected with greater longevity. In fact, this is the case with most of the vertebrated animals. In many invertebrated animals the appearance of the generative functions is, on the contrary, the forerunner of death. Numbers of insects die immediately after having procreated. The male butterflies die even sometimes in accomplishing the act of generation. The ephemerides live one or two years in the state of larvæ, and only a few hours as perfect insects.

In plants, fructification is also a supreme act, presaging a complete and speedy death in herbaceous plants, a partial death in the perennial vegetals. The celebrated saying of Proudhon, "Love is death," is then in this case an exact expression of the truth. So true is it, that we may sometimes abridge or lengthen life by accelerating or retarding the moment of reproduction. If, by means of a rich manure, we cause biennial plants to fructify during the first year of their existence, they die that same year. On the contrary, the mignonette is rendered ligneous and long-lived by cutting its flowers before the formation of the seed.[1]

Insects themselves live much longer when they are prevented from pairing.

Death may be general or partial. This latter is frequent in all organisms with a feeble physiological centralisation, in vegetals and in many lower animals. It is even not uncommon in the mammifers. Inversely, in these latter, the partial life of certain elements, especially of the epitheliums, often continues after general death, characterised by the cessation of the three primordial functions, circulation, respiration, and innervation.

Death may ensue through simple old age, in consequence of an extremely gradual slackening of the molecular movements of nutrition. It then takes place without pain, without illness, without agony, sometimes without consciousness, and may even be accompanied by a certain feeling of comfort; it is then the *euthanasia* of Plato.

Pinel has observed that at the Salpêtrière most of the nonagenarian women died without any shock, and often in their sleep.

Death is, in fact, only a final cessation of nutritive exchanges. Every being, every anatomical element which ceases to assimilate and disassimilate, returns consequently to the mineral world. The materials which constituted it then undergo purely chemical unfoldings, decompositions, and disaggregations.

During the life of a complex organism, many of its histological elements perish, without their death being in any way prejudicial to the life of the whole. Sometimes the elements of certain tissues become mutually compressed, and thus, by simple pressure, cause the gradual disappearance of some of them. Sometimes the

[1] De Candolle, *Organographie Végétale*, t. II.

anatomical elements liquefy, as the embryonary cells do normally : this is what is called ulceration. On the surface of the skin and of the mucous membranes and in the glands, thousands of epithelial cells incessantly detach themselves, fall, and dissolve.

Even in the web of the deep tissues, a number of histological elements disappear by simple resorption, and are either replaced or not by elements of new formation.

All that has gone before shows clearly that life is a thing modifiable and variable in intensity and duration. In certain individuals it is prolonged two or three times beyond the average life of their species. Thomas Parr was married at a hundred and forty-two years, and was still fit to accomplish the act of generation. He died at the age of a hundred and fifty-two; and Harvey, who made the post-mortem examination, found his muscles still full and well developed, the viscera in a good state, and no ossification of the cartilages.[1]

A large number of examples are on record of partial rejuvenescence in old men. White hair has become again black, and new teeth have appeared. In other cases, with aged women, the menses and the aptitude for fecundation have reappeared. We have ourselves seen once, in the case of a woman more than sixty, white hair replaced by black, after erysipelas of the hairy scalp.

In truth, death is a necessity to all the organised beings of our planet; but it is not a fatality. As Ch. Robin says,[2] "No scientific contradiction would hinder our conception of a perfect equilibrium between assimilation and disassimilation indefinitely repeated in all existing beings, without interrupting the continuity of that molecular renovation, and without a decomposition of the organised substance ensuing . . . . The anatomical element or organism, once produced, once born, may be supposed to present a perfect equilibrium of *indefinite* duration between the

[1] *Philosophical Transactions*, 1669.

[2] Ch. Robin, *Elém. Anat.*, p. 96.

act of assimilation and that of disassimilation." Condorcet had already written: "Would it be absurd now to suppose that the perfectionment of the human species may be regarded as susceptible of indefinite progress, that a time must come when death will only be the effect either of extraordinary accidents, or of the more or less gradual destruction of the vital forces, and that finally, the duration of the average interval between birth and this destruction has no assignable term? Doubtless, man will not become immortal; but may not the distance between the moment when he begins to live and the ordinary epoch when naturally, without illness, without accident, he experiences the difficulty of existing, be ceaselessly increased?"[1]

To dare now to assert that it is not impossible to conquer death, the great enemy, is to expose ourselves to an accusation of madness. The animist and vitalist doctrines fail; they have lost all credit with science; but a yoke borne for a long time always leaves a permanent impress, and, in the domain of opinion, the effect often long survives the cause. For centuries life has been considered as a mysterious, miraculous fact, beyond all investigation. Each organism was regarded as a monarchy despotically governed by a metaphysical entity. It was believed that the problem of life must eternally defy the power of human science. It was a *fatum*, against which it was useless to struggle. Such is still the prevailing opinion; but it exists only by force of habit. The phenomenon of life has been analysed. We know that it is the result of simple molecular exchanges, comparable to those that take place in an electric pile. That there is in the vital phenomena something immutable, predestined, no one can now maintain. Every living being conserves itself as long as there is in it a certain nutritive equilibrium, as long as assimilation and disassimilation are almost equally balanced. Now it is certain that the duration of this equilibrium depends upon an infinity of causes, internal and external. Of two children born, one may live ar hour, the other half a century. There is

[1] Condorcet, *Progrès de l'Esprit Humain.*

neither law nor rule when the course of life is abandoned to the hazard of events, as always happens. *A priori*, it is surely not impossible, an organised being given, to maintain indefinitely in it the tide of life at a constant watermark, and it seems to us that science is now sufficiently armed to attack boldly this great problem. It would be necessary to bring observation and experiment to bear first of all upon very simple organisms, whose normal life is very short. We should commence by determining the average duration of that life when the organism is abandoned to itself. Then, being guided as much as possible by scientific facts already acquired, we should vary in a hundred ways the light, the temperature, the alimentation, the composition of the atmosphere, &c., &c., noting carefully the effect of each new factor, of each variation of medium.

Furthermore, we should scrutinise the conditions of life and of organism in the species remarkable for their longevity, then in individuals whose duration is exceptionally long. Evidently the investigation itself would suggest new modes of research. We should fail, or we should succeed. In any case, something useful would result from this labour.

In fact, biology will not be a complete science till it has learned to comprehend life. A man of science, who cannot without injustice be accused of philosophical temerity, has already proclaimed this: "The physico-chemical actions, which manifest and regulate the phenomena belonging to living beings, are included in the ordinary laws of general physics and chemistry" (p. 4).

" . . . . There is only one system of mechanical philosophy, only one system of physics and of chemistry, which comprehend in their laws all the phenomena which take place around us, either in machines living or machines dead. Under the physico-mechanical relation, life is only a modality of the general phenomena of nature; it engenders nothing, it borrows its forces from the outer world, and only varies its manifestations in thousands and thousands of ways" (p. 135).

"By modifying the inner nutritive and evolutive mediums, and by taking the organised matter in some degree from its birth, we may hope to change its evolutive direction, and consequently its final organic expression. In a word, I think that we can produce scientifically new organised species in the same way that we create new mineral species, that is to say, that we can cause the appearance of organised forms, which exist virtually in the organogenic laws, but which nature has never yet realised" (p. 113).

"Life is extinguished, and natural death takes place, only because the production of the plastic element stops, and because then the passive tissues become impregnated and incrusted with mineral and other matters, which cramp their functions, and lessen more and more the nutrition, or the genesic formation of the active histological elements" (p. 126).

"To sum up, what the physiologist wants, is to be able experimentally to direct the evolutive phenomena in such a way as to modify the nutrition of the organised matter, in order thereby to change more or less the duration, the intensity, or even the nature of its vital properties" (p. 129).

"Up to the present time, physiology has been struggling with transitory ideas, which will disappear in proportion as science advances. With regard to physiology, we are just at the point where alchemy was before the foundation of chemistry. The views which may now be expressed with relation to the modes of action of the physiological experimentalist will only be the result of gropings more or less vague; but, nevertheless, these acts will not be less positive, and the scientific principle of general physiology cannot remain doubtful or uncertain. Physiology, like all the terrestrial sciences whose phenomena are within our reach, must become in time an active experimental science upon the phenomena of life" (p. 219).

"When the progress of general physiology shall have shown the experimentalist the special organic elements upon which he acts, and he shall have learned to master the conditions of their activity, then he will have acquired the power of scientifically modifying and regulating the phenomena of life. He will extend his dominion over living nature, as the natural philosopher and the chemist have acquired their power over the phenomena of inert nature."[1]

[1] Cl. Bernard, *Rapport sur les Progrès de la Physiologie Générale en France.*

NOTE.

The reader is anew reminded that when degrees of heat are mentioned the Centigrade scale is always implied, so that thirty and thirty-five degrees Centigrade correspond respectively to eighty-six and ninety-five degrees Fahrenheit.—*Translator.*

BOOK IV.

OF GENERATION.

CHAPTER I.

OF THE ORIGIN OF ORGANISED BEINGS.

WITHOUT remounting to the cosmogonies so fascinating and so probable of Kant and of Laplace, we must admit, with contemporary geology, that the earth was formerly in the state of incandescent globe; that during numerous cycles it was absolutely uninhabitable for the organised world we now know. We are compelled therefore to admit that the first living beings spontaneously organised themselves at the expense of mineral matter. The first inhabitants of the earth were, we know, of an extremely simple structure. The monera of Haeckel, some types of infusoria, the rhizopods perhaps—such are the existing organised beings which best recall to us those primitive ancestors of the organic world. But the Darwinian doctrine, which results with such evidence from palæontology, from embryology, from the well hierarchised classification of the organisms, demands as its indispensable complement spontaneous formation, without germs, without parents, of the first examples of the living world.

In the scientific domain any logical and necessary deduction or induction ought to be admitted without contest, though it may shock old ideas and shatter old dogmas. This is far, however, from being the case. The same religious and metaphysical prejudices which have been so deeply disquieted by the doctrine of organic evolution are still more alarmed and annoyed by the idea of spontaneous generation of any kind. It is only step by step that those antiquated theories yield the ground to scientific demonstrations. First of all it was denied that organised matter had in itself the faculty of living. The equilibrium was only maintained, it was thought, in every organised being through the perpetual intervention of an entity, of an archeus, of a vital principle, of a soul, and so on, guiding and governing the vital phenomena as a charioteer conducts a chariot, to use the expression of Tertullian. It becomes, however, unavoidable to admit that there is no metaphysical dualism in the plant and in the animal, but that they both live because they both combine in themselves the conditions necessary and sufficient for an incessant nutritive renovation. The doctrine of evolution has had the same destiny. Not many years ago all naturalists, or almost all, believed in the perfect immutability of the organised species, and, as every epoch had its special fauna and flora, it was necessary to recognise, with Cuvier, as in effect was done, a series of successive creations, of visible or organic changes. When God, irreverently compared to the machinist of an opera, whistled once, an implacable cataclysm annihilated all the living world; when he whistled a second time, but creatively, a new fauna and a new flora rose to life. Thus had things to go on at every geological epoch. From the tribolite to the mammoth every species had thus to be formed by magical crystallisation. Assuredly there was here a spontaneous generation of the most astonishing kind, but it shocked no one, because it was in more or less tacit accordance with metaphysical and religious ideas. But little by little the idea of miracle has been driven from this domain as from so many others. It became inevitable to confess that the geological epochs had not been separated by abysses, that the cataclysms, where there had been any, had only been partial; that the modifications and transformations of the soil had been produced little by little, slowly, by this patient labour of accumulated ages. But this new geological doctrine was incompatible with the sudden destructions and creations of the organised world. If the habitat had been slowly modified, it was necessary to believe that the inhabitant had been slowly modified too. The grand doctrine of organic revolution created by Lamarck, completed by Darwin, has come then to demonstrate the mutability of the organised species, and to furnish the genealogy thereof. It was a real revolution, and many naturalists have not decided for or against it; but nevertheless, the tide is gradually rising. The doctrine of evolution is already almost triumphant. There scarcely remains for the recalcitrants any other resources than to demonstrate its perfect agreement with the dogmas they are not willing to abandon. The thing is in process of execution. The interpreters are skilful, the sacred texts obliging, the metaphysical theories ductile, malleable, and flexible. Courage! We must be very narrow-minded indeed not to recognise in the first chapter of Genesis a succinct exposition of the Darwinian theory.

But if the doctrine of evolution makes us descend step by step to the most rudimentary organised beings, it does not go further; and we must admit, at least here, the intervention of a spontaneous generation. This consequence is revolting to many men of science who accept all the rest. They reject the notion that matter has in itself the power of self-organisation in certain given conditions. Yet many of these rebels formerly believed in the spontaneous generation of the mammoth at a time when the doctrine of successive creations was in vogue. But now, rather than resign themselves to the cruel necessity of admitting, even as a simple possibility, spontaneous generation, they prefer resuscitating the old theory of errant germs. The interplanetary spaces are, they think, bestrewn with germs, born we know not how, coming we know not whence, and waiting we know not how long, for a passage to a planet sufficiently mature to serve them as nourishing receptacle, as matrix.

Such puerilities as these are surely far more difficult to admit than the spontaneous formation of some very simple living types at the outset of the organic world, even though many observations and experiments do not plead very eloquently in favour of spontaneous generation even at our epoch. Certain vegetal parasites develop themselves under the epidermis of living plants. Whence can come the seeds of these entophytes which appear even in plants destitute of stomata?[1] Microscopic mushrooms spring up and live in the citrons. The forester Hartig found some in the cavities of the ligneous part of trees, under numerous sound annual layers. Marklin has seen the white of a hen's egg converted into *sporotrichum*. Very recently vibrions were met with in the pus of a closed abscess. In his *Histoire des Helminthes*, M. Dujardin speaks of the *rhabditis aceti*, which dwells exclusively in wine vinegar, and is found neither in the vine nor in grapes. Whence then does it come? And especially where were its germs when man made neither wine nor vinegar?

The cause of generation spontaneous, equivocal, heterogenical has moreover never ceased to have partisans. No one has defended it with more force, perseverance, and talent than F.-A. Pouchet, from whom we now propose to borrow some arguments.

Most of the organic macerations abandoned to themselves, at a suitable temperature, at the end of a time which varies, but is tolerably short, are peopled by vegetal and animal proto-organisms. According to the adversaries of heterogenesis, these myriads of new beings proceed from germs floating in the air. But whoso says germ says ovulum, that is to say, cell of a diameter perfectly appreciable in the miscroscope, for the ovulum of certain ciliated microzoa varies from $0^{mm}0028$ to $0^{mm},0420$ in diameter.

Now, if by the aid of the aëroscope of Pouchet, we examine microscopically one cubic décimètre of air, by making it pass through an orifice of a quarter of millimètre of section, that is to say by drawing it out to a length of 4000 metres, it is only

[1] F.-A. Pouchet, *Nouvelles Expériences sur la Génération Spontanée et sur la Résistance Vitale*, p. 117.

very exceptionally that we find the ovum of a ciliated microzoon or the spore of a mucedinate.

If, by the help of an eight-horse motive power, we project on diverse macerations of plants 6 million litres of that atmospheric air which is said to be full of germs, we see that the macerations, exposed to this torrent of ovula, do not become richer in infusoria than those which are imprisoned in a single cubic décimètre of air.

We obtain spontaneous generations by making use of artificial water and air, and even of oxygen, instead of air.

When we mingle together two different fermentiscible liquids, we find, in the mixture, organised beings different from those which the liquids when separate contain.

In experimenting simultaneously with calcined air, a body heated 200 degrees, and water which has undergone ebullition, we obtain animal and vegetal proto-organisms. Now not one of the reproductory bodies resists boiling water. But the following experiment is assuredly more convincing still.

A vessel of crystal $0^{m},30$ in diameter is washed with sulphuric acid and filled with distilled boiling water. Then we plunge therein 10 grammes of filaments of flax heated for two hours to 150 degrees. We cover it afterwards with a receiver, and we place it in the centre of another large vessel $0^{m},50$ in diameter, filled with distilled water. We keep it at a temperature of 28 degrees, and at the end of four days the maceration is filled with paramæcia, while we do not find one of these animals, and not even one of their ova, in the large vessel.[1]

But why do the so-called floating germs persist so obstinately in being invisible? In effect, atmospheric micrography finds in the air nothing but particles of fecula of wheat or some very small particles of silex. The ovula, the spores are so rare that habitually we do not meet with a single one in a cubic décimètre of air.

Aërian germs being so rare can only very slowly people infu-

[1] F.-A. Pouchet, *loc. cit.*, p. 122.

sions. Some ovula may fall thereinto first of all, be evolved, and give birth to organisms which may multiply little by little. Now in a fermentiscible liquid, we see at the outset a very thin mucous pellicle forming on the surface. Then suddenly appear in this pellicle many small lines, pale, immobile, ranged side by side in a certain disorder. These lines have the form and the diameter of *bacteriums*, and in effect, at the end of some hours we see them take animation and become living *bacteriums*, moving rapidly in a straight line.

Gérard had noted, at the time of the appearance of proto-organisms in the macerations, a certain evolutive order. In the maceration of hay we see, the second day, specimens of the *bacterium termo* simple, whose articulations augment little by little. Then come *monads*, and at the end of fifteen days we find *trichods*, *colpods*, the *proteus* tribe closing the series.[1]

F.-A. Pouchet has likewise demonstrated this evolution, and has drawn from it a whole theory.[2] According to him there appears, first of all in the macerations, an ephemeral population of vibrions and monads. These proto-organisms die. Their *débris* and their dead bodies mount to the surface, fall asunder, dissolve more or less, and all these detritus form a sort of membrane, which organises itself anew, engendering ovula of superior infusoria, microzoa with vibratile cilia. At certain points of this pellicle, which Pouchet calls *proligerous membrane*, we see granulations accumulate, mass themselves into a kind of spheroidal nebulæ. Then this nebulous substance becomes a true ovular cell, surrounding itself with a translucid membrane, with a bright zone. Finally this ovulum is evolved. We observe therein the gyration of the content or vitellus, the formation of the embryon, and the fifth day comes forth a rameciate.

The same maceration gives, or does not give, birth to ciliated microzoa, according as its proligerous pellicle is more or less thick.

[1] Gérard, *Dictionnaire d'Histoire Naturelle*, art. Génération.

[2] F.-A. Pouchet, *loc. cit.*, p. 110.

If we place the half of a given liquid in a vessel with a narrow surface, and the other half in a vessel with a broad surface, the proligerous pellicle of the narrow vessel is much thicker than that of the broad vessel. For in the one vessel and in the other has been produced an equivalent generation of vibrions and of monads; but in the broad vessel the residua have been obliged to spread themselves over a larger surface. They have not, for this reason, been able to form a proligerous membrane sufficiently compact, and no ciliated infusorium is created.

Certainly these are very eloquent facts. A last fact let us cite. There was poured into a porcelain basin with a flat bottom the paste of boiling flour, about a centimètre in thickness. Then, when this paste commenced to congeal, there was written on its surface, with a pencil, imbibed with a strong maceration of gall nut powder, previously examined in the microscope and filtrated, these two words:—*generatio spontanea*. The basin was then covered with a sheet of glass, and left to itself for four days. The temperature was 24 degrees on an average, and the pression 0,76 during this space of time, at the end of which the words *generatio spontanea* were seen traced in black. These characters were formed by crowded tufts of a microscopic mushroom *absolutely unknown*, with stalklets simple, cylindrical, not articulated, and with capitula of a beautiful black. M. G. Pennetier proposes to call this new organic species *Aspergillus primigenius*,[1] Many other topical facts can be found in the publications of F.-A. Pouchet, and of his rivals MM. G. Pennetier and Mantegazza, Joly and Musset, to which we must content ourselves with referring. The short extracts we have just given suffice to show that the doctrine of spontaneous generation does not merit the vulgar disdain with which it is assailed. Its partisans are able to rely on serious arguments drawn from observation and experiment. It is evidently not enough to oppose to them dogmatic contradictions and some questionable chemical experiments.

[1] G Pennetier, *Origines de la Vie.*

The facts observed by F.-A. Pouchet render moreover spontaneous generation much more intelligible. In effect, the birth of infusoria, at least of the more complex infusoria, seems to be preceded by the formation of a sort of living blastema, in the heart of which the ovula originate, exactly as they originate by spontaneous genesis in the blastemas of animals and plants.

What radical difference is there between the formation of the ovula in the proligerous pellicle of F.-A. Pouchet and the formation of cellular elements under the microscope in saccharine serum? It is one of the adversaries of spontaneous generation, M. Cl. Bernard, who affirms that he had observed this last fact of veritable heterogenesis.[1] We must allow him to speak himself:—"I have observed that in saccharine serum, under the influence of a mild temperature, there are developed amyloidal productions perfectly analogous to the white globules. . . . . In a drop of saccharine serum, completely transparent, and in which the microscope reveals to us nothing, there are soon formed leucocytes, or globules of beer yeast."

[1] Cl. Bernard, *Rapport sur les Progrès de la Physiologie*, &c., pp. 210, 217.

CHAPTER II.

OF GENERATION IN THE TWO KINGDOMS.

Growth is only an excess of nutrition, and generation is only an excess of growth. Growth and generation have for cause a superabundance of nutritive materials. This superabundance has for effect first of all to carry the anatomical elements to their maximum volume, then to provoke the formation of new elements. As long as the individual has not attained all the development compatible with the plan of his being, the elements, newly born, remain aggregated to the pre-existent elements. When the limit of growth is attained, when there is no longer room in the organised individual for a new adjunction of histological elements, the newcomers detach themselves from their organic stem and constitute independent individuals which evolve in their turn.

Generation is so much a continuous growth, that its processes are identical with or analogous to those of growth. Definitively, in growth as in generation, those processes reduce themselves to two: the process of segmentation and that of genesis, which however can aid each other and combine their action.

Division (scissiparity, fissiparity, and so on) is observable in the majority of the lowest representatives of the two kingdoms. It is habitually the division into two parts which is then in use, bipartition, sometimes longitudinal as in the vorticels, sometimes transversal, as in the hydræ, the acalephans (Fig. 43).

It is by bipartition that the polypiers are produced and the humblest seaweeds and mushrooms multiply.

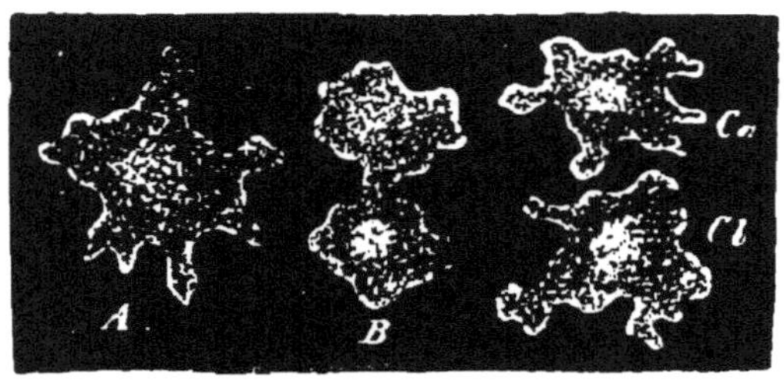

FIG. 43.

Reproduction by segmentation of an elementary organism, a moneron: *A*, entire moneron (protamœba); *B*, the same moneron divided into two halves by a median fissure; *C*, the two halves have separated, and constitute now independent individuals.

At a higher degree of organisation it is no longer the whole organism which is divided. The function is localised. There is formed, often by genesis, a special cell, the *ovulum*, charged to reproduce a new organic individual. This cell always commences its labour of organic construction with a series of bipartitions, with a segmentation; whence result numbers of new cells, of bricks, which are the first materials of the future edifice. But usually to be fit to traverse its whole evolutive cycle, the ovular cell needs a special impulsion. It has to blend with another anatomical element, likewise special. The element which segmentises and multiplies itself is called *female*. That which impresses on it the evolutive impulsion is called *male element*, and the process of union is one of the simplest: it is the process of conjugation of which we have previously given an example. The two cells come into contact: the female element absorbs the male element, *impregnates itself therewith*, and from that moment it is *fecundated*, and pursues the course of its formative labour. In the two kingdoms everywhere there are sexuality and fecundation; whatever may be the complication of accessory organic apparatus, the fundamental fact reduces itself always and everywhere to the conjugation of two cells, and the absorption of one by the other. The names change, but the phenomenon is essentially the same, whether we contemplate the oosphere and the antherozoids of the algæ, the embryonary sac and the pollen of the phanerogams, the ovulum and the spermatozoaries of the animals. The oosphere, the embryonary sac, the ovulum are simple varieties of the female element, as the antherozoids, the pollinical cells, the spermatozoaries are varieties of the male element.

Between the organisms in which reproduction is effected by simple bipartition, and those in which it is only possible after the fusion of a female cell and a male cell, that is to say, after a fecundation we can place the cases of parthenogenesis, the most celebrated example of which is that of the aphides. Here the intervention of the male cell is only necessary after long intervals. If it has once taken place it suffices for the formation of a series of ovula, to which it is no longer indispensable, and the female can afterwards engender, without the co-operation of the male, a whole line of young. But little by little the field of evolution is abridged for each new ovulum: the products are more and more imperfect, more and more incomplete: finally the evolutive force of the ovula is extinguished, is exhausted, and their revivification by a new sexual, impregnation becomes the very condition itself of generation.[1]

In most plants, in a great number of inferior bisexuate animals, the male and female apparatus are united in the same individual. Then autofecundation is often possible: sometimes also it is impossible. But even in the first case the diversity of origin of the male and female sexual elements is a condition most frequently favourable, sometimes indispensable to fecundation. The observations and experiments of Sprengel and Darwin have demonstrated that nearly all hermaphrodite plants need, to fructify, a cross fecundation. To use the expression of Darwin,

[1] See Cl. Bernard, *Des Phénomènes de la Vie communs aux Animaux et aux Végétaux* (*Revue Scientifique*, 1874).

"Nature tells us in the most evident manner that she has horror of a perpetual autofecundation."

The female ovulum can be fecundated at various moments of its evolution, at various degrees of maturity, and it is probable that the epoch of fecundation has an influence on the future direction of this evolution, which has not an immutable character. We shall see in effect, further on, that disturbing causes, even though slight, acting on the fecundated ovum, can make the embryonary development vary in a very large measure. That even the production of the one sex or the other may be connected with the degree of maturity of the fecundated ovulum is not, *a priori*, an inadmissible supposition.

Girou de Buzareingues demonstrated in the female slips of phanerogamous dioic plants with the flowers in clusters, that the male or pollinical cells, when they fell on the whole cluster and fecundated all the flowers thereof, produced very various results according to the degree of maturity of the ovula. At the lower part of the cluster, where the most advanced flowers are found, the fructification gave male seeds, while the flowers less ripe, those at the top of the cluster, produced female seeds.

Starting from this idea, and supposing that the complete maturity of a female ovulum might be very favourable to the production of the male sex, and inversely, M. Thury, of Geneva (1863), caused cows to be impregnated, sometimes at the commencement, sometimes at the end, of the rutting period. In the first case he obtained female calves; in the second, male calves. The experiment was recommenced by a Swiss agriculturist, M. Cornay, who, twenty-nine times in twenty-nine cases, succeeded in producing at will such or such a sex.[1] These are positive facts which are not weakened in any considerable degree by contradictory observations, almost all susceptible of different interpretations; and assuredly the question deserves to be taken up afresh and seriously studied.

In the preceding fact an observation made on ovular develop-

[1] Cl. Bernard, *loc. cit.*

ment in the vegetal kingdom has been extended and applied to the animal kingdom. In truth, if we remount to the primordial phenomenon of sexual generation, we find it nearly identical in the two kingdoms. The animal ovulum and the vegetal ovulum are rigorously comparable.

The animal ovulum is a cell, which, at first, is not distinguished from many other histological elements by any observable characteristic. This cell is composed essentially of an enveloping wall, of a protoplasmic content, of a nucleus. It has been a mistake therefore to give to these diverse plants of the ovular cell special names, tending to give the notion that they represent things without analogy, when in reality the differences are purely virtual. The faculty of reproduction, of generation, considered in a general manner, by no means specially appertains to the ovular cell. It is merely developed more and more with a longer range in the ovulum than in the other cells.

Let this be as it may, it is indispensable to know the names given by the embryologists to the diverse parts of the ovular cell. The enveloping membrane, which is hyaline and transparent, and whose projection has, in the microscope, the appearance of a ring, has been called *transparent zone, zona pellucida*. The content or cellular protoplasm of this envelopment, a substance more or less viscous and granulous, has been called *vitellus*. The ovular protoplasm or vitellus contains a nucleus which hollows for itself a way from a cavity and then bears the names of *germinative vesicle, vesicle of Purkinje*. Finally, this nucleus contains a nucleole, which is the *germinative spot*.

In the most perfect of the sexual vegetals, the phanerogams, that which habitually is called *ovulum* is nearly equivalent to the ovarium of animals. In the cellular tissue with polyhedrical elements which this ovary contains, and which is called *nucula*, is found the real ovulum, which has been called the *embryonary sac*. It is an oval cell, whose wall represents the vitelline membrane of the animal ovulum. Its content is a mucous, granulous, and greyish protoplasm, manifestly equivalent to the vitellus. Finally, the nucleus or *embryonary vesicle* of the botanists is comparable with the germinative vesicle. We distinguish besides therein a nucleole, which seems thoroughly to correspond to the germinative spot. The differences between the embryonary sac of plants and the ovulum of animals bear only on the dimensions or the number of the nuclei. The embryonary sac has in effect dimensions relatively considerable, and it contains sometimes two or three nuclei or embryonary cells.

With regard to generation as with regard to all the grand biological facts, we see evermore revealed the unity of life. We shall have to signalise many other analogies in the pages which have to follow and which are to be consecrated to a description of the principal phenomena of generation in the two kingdoms.

CHAPTER III.

OF VEGETAL GENERATION.

Asexuate generation is observable in certain algæ, especially the confervæ. It is accomplished by endogenesis. The protoplasm of certain cells contracts first of all, by separating from the wall, and usually divides afterwards, segmentises into many new cells. Then the enveloping membrane of the mother cell opens and is reabsorbed. The daughter cells escape and swim in the water, eddying, thanks to the movements of two or more vibratile cilia. These *zoospores*, as they are thus called, end by fixing themselves and germinating (Fig. 44).

Other algæ are endowed with sexuate generation, either in its simplest mode, which is conjugation, or in its complete mode, with male cell, female cell, and fecundation. In the first case (*ulothrix, chlamydococcus, pandorina*) two cells analogous to the zoospores in movement meet and melt into each other. When there is sexuality, the phenomena still offer great simplicity The female element is represented by a cell, whose protoplasm contracts while separating from the cellular wall. This small protoplasmic mass, thus modified, and ready to evolve, is called *oosphere*, and the envelopment is called *oogon*. In other cells the protoplasm transforms itself into mobile elements very analogous to the spermatozoaries of animals. These fecundatory elements are denominated *antherozoids*, and the cell which contains them and has engendered them is called antheridion (Fig. 45). Once set at liberty, the antherozoids penetrate into the *oogon*, by a special opening, and confound themselves with the mass of the oosphere which they impregnate.[1]

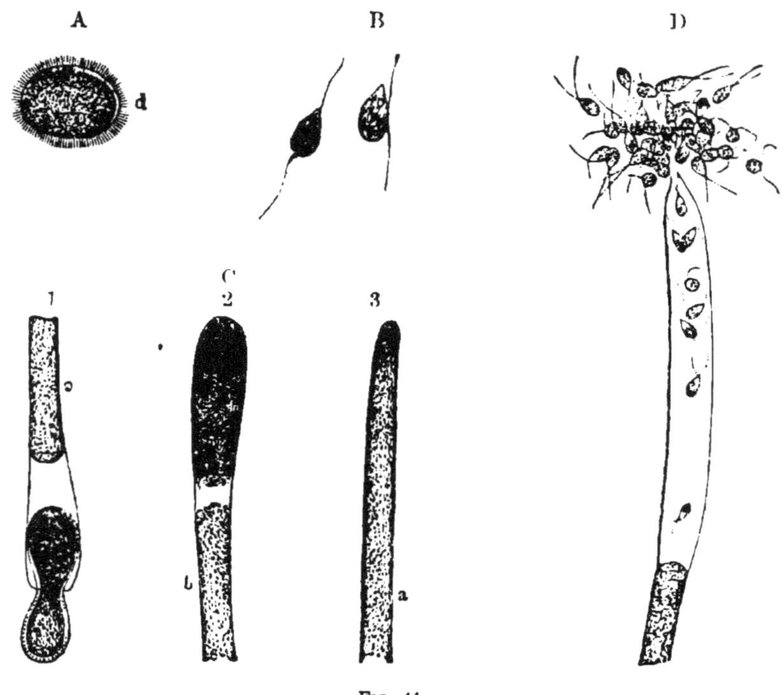

Fig. 44.

A, zoospore covered with vibratile cilia.
B, zoospores furnished only with two vibratile cilia.
C, evolution of a zoospore of Vaucheria : *a*, filament before the fructification ; *b*, at the extremity of the filament an accumulation of protoplasm has formed, whence is to proceed the zoospore ; *c*, zoospore, almost completed, and already furnished with its enveloping membrane, projects beyond the filament.
D, Numerous zoospores set at liberty.

Apart from the abundance of organs, of apparatus, which in the highest animals are the means of generation, the same description of fecundation could serve equally for the algæ and

[1] Sachs, *Traité de Botanique*. p. 285.

for man. So great is essentially the simplicity of biological phenomena of the first order.

When the oosphere has absorbed some antherozoaries, generally of a volume much inferior to its own, it has become capable of pursuing the course of its evolution : it is fecundated. We see it then envelop itself with a solid membrane, fix itself, and germinate. It has, in a word, become an oospore.

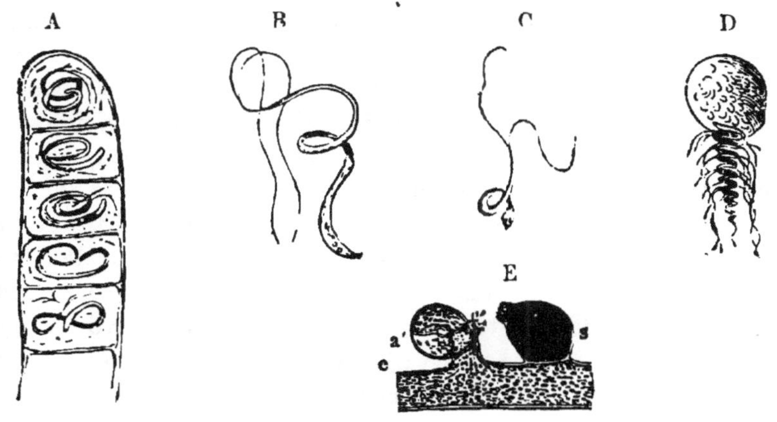

FIG. 45.

A, extremity of an antheridion filament formed of cells juxtaposed and containing each an antherozoid.
B and C, antherozoids in the state of liberty.
D, antherozoid of an aquatic fern.
E, fragment of an antherozoid of *Vaucheria* twisted spirally (a'). From the extremity of this antheridion come forth antherozoids having the form of short rods. By the side of the antheridion is found the oogonion (s), or saclike cavity full of ovular corpuscles. At a point of the oogonion situated in face of the extremity of the antheridion is formed an orifice by which penetrate these antherozoids. Then these antherozoids impregnate the ovula, which evolve or cover themselves with an enveloping envelope; in short become oospores.

The modes of reproduction of mushrooms are very varied, as regards the accessory apparatus: but essentially the processes of generation are perfectly comparable with those of which we have just spoken. Sometimes the reproductory bodies, or spores, are formed without fecundation; sometimes fecundation is necessary. These spores are sometimes immobile, sometimes mobile, and furnished with vibratile cilia; consequently they are in every respect similar to the zoospores of the algæ. But definitively, the spores are formed at the expense of certain cells, and if they are multiple they result from a repeated bipartition of this protoplasm. What matters it, after this, that they are produced or not on the filaments of a mycelium, that they are gathered together or not in the cavity of a receptacle, or *sporangium*, of this form or of that?

The system of reproduction by oospheres and mobile antherozoids exists also in the characeæ, the muscineæ, and the ferns. The form of the antherozoid varies, but usually it is a filament more or less elongated, mobile, furnished or not with vibratile cilia. Always the antherozoid is formed at the expense of the protoplasm of special cells or antheridia whose wall is torn or reabsorbed. Before the dehiscence or disappearance of this wall, each antherozoid, habitually alone in the mother cell, is rolled two or three times on itself (Fig. 45). As to the apparatus containing the female corpuscules, they vary much in form, and have for that reason received diverse names; but generally the antherozoids do not penetrate into the receptacles of the oospheres (archegons) except through a soft mucilaginous substance. Once in contact with the oosphere or oospore, the antherozoid confounds itself with it, and as happens usually after all fecundation, the oospore segmentises itself by bipartition, and the embryonary development follows its course.[1]

The phenomena of reproduction and fecundation are manifestly the same in the phanerogams. We have already spoken of the embryonary sac, of the embryonary cells which it contains. The fecundating agent is no longer here an antherozoid; it is the granulated viscous substance (*favilla*) contained in the particles of pollen. These particles are of very varied form, but have all a double enveloping membrane (endhymenine, exhymenine). It is well known that from contact with the humidity of the stigma,

[1] J. Sachs, *loc. cit.*, pp. 391, 318, 401, 439.

the particle of pollen swells by endosmosis, that its internal membrane forms a hernia filled with favilla (pollinised content), and penetrates thus to the embryonary cells. The fecundation is made by the intense blending of the *favilla*, and of the contact of the embryonary cell. The fecundated cell always develops itself, and forms by division a mass of spherical cells, whence proceed the cotyledons and the embryon.

CHAPTER IV.

OF ANIMAL GENERATION.

The two living kingdoms were probably identical originally, as the naturalists, partisans of evolution, try to demonstrate by patient researches and ingenious pictures of affinity. Assuredly there still exist many points of union between the two grand divisions of the organic world, and there is no differential characteristic constant and applicable to all animal and vegetal species. We must therefore expect to meet with a great analogy between the principal phenomena and even the grand processes of reproduction in the two kingdoms.

At the lowest degrees of the animal hierarchy, in the beings little or not at all differentiated, for example in the polypi, the morphological centre of the whole creature seems to exist in each of the cells, and any portion whatsoever of the body can reproduce the whole animal. It is in the species thus endowed with an enormous power of reparation, of regeneration, that fissiparous reproduction is especially observable. The individual, on attaining his maximum of growth, overflows, so to speak, beyond his natural limits; he unfolds himself, and forms, by simple division, a new individual (Fig. 46). This process of multiplication acquires sometimes an extreme energy, if it is true, as M. Balbiani affirms, that in forty-two days a paramecium can produce, by fissiparity, 1,384,116 new individuals, that is to say, that a single animal, $0^{mm},2$ long grows 277 mètres in bulk. We see then the body of the infusorium lengthen, thereupon narrow toward the middle, and form thus a couple conducted by the anterior individual.[1] Finally the separation is effected, and each of the halves completes itself. The naïds divide themselves in the same manner transversally into two parts, of which

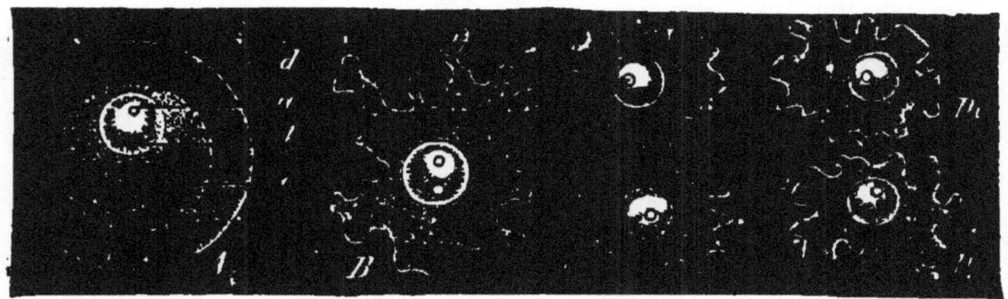

FIG. 46.

Amœba sphærococcus, in the different stages of its evolution. *A*, encysted amœba. Protoplasmic mass (*c*), containing nucleus (*b*), nucleole (*a*), and enveloping membrane (*d*). *B*, amœba freed from the enveloping membrane; *C*, amœba commencing to divide itself; *Da* and *Db*, amœba totally divided into two independent parts.

the anterior remakes itself into a head, and the posterior into a tail.

It is by germination or gemmiparity, that is to say, by a variety of fissiparity, that are formed the aggregations, the colonies of the zoantharies. Sometimes the bud detaches itself, playing the part of a vegetal spore or bulbille, and goes to form an independent individual.

Many monadaries seem to reproduce themselves solely by gemmiparity. This mode of generation exists sometimes alone, sometimes associated with sexuality, sometimes alternating with this last in many inferior animals. In the hydras or fresh-water polypi we easily see the young hydras budding on the mother hydra. Their body at first communicates with the cavity of that of the mother, and is nourished at the expense of the same until the moment when the new individual is in its turn furnished with prehensile tentacles capable of seizing its prey.

[1] Dugès, *Physiologie Comparée*, t. I. p. 213.

Internal gemmiparity, *endogenesis*, is observable in the olvoces, the acephalocystes, and so on. The cavity of the animal fills itself with vesicular animals similar to the mother vesicle. Certain of these newcomers, those the most developed, contain themselves other individuals, which in their turn contain others; so that three generations are thus inclosed one within the other.

According to M. Balbiani many fissiparous or gemmiparous species are at the same time capable of sexuality, but of an alternant sexuality.[1] In the green paramecium (*Paramecium bursaria*) the sexuate cells pre-exist; they enlarge at a given moment. In the largest of these cells ovula with a nucleus and a nucleole are formed. In the smallest the intercellular substance divides into small bacilla, which are spermatozoaries charged with fecundation.

This hermaphrodite sexuality enters into action at a determinate epoch. The scissiparous process would exhaust itself little by little, would end by no longer producing any but puny, imperfect individuals, and would lead to absolute sterility, if sexuality did not come to give to the reproductive property a new vigour.

The lowest mode of sexual reproduction, conjugation, is observable in the colpods. Two individuals unite, encyst themselves, and blend into one common mass. Then this common mass reproduces by fractionment new individuals.

The mode of copulation of the green paramecium recalls still, though faintly, conjugation. In its sexual phase, each paramecium is provided with reproductory bodies male and female, as we have recently seen; but nevertheless,—a thing, however, frequent among the hermaphrodites,—the fecundation is mutual. Two individuals juxtapose themselves mouth to mouth during a duration of many days, and they exchange their spermatozoids. It is an example of the horror which Nature has of autofecundation, to use Darwin's expression. Many hermaphroditical

[1] Balbiani, *Sur l'Existence d'une Génération Sexuelle chez les Infusoires* (*Comptes Rendus de l'Académie des Sciences*, t. XLVI).

invertebrates are provided with the genital apparatus of the two sexes; but these apparatus are so arranged that autofecundation is almost impossible.

Sometimes, however, especially in the bivalvous mollusks, for the most part fixed and immovable, there is autofecundatory hermaphrodism.

On the whole, a general fact results from the examination of generation among the inferior animals,—the great diversity of processes, and also a sort of confusion in their employment. The species are still ill differentiated; there is in the plan of their organism a sort of indecision. Nature, to use the language of Darwin and of many others, seems to hesitate, to grope; she has not yet found the best way, and tries all ways simultaneously. But when the sexuality is well marked, when there is formation of a male cell and a female cell, the primary acts of reproduction and fecundation assume an almost uniform character.

In all animals, the female cell, the ovulum, is nearly identical. At first the ovulum differs in nothing from an ordinary cell, if we do not allow ourselves to be led astray by the special names invented by the embryologists. The complete ovulum is composed in effect of an enveloping membrane, the *vitelline membrane*, or *zona pellucida*, of a content or *protoplasm*, the *vitellus*, of a nucleus or *germinative vesicle*, of a nucleole or *germinative spot* (Fig. 47).

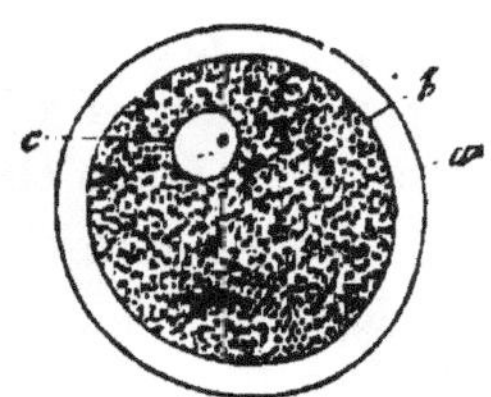

FIG. 47.

Ovulum of woman magnified 250 times: *a*, pellucid zone; *b*, vitellus; *c*, germinative vesicle containing the germinative spot.

According to M. Van Beneden, there are in the ovulum an accessory part and an essential part. The last, which he calls *cell-ovum*, has in all the animal kingdom an identical evolution, and is represented solely by the nucleus of the ovulum, by the *germinative vesicle*, and a small part of the vitellus which surrounds it.[1] It is certain, at all events, that the germinative

[1] Cl. Bernard, *loc. cit.*

vesicle is the least variable part of the ovum. It can be double, and then gemelliparity seems constantly to result therefrom. This is what takes place at least in certain species, especially in the *Vortex balticus*, in which this duplicity of the germinative vesicle is the rule.[1]

The cellular content or vitellus is likewise called *yellow*, very improperly, however, for its colour is very variable. In effect, it is, according to species, white, yellow, red, brown, green, violet, and so on, and is usually composed of a liquid more or less viscous, holding in suspension granules, and often fatty globules.

The vitelline or enveloping membrane is also exceedingly diverse. Often it is a transparent, anhysted membrane; sometimes it is a simple albuminous layer. In certain species it is covered with designs, with grooves, with reliefs. At other times, as happens in the ovulum of fishes, it is pierced with holes, with micropylæ, which facilitate the impregnation of the ovulum by the spermatozoaries.[2]

We must then consider the nucleus of the ovulum as a germ englobed in a mass purely nutritive, which is the vitellus.

The ovulum of viviparous animals differs very little as to volume, whatever the size of the animals may be. That of the oviparous animals is much larger, is sometimes indeed very voluminous; but the increase bears on the nutritive portion of the ovulum, the vitellus, which in birds acquires an enormous development.

What distinguishes the ovulum from every other cell, even before fecundation, is the rapidity of its evolution. There is a whole series of phases, of modifications, through which it passes in a very short space of time, without even undergoing the fecundating impulsion. Soon, in the whole animal series, the ovulum loses its first differentiation; the nucleus and the nucleole (vesicle and germinative spot) disappear, and seem to melt into the vitellus. The ovulum is then nothing more than

[1] Leydig, *Traité d'Histologie*, p. 620.
[2] Leydig, *loc. cit.*, pp. 616, 617.

an imperfect cell composed of an enveloping membrane and of a mass of granulations.

At the same time the vitellus is drawn back toward the centre of the cell, and a transparent zone then separates the cellular membrane from the granulous content (Fig 48 A). Another phenomenon, the utility of which is not yet well known, is afterwards produced. It can take place also independently of fecundation. It is the emergence of the *polar globules.* According to M. Ch. Robin, these globules are formed by a sort of vitelline germination. We see successively appear on the surface of the vitellus two

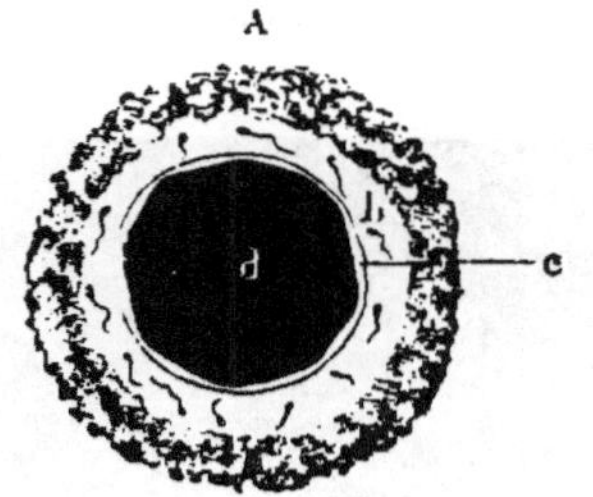

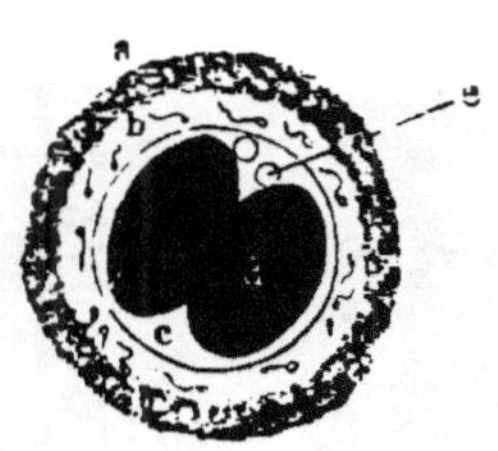

FIG. 48.

A, ovum of she dog immediately before the commencement of segmentation. The contraction has given to the vitellus a polyhedrical form.—*a*, epithelial cells; *c*, space between the contracted vitellus (*d*) and the vitelline membrane or pellucid zone.

B, ovum of she dog some hours later. The epithelial cells are further diminished; the vitellus separated into two segments or spheres of fractionment. Between these two halves are seen the bright vesicles (*c*) called vesicles *of direction.* The letters as before.

or three projections, which take the hemispherical form, and separate by division from the mass which has engendered them, remaining interposed between the cellular membrane and the vitellus. These are the *polar globules* (Fig. 49, A e).

All the evolutive phenomena which we have just signalised can be accomplished before fecundation; they can also be posterior thereto. But in certain species the spontaneous ovular evolution goes much further, since, in the cases of parthenogenesis the whole development of the embryon is effected without the succour of fecundation. This ovular generation, without males, this parthenogenesis, which the doctors of Catholic theology are sure to invoke one day in support of their mysteries the most important and the most improbable, is not rare in the animal kingdom. Many examples thereof can be cited, of which the most celebrated is the parthenogenesis of the pucerons, signalised first of all by Bonnet.

But for the most part the spontaneous evolution of the ovulum does not go further than the emergence of the polar globules. If then the fecundation does not intervene, the ovulum withers and dissolves. In the contrary case, it undergoes the important phenomenon of fractionment, that is to say, a series of fissiparous divisions, of bipartitions. Fissiparity has certainly

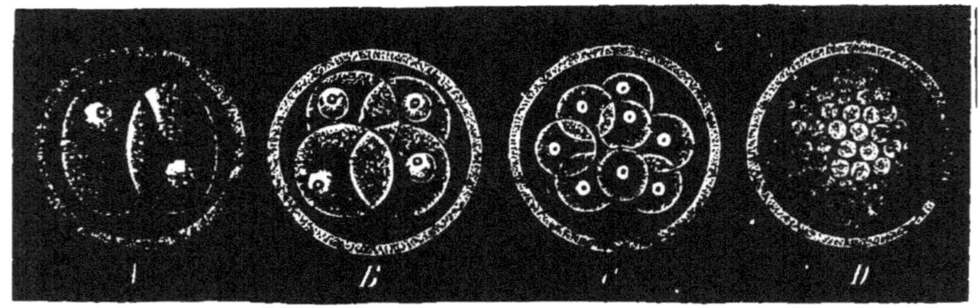

Fig. 49.

First stadium of the evolution of a mammifer, "segmentation of the ovum," multiplication of the cells by reiterated scissions: *A*, the ovum divides by a first fissure into two cells; *B*, the two cells divide into four cells; *C*, these last divide into eight cells; *D*, the segmentation, indefinitely reiterated, has produced a spherical mass of numerous cells.

been the first process of reproduction of the rudimentary organisms, and it still exists in the state of transitory phase, of organic tradition, in sexuate generation. The phenomenon is one of the simplest. After the disappearance of the vesicle, the retraction of the vitellus, the emergence of the polar globules, the vitellus narrows in the middle, and is thus severed into two granulous masses, of which each is divided in its turn into two halves, and so on in succession, till the moment when the whole vitelline mass is transformed into a heap of globules, spherical, granulous, destitute at first of enveloping membranes and of nuclei, then completing themselves little by little. They are then *vitelline cells*, the agglomeration of which gives to the ovular surface the appearance of a mulberry (Figs. 49 and 50).

Soon after their appearance the vitelline cells displace themselves; they come and fix themselves on the internal face of the ovular membrane (vitelline), and thus form a membrane, limiting a hollow sphere full of an albuminous liquid. This membrane has been called *blastoderm*, and the cells which compose it become the *blastodermic cells* (Fig. 50). It is at their expense that the embryon forms itself.

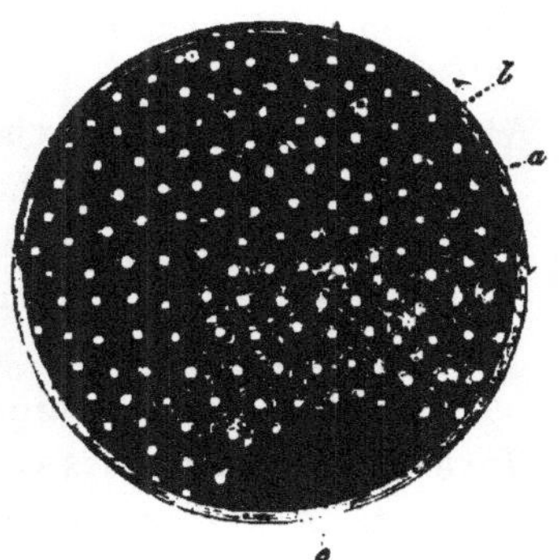

FIG. 50.

Ovum of rabbit taken in the uterus.—*a*, pellucid zone; *b*, blastodermic vesicle formed of hexagonal cells, in symmetrical conjunction like the bricks [of a pavement]; *c*, accumulation of internal cells.

The vitelline fractionment is a general phase of ovular evolution; but it is not always accomplished uniformly. In birds, in fishes, in testaceous reptiles, in the cephalopodal mollusks, it is partial; a portion only of the vitellus takes part therein. Then the blastodermic membrane, instead of forming a hollow sphere, becomes a simple spherical calotte, and the portion of the vitellus not employed plays merely the part of a nutritive substance.

In the ovum of insects, of the arachnida, of certain crustaceans, there is produced, instead of a regular series of bipartitions, an irregular cleaving, whence results the formation of unequal fragments. But these fragments end by grouping themselves into polyhedrical masses.

Whatever besides may be the form and the extent of the blastodermic membrane, it is not long in unfolding into two superposed membranes. This division is especially perceptible at the point of the blastoderm, whither go to form themselves the first lineaments of the embryon. The external blastodermic vestment has been called *serous* or *animal vestment*, because it is the rudiment of most of the apparatus and organs of the animal life. The internal vestment is called *mucous* or *vegetative vestment*. From it proceed the nutritive apparatus.

Finally the cells accumulate at a point of the blastoderm, and form there a sort of spot called the *embryonic spot*. It is there that the embryon develops itself.

In their ensemble the phenomena we have just described are common to all vertebrated and invertebrated animals, endowed with sexuate ovulation. But, starting from the formation of the embryon, differences arise corresponding first of all to the two great branches of the animal kingdom. To follow step by step these differences would be evidently to go beyond the limits assigned to us. We must content ourselves with indicating some general facts.

The division of the animal kingdom into the vertebrated and invertebrated branches is of all the most general, the most important. Also it is marked at the very outset of the embryonary evolution. In the vertebrated ovulum the embryonary spot elongates into an ellipse, rises in the middle after the fashion of a buckler, and at this point the serous vestment hollows out for itself a rectilinear furrow (primitive line, *nota primitiva*). The edges of the furrow rise more and more, and in the depth is formed a solid cellular cord, a sort of provisional scaffolding, round which the vertebral column is to be formed. This cord is the notocord or dorsal cord (Figs. 51 and 52).

It is to special treatises that the reader must go to follow the ulterior phenomena of development, the formation of the *umbilical vesicle*, of which the content serves still for the nutrition of the young being, that of the amnios, which envelopes it in an aqueous medium, the origin of the placenta or of the vascular organ, which, in the superior viviparous vertebrates, establishes a communication between the circulatory system of the mother and that of the embryon, when this last has exhausted the nutritive resources of the vitellus.

A very important general fact, of which the partisans of the evolutive or Darwinian doctrines make great use, and with justice, is the identity of the embryonary forms at the outset in all the vertebrates, and the appearance in a regular and

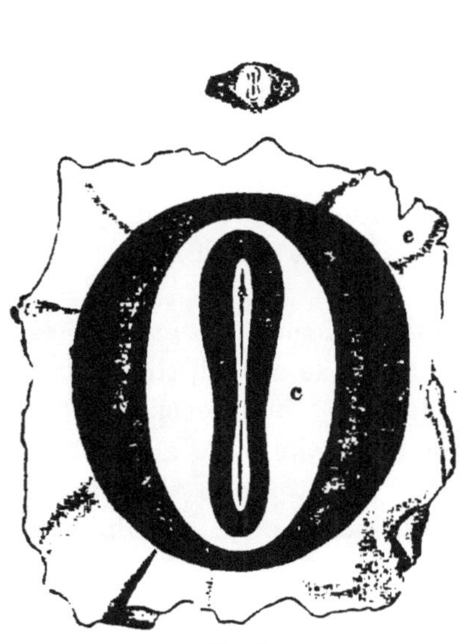

FIG. 51.

Ovum of dog. The embryon in form of shoe sole is outlined.
a, dorsal fissure;
b, dorsal plates;
c, clear area;
d, opaque germinative area;
e, membrane of the germinative vesicle.
The small superior figure is of natural size.
The inferior figure is magnified.

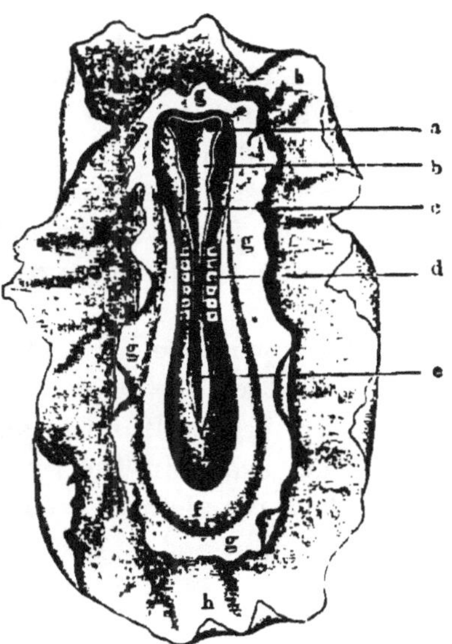

FIG. 52.

Embryon of a dog's ovum, twenty days old. Dorsal fissure widely open and surrounded everywhere with a bright edge, which indicates the first deposit of the nervous substance on the depth and the walls of the fissure. In the depth is seen on the median line the *chorda dorsalis* represented by a darker stripe.—*a*, *b*, *c*, rudiments of the cerebral vesicles; *e*, posterior rhomboidal sinus; *d*, body of primordial vertebræ; *f*, lateral plates; *g*, middle and external vestments of the blastodermic vesicle, still united; *h*, mucous vestment; *i*, body of the embryon.

successive order of the characteristics of each type. We may compare the embryons of the vertebrates to a group of travellers starting from the same place, and entering first of all a vast general route, which they forsake to enter roads of more and more secondary importance, and diverging less and less. It is first of all fishes which differenciate themselves; then reptiles, then birds, then mammifers; finally the diverse mammiferous types. The less individuals are to differ in the adult state, the more tardily their embryons differenciate themselves. It is at the last period of development, for instance, that the embryon of a monkey can be distinguished from the embryon of a man, and there is a moment when both do not differ more from the embryons of a fowl or of a tortoise than they differ from each other.[1] It is difficult not to see here, as the partisans of evolution are inclined to see, a sort of living palæontology, an abridged picture of the formation of diverse animal types, such as has been accomplished in the course of the cycles which have rolled away.

The development of the fecundating element is at first so analogous to that of the ovulum, properly so named, that it has not unjustly been called the *male ovulum*. Wherever sexuate reproduction exists in the two kingdoms, the male element is primitively a complete cell, containing a protoplasm comparable with the vitellus and a nucleus analogous to the germinative vesicle.

These male ovula, according to the testimony of many observers, seem usually to be formed by spontaneous genesis, as well in the anther of phanerogamic glants, as in the special apparatus of animals.

Once the male cell formed and arrived at a certain degree of maturity, its granulous protoplasmic content undergoes, spontaneously, a phenomenon, which vividly recalls the fractionment of the fecundated female ovulum. The last elements of that fractionment, which are generally in an even number, resemble each other much in most of the cryptogamic plants, and in all the animal kingdom. They are cells, whence comes forth finally a mobile corpuscule, called *spermatozoid* in the vegetal kingdom, and *spermatozoary*, or *zoosperm*, in the animal kingdom. The cells of fractionment, each of which produces, as a male, only one of

[1] Huxley, *Man's Place in Nature.*

these fecundating corpuscles, have been called, always by analogy, *male embryonary cells.*

In the phanerogamic plants the male ovula are the great cells, which are called *pollinical utricles,* and the products of their fractionment are *pollinical cells,* each of which ends by representing a particle of pollen. It is the content of these particles of pollen, or *favilla,* which is the fecundating substance. The evolution is here less complete than that of the male embryonary cells. In effect, these last resemble at first, almost identically, the pollinical cells. But arrived at a higher degree of maturity, they differenciate themselves more, and their content becomes spermatozoary, or spermatozoid. If the content of the particles of pollen does not organise itself into corpuscles, endowed with a movement of totality, it is not, nevertheless, completely immobile, and the granulations observable therein are animated by movements comparable besides with those which are effected in many other vegetal cells. Whatever, moreover, may be the apparent dissemblance between the favilla of the particle of pollen and the mobile fecundating corpuscle, spermatozoary or spermatozoid, the phenomenon of fecundation does not, on that account, change in its essence. On the whole, the fecundating substance penetrates always into the ovulum, and when once it has arrived there its molecules sever and untimately blend with those of the female ovulum.

The form of the spermatozoaries and of the spermatozoids is somewhat variable. These fecundating elements are globulous in some vegetals ; they are also globulous in the myriapods and some crustaceans. But in general they are more or less filiform, and at one point swelled out. They have already this form in the male embryonary cells, when, before the rupture of those cells, they are seen to be rolled spirally (Fig. 53).

As a rule the enlarged part of the spermatozoary is at one of its extremities. This part is of variable form, cylindrical, spherical, oval, elongated to a point, and so on. The spermatozoaries of the mammifers have a somewhat short head and a caudal filament long and thin. But the form of the spermatozoary seems to be secondary, for certain animals have simultaneously zoosperms of two different forms, (*Paludina vivipara.*)[1] The filiform part is the locomotory organ of the spermatozoary. The movements produced by this sort of vibratile cilium are moreover very varied. They are sometimes movements of reptation, sometimes movements of torsion, sometimes flutterings, sometimes spirally penetrating movements.[2] There are no immobile spermatozoaries among the vertebrates. They seem, however, to be very numerous among certain invertebrates, especially the crustaceans. But probably their rigidity is only apparent, and ceases as soon as they penetrate into the body of the female. It is certain that even the mobile spermatozoaries of man and of the mammifers move with much more velocity when they enter the mucus of the uterine neck or body. Their rate of progression however, is infinitely less than it appears in the microscope According to Henle it is only eighteen hundredths of a millimètre a second in man. The rate of progression varies much with the nature of the ambient medium. It increases in an alkaline solution,—diminishes, on the contrary, in an acid medium.

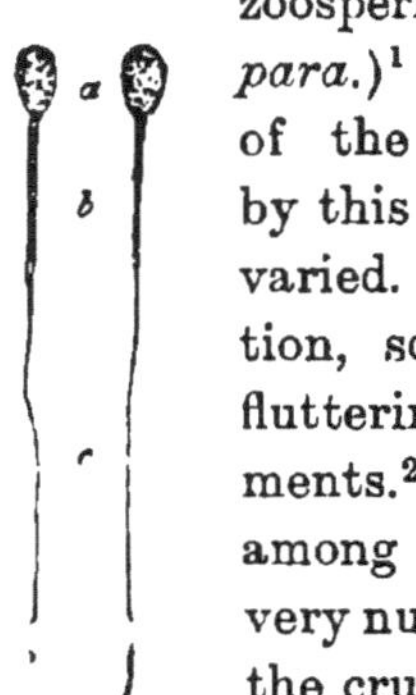

Fig. 53. Spermatozoaries of man: *a*, head; *b*, body; *c*, tail.

Since the discovery of the spermatozoaries, by Leuwenhoeck the question has often been asked whether the spermatozoaries were or were not animals. Certain micrographers, partisans of the first opinion, have gone so far as to assert that the spermatozoaries of the mammifers are differentiated animals. Some have even thought that they had testicles. Judged by our present instruments of observation, the spermatozoaries are absolutely without structure, are very inferior in this respect to the vibratile epithelial cells, with which they have been compared. We must consider them special anatomical elements, superior to most o

[1] Leydig, *loc. cit.*, p. 600. [2] *Ibid.*, p. 554.

the other histological elements by their motility, and especially by their fecundating property, very inferior on the other hand by their structure.

All the particles of pollen, all the spermatozoids and spermatozoaries are endowed with the fecundating property. But it seems established by experiments of artificial fecundation, that a single particle of pollen, a single spermatozoary cannot alone fecundate an ovulum. The ovum needs veritably and literally to be impregnated with fecundating molecules. There is a certain minimum which does not suffice to give the necessary impulsion.

We have seen what great analogy exists between the male ovulum and the female ovulum. In the inferior animals and even in the superior animals, at a certain period of embryonary evolution, the analogy extends to the whole male and female generative apparatus. At a particular moment of the intra-uterine life it is impossible to distinguish the male embryon from the female embryon, and in many invertebrates the difficulty persists during the whole duration of life, to such a point indeed that men like Cuvier, Carus, and Blainville never succeeded in distinguishing from each other the ovarium and the testicle of the gasteropods.[1]

The ovulum and the spermatozoaries have another property in common, the property of reviviscence. Spermatozoaries desiccated can get back their motility after a very long time, when they are humected anew; and M. Davaine was able to keep for five years, without killing them, in a solution of chromic acid, at 2 in 100, the eggs of the lumbricoid ascarides of the Greek tortoise. As for the vegetal seeds, we know that some of them can still germinate after thousands of years.

[1] Dugès, *loc. cit.*, t. I., p. 231.

CHAPTER V.

OF REGENERATION.

It is impossible to terminate an exposition of generation without saying some words on regeneration, which essentially does not differ from it. We must satisfy ourselves with mentioning the general laws of regeneration, without describing the innumerable series of particular facts which everybody knows.

It seems to result from modern micrographical observations that there is identity between primordial phenomena of the first generation of the tissues in the embryon, and those of their regeneration in the adult. In both cases the elements of the newborn tissues are formed by spontaneous genesis in the midst of an organisable blastema. In the animal embryon this formative blastema results from the liquefaction of the embryonary cells : in the adult animal it is formed by the anatomical elements of the conservated tissues. But in both cases the spinal tissues do not appear all at once : they are first of all preceded by a transitory generation of nuclei and of cells called *embryoplastic*, to which at last succeed the special histological elements.

The lower an animal is placed in the hierarchy, the greater the degree of regenerative faculty he possesses. In this case there is still physiological confusion. There are no elements specially charged with reproduction. Every histological element possesses then in the state of indivision, the fundamental properties of nutrition, of development, of reproduction. All the world knows that in the inferior animals we can make with full success artificial fissifarity. The *Polypus brachiatus*, the earthworms, the naïds, the planaries can be sectionised with impunity. Each isolated fragment lives its own life, completes itself, and becomes an entire animal.

We can even thus create monsters. For instance, by sectionising longitudinally a planary to the half of its length, Dugès was able at the end of a few days to obtain a bicephalous planary.

When an inferior animal is sectionised transversally into two parts, it is the more differentiated, the more centralised the part which completes itself the first. In a *Polypus brachiatus* it is the buccal portion which first renews itself. In the earthworm the anterior fragment remakes for itself a tail before the posterior part succeeds in refashioning a head.

In all cases the organs of new formation are regenerated at the expense of the blastemas of the portion which they complete. When a snail remakes a head, when the tail of a lizard grows again, when, as Spallanzani has observed, a salamander regenerates a new tail with the nerves, the muscles, the vertebræ, the vessels, the skin thereof—when it creates anew claws, the lower jaw, and so on, all this labour of reconstruction is achieved without augmenting the weight of the trunk which has remained alive. The animal has become again entire and complete, but it is enfeebled, diminished. This is why it is impossible to renew indefinitely the operation without causing death. The head of the earthworm, for example, cannot be reproduced more than twice or thrice. This is also the reason why the regenerated is often smaller than the old and more imperfectly fashioned part.

By an inexplicable singularity regeneration is sometimes the more certain, the larger the portion taken away is. When Réaumur merely broke one or two joins of the anterior claw of a crab, the loss was not repaired. On the other hand, the whole claw was regenerated when it had been entirely removed.[1]

[1] Réaumur, *Mémoires de l'Académie des Sciences*, 1712.

Regeneration being simply, like generation, an exaggeration of the property of nutrition, it is natural for it to be the prompter and easier the younger the individual is. It is operated, in effect, more easily in youth than in the period of maturity, and more easily in the larva than in the complete animal. The young salamanders reproduce their claws more easily and more completely than the old. The sectionised tail of the tadpole grows again the faster the younger the animal is. The perfect insect does not reproduce its antennæ, while this regeneration is accomplished with facility in its larva; and so on.

The faculty of regeneration—and this is a very singular fact at the very outset—seems to be almost peculiar to the animal kingdom. It exists little, or not at all, in plants. A leaf or a branch when cut is not remade. This fact comes in support of the theory which considers every complex vegetal as an aggregation, a sort of colony of polypiers, of which the diverse parts have their own, their individual life. Every leaf, every bud lives on its own account, and the diverse parts of the vegetal are connected with each other by a very feeble solidarity.

In the superior animals and in man we do not see, as in the inferior animals, an entire organ, complex, composed of diverse tissues, reproduce itself in totality. But all the tissues, except perhaps the striated muscular fibre, can be reproduced isolately and in small portions.

It is a fact of vulgar notoriety and often utilised in surgical therapeutics, that great quantities of osseous tissue can be regenerated, and even very promptly, provided the vascular membrane of the bone, the periosteum is uninjured.

The regeneration of a tissue is the easier the more this tissue is differentiated, the more delicate and noble are the functions to which it appertaineth. Yet, in conflict with this general proposition, the muscular striated tissue can be little if at all regenerated, while the reproduction of the nervous tissue is not rare. A sectionised or even a resectionised nerve remakes itself from the injured point to its furthest ramifications, and according to M. Ranvier it regenerates itself by traversing all the phases through which it has already passed from the beginning of its fœtal development. The nervous centres, the central hemispheres even, seem to be able to restore themselves in a certain degree. How otherwise explain a large number of incontestable cases of complete re-establishment of the intellectual faculties after serious lesions, and even after notable losses of substance of the encephalon?

No doubt, the biological facts, like all the phenomena of the universe, obey laws, true laws, inflexible, and without exception. But those facts are so complex, so intermingled, the *ensemble* of their causes forms a skein so entangled, that the most of our pretended biological laws are simply imperfect generalisations, suffering many exceptions and always subject to revision.

BOOK V.

OF MOTILITY.

CHAPTER I.

OF THE BROWNIAN MOVEMENTS.

THE general properties of organised matter, with which we have so far been occupied, are closely connected with the truly fundamental biological property, nutrition; to speak correctly, they are only its dependencies. It is otherwise with that which we have now to examine. Doubtless, the faculty of movement is subordinate to nutrition, which is the very essence of life, but it differs from it; it is, as it were, superadded to it.

Before proceeding further, it is important to distinguish carefully between motility, an organic property, and movements, essential attributes of every organic and inorganic matter. At the present time, most men of science, and a certain number of thinkers, have finally broken with the old metaphysical distinction, which, with one stroke, cut the universe into two parts, having between them only relations of contingency. We have ceased to abstract the active qualities of the material substance from the substance itself; we no longer figure the world to ourselves as constituted by an extended substance, but in itself perfectly inert, ceaselessly incited and put in motion by an impalpable agent called force. The ancient divorce no longer exists. Force is espoused anew to matter, from which moreover it was never separated, except in the brain of metaphysicians. There is no longer inertia in the universe. Behind all these phenomena is concealed an extended substance, never immobile, and which is itself its own mover. If we still often speak of forces, it is from pure habit, or by pure artifice of language. In final analysis, the idea of force may always be brought back to movements of an active material substance.

Without speaking even of the planetary movements, which carry everything along, we know that those bodies most solid and immovable in appearance resolve themselves into atoms and molecules, ceaselessly animated with rapid movements. Naturally, these molecular movements also take place in organized bodies. There, they are varied very differently, since these bodies are nourished, that is, are the seat of incessant molecular exchanges, and are decomposed without intermission to be recomposed.

The molecular movements are, then, effectuated in every living substance, but all living bodies are not endowed with motility. Motility may be defined as the property which certain organized bodies have, either of totally changing their position in space, or of contracting, or of becoming shorter, or of modifying for the moment their form, spontaneously and independently of any mechanical action of exterior origin.

Definitions, with whatever precision we may endeavour to give them, are always adjusted with difficulty to all the peculiar facts which they embrace. There are certain movements, for example, which we do not very well know how to classify; we wish to speak of movements still incompletely studied, and known under the name of Brownian movements. The botanist, Robert Brown, was the first to prove that the fine particles of mineral dust, well pounded, moved, apparently spontaneously, when held in suspension in a liquid medium. In fact, all corpuscles having from three to four thousandths of a millimètre in diameter, execute in liquids, under the microscope, a kind of oscillation, with an alteration in position of from four to five times their diameter. It is difficult to attribute to simple liquid currents, as has often been done, these movements, which seem rather due to phenomena of attraction at a short distance, and must be brought back to the domain of capillarity. Now almost similar movements are also observed in certain organic substances. Thus, when leucocytes or infusoria are disaggregated, they resolve themselves into granulations animated with very lively Brownian movements. Must we consider these movements as organic or inorganic? We scarcely know; and the distinction becomes still more embarrassing with regard to the movements of the pollinical *favilla*, of which we have already said a few words.

A number of observers, Needham, Amici, Guillemin, &c., &c., have occupied themselves with the movements of the favilla, or rather of its corpuscles. The last of the observers whom we have just named has even attempted to compare them with the movements of the spermatic animalcules. Robert Brown had already observed these movements, and he had also seen them in pollinical particles, especially in the ovoid and transparent particles of the pollen of gramineous plants. These movements have even a singular persistency, since R. Brown said that he found them still in particles of pollen preserved in a herbarium for twenty-five years. Moreover, Alex. Brongniart has noticed some interesting peculiarities regarding them. In one case, particles of pollen having burst in water, the granulations of the favilla of the *pepo* were still animated with their ordinary oscillatory movements, and were displaced. The pollinical granulations of several other species (*hibiscus palustris* and *syriacus, rosa bractea,* &c.), *curved inwards and changed their form.*

Doubtless, there is nothing organic in the Brownian movements of mineral particles, but they are nearly similar to the Brownian movements of the *débris* of white globules, and of disaggregated infusoria, and these last resemble the oscillations of the *favilla*, which, in certain species mentioned by Brongniart, have truly the appearance of spontaneous and organic movements, since the granulations not only incurve and change in form, but are, besides, paralysed by contact with alcohol and various other substances, which never happens to the mineral particles.

It seems, then, that there may be, in the phenomena which we have just mentioned, a kind of gradual transition from inorganic movement to organic motility. But many purely organic movements take place in the two living kingdoms, and it remains for us to enumerate them, and to indicate, as far as possible, their conditions and laws.

CHAPTER II.

OF MOVEMENTS IN THE VEGETAL KINGDOM.

We shall first of all set aside, as foreign to our subject, the slow insensible movements of the stems and roots, which depend upon development, growth, &c., and whence result, for example, the direction of the stems towards the sky, of the roots into the soil, the spiral twisting of certain stems, &c.

On the contrary, the movements of certain species of confervæ strongly recall those of animals. In fact their filaments balance, lower, raise themselves, twist themselves, and oscillate; and so on. Heat and light accelerate them; cold and darkness retard them. Acids, alkalis, alcohol, &c., suppress them.

We have already pointed out the movements of the spores and of the antherozoids. In a number of phanerogamic species, the various parts of the flower execute extended movements. Certain flowers open in the day and shut in the night, or inversely. This phenomenon even takes place habitually, in each species, at a certain hour, so that by choosing suitable species, Linnæus could form what he called *Flora's clock*.

The movements of the petals depend much upon heat and light. It is only between 8 and 28 degrees that the flower of the *crocus* opens. Below this it does not open at all; above, it shuts. At a constant temperature, any abrupt variation of light tends to decide the blowing of the flowers of the crocus, the tulip, and of compound flowers; every abrupt diminution tends to close them.

In cellars, lighted at night by lamps, and kept in darkness

during the day, De Candolle has seen flowers, habitually diurnal, open in the evening and shut in the morning. Flowers of nocturnal habits changed also inversely.[1]

Movements of the stamens are also common and generally more rapid than those of the petals. The best known are those of the stamens of the *berberis vulgaris*. When irritated by any kind of contact, these stamens bend towards the pistil, and scatter their pollen upon the stigma; after which they stand upright again, and approach the petals. The movement of the stamens is sometimes spontaneous and successive; thus Humboldt has seen the stamens of the *Parnassia palustris* approach each other one by one, and by jerks of the pistil, scatter their pollen three times over the stigma, and resume their first position. Certain venomous substances abolish these movements of flowers. Flowering branches of species having petals or mobile stamens are paralysed, when their stems are plunged into various solutions (prussic acid, water of bitter almonds, alcohol, ether, acetic acid, ammonia, &c.). The movements stop as soon as the venomous liquid, by absorption, reaches the mobile organs.[2]

When once fecundation is accomplished, the vitality of the flower diminishes, and the movements cease.

Many plants have leaves which execute each day a regular and periodical movement. They have a diurnal position, a nocturnal position, and pass slowly from one to the other. Moreover, certain of them are excitable by various contacts, mechanical agents, light, &c., so that accidental movements are superadded to the diurnal movements. It is peculiarly in the leguminous plants, especially the mimosas and oxalideæ, that the slow movements of wakefulness and slumber are most easily observed. They are caused by the variations of the luminous intensity, and are specially due to the action of the most refrangible rays.[3]

The species most celebrated for the mobility of their leaves

[1] Mémoires présentés à l'Institut.

[2] Goeppert, *De acidi hydrocyanici vi in plantis commentatio*. Breslau, 1827.

[3] J. Sach's, *loc. cit.*, p. 1029–1031.

are the *Dionæa muscipula*, the *Oxalis sensativa* of Java, and the *Mimosa pudica*. The leaves of the dionæa are furnished with true darts, which transpierce the insects as soon as they come in contact with them. As to the *Oxalis sensitiva*, its mode of action is nearly like that of the *Mimosa pudica*, which we shall take as a type.

The periodical movement is very marked in the *Mimosa pudica*. According to the observations of MM. P. Bert and Millardet[1] it is effected thus. In the evening the petiole of the compound leaf is much lowered; then it begins to rise again towards midnight, consequently without any luminous excitation, and before sunrise it has attained its maximum of uprightness. At sunrise it begins to lower itself very rapidly, whilst the folioles separate and spread themselves out. At the commencement of the night, the petiolary depression is at its maximum, and the folioles are close to the petiole. But, in addition to this slow periodical movement, the leaf has extreme sensitiveness, and all contact, all mechanical or chemical irritation, causes it instantly to assume its nocturnal position.

It is towards 30 degrees that the sensitiveness of the *mimosa pudica* attains its maximum. Below 15 degrees it is null. At 40 degrees the folioles become rigid in an hour, at 45 in half-an-hour, from 48 to 50 in a few minutes. Up to this point they may still recover their sensitiveness in a lower temperature, but at 52 degrees they lose it finally. Below 15 degrees, or after having remained some days in darkness, the folioles also become rigid, but transitorily.[2]

Ingenhouz and Humboldt have seen the motility of the sensitive plant disappear when the plant was plunged into carbonic acid, azote, &c., but this was only because it could no longer respire. On the contrary, they have seen it naturally persist in oxygen. An electric current causes the depression of the leaves. On the contrary, chloroform and ether abolish the motility.

[1] Bert, *Recherches sur les Mouvements de la Sensitive.* Paris, 1867.

[2] Sachs, *loc. cit.*, p. 855, 1035, 1037; and Dutrochet, *Mémoires pour servir*, &c., t. I. p. 524.

Lindsay, Dutrochet, and, after them, many other observers have proved that the principal agent in the movements of the sensitive plant is the bourrelet at the base of the petiole. According to Dutrochet, this bourrelet acts as if it were composed of two springs, an upper and a lower. The first lowers the petiole, the second raises it again. The action of the bourrelet would coincide with its distension, its turgescence, and would be due, according to certain botanists, to an afflux of sap. But how is this afflux of sap produced? How can the movement of the folioles be explained? Above all, how can we comprehend that an excitation may be transmitted from one leaf to another, sometimes even to a distant leaf, missing the nearer leaves, sometimes across the entire plant, since all the leaves contract successively, when a drop of sulphuric acid is sprinkled upon the roots.[1]

We must admit here the intervention of the anatomical elements of contractile tissues, analogous to those which exist in animals. In effect, M. Vulpian has found, at the junctions of the folioles, and at the base of the petioles, cells containing a finely granulated jelly, which he compares to the substance of muscular fibres, and which he says he has seen contract under various excitive influences. M. Cohn, of Breslau, has found analogous histological elements in the filaments of the anthers of the cynareæ; that is to say, elongated cells, longitudinally striated when at rest, contracting under the influence of various excitants, and then presenting transverse striæ, of a very vivid kind.

These anatomical elements may possibly be the agents of the local movements, directly provoked. As to the indirect movements, transmitted to leaves non-excited, we cannot account for them without admitting a certain contractility of the sap vessels, along which the stimulus is transmitted from one to the next. This kind of vascular contractility is rather animal

[1] Vulpian, *Leçons sur la Physiologie Comparée du Système Nerveux*, p. 31, 32.

than vegetal; nevertheless it exists in a large number of vegetals, if the facts observed by Goeppert are exact. According to what this observer says, branches cut from plants rich in milky juices (*Chelidonium majus, Lactuca perennis, Euphorbia esula,* &c.) no longer pour forth liquids when plunged into prussic acid.

Here then we have a property common to the two kingdoms, which, moreover, touch at many other points. The manner in which the spores and the antherozoids of the cryptogams move, forms a new feature of union. Here is no longer analogy, but identity. We know that many infusoria alter their position by means of vibrations, oscillations of one or many filiform appendages called *vibratile* cilia, which are also found in animals on a variety of epithelial cells, hence called *vibratile epithelium.* Now these animal vibratile cilia in no wise differ from those which serve as propulsive agents to the spores and antherozoaries of sea-weeds and of certain mushrooms, &c. Our classifications, with their arbitrary divisions, are never the exact expression of nature. Everything is connected; everywhere there are transitions and shadings.

CHAPTER III.

OF MOVEMENTS IN THE ANIMAL KINGDOM.

WITHOUT, as we have seen, specially appertaining to the animal kingdom, motility is nevertheless a property infinitely more spread and developed in the animal kingdom than in the vegetal kingdom. There is scarcely any animal species which is not more or less endowed therewith : but it is not a property inherent in organised matter : for many histological elements are destitute thereof, and when the animal is perfectionated, differentiated in some small degree, motility is the attribute and the function of a special tissue, at least in its most perfect mode.

At the lowest degree of animality, when all is still confused in the living substance, it is the whole body of the animal which is constituted by a substance, contractile, homogeneous, changing form perpetually, emitting and retracting expansions unceasingly. This contractile, amorphous substance has been called *sarcode*. It is sarcode which forms exclusively the body of the amœbæ of the lowest monerians (*Bathybius haeckelii*), also of the rhizopods which move in emitting and retracting sarcodical expansions.

The first effort of differentiation seems to be the formation of vibratile cilia. Here the mobile expansions are no longer transitory. They have a fixed, definite form. They are persistent organs, constituting, as we have seen, the principal agent of locomotion in the infusoria.

In the hydroids, the hydra properly so-called excepted, there is already a certain differentiation of tissue and a thin contractile layer exists, on the periphery of the body, beneath the epithelial lining. This tissue is already composed of long and fine fibres.[1]

On the contrary, the substance of sponges is formed of corpuscles, amibiform, amorphous, and contractile.

In many animals, especially in numbers of worms, in the first period of life, the only locomotory organs are still vibratile cilia : but at the adult age there is developed under the external tegument a muscular layer constituted by fibres confusedly interlaced (Fig. 54).

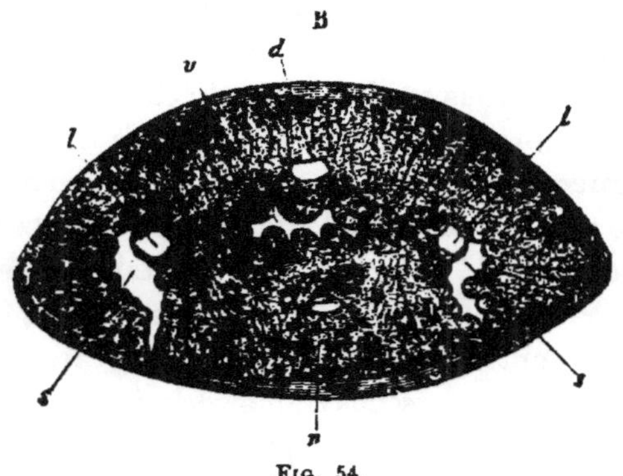

FIG. 54.

Section of *Hirudo*. *c*, cuticular layer; *m*, muscular layer; *r*, lateral line with the excretory organ; *p*, *p*, median lines, superior and inferior; *q'*, oblique fibres; *τ*, intestine; *d*, dorsal vessel; *l*, lateral vessel; *s*, vesicle of the excretory organ; *n*, ventral ganglionary chain.

Already in the echinoderms the case is different. Here the contractile apparatus always forms a subtegumentary tube : but the fibres which compose it are regularly grouped.[2]

In respect to the muscular apparatus, the mollusks represent a well-marked link of transition. During the first phases of their development they often move by means of vibratile cilia. At the adult age they are furnished alike with a superficial

[1] Gegenbaur, *Manuel d'Anatomie Comparée*, p. 118

[2] *Ibid.*, p. 169, 170.

muscular envelopement and distinct muscles, traversing the cavity of the body and serving to open and to shut the valves. The most of the muscular fibres constituting the motory apparatus have still as in worms, the form of long filaments a little flattened, which sometimes are subdivided into fibrils. A certain number of those fibres are however slightly striated transversally.[1] We perceive therein also some nuclei, which are probably the last vestiges of the primitive embryoplastic cells, at the expense of which the fibre has been developed.

A new and decisive progress is accomplished in the arthropods. No longer any contractile envelopement. The muscular fibres are all grouped into distinct masses, having a definite form, into muscles, which are inserted into such and such a part of the body by means of fibrous tendons. Lastly those muscles are almost solely formed of fibres bearing very numerous transversal striae.[2] Under the microscope they scarcely differ from those which constitute the muscles of animal life in the vertebrates, that is to say, muscles charged to excite the voluntary movements.

The voluntary muscles of the vertebrates represent the most perfect form of contractile substance. They are supported by the diverse pieces of the osseous framework, into which they insert themselves, and on which they impress the movements which the will commands. The cutaneal muscular apparatus disappears, unless we are disposed to see vestiges thereof in the small muscles which put in movement the scales of serpents, the pilose bulbs, and the hair of the mammifers, the feathers of birds, in those which act on the skin of the human face, and so on. Yet according to Gegenbaur, these dermic muscles are completely lacking in fishes, and do not at all represent the contractile tube of the inferior animals.

From the point of view of elementary form the contractile substance which may be called muscular presents itself under many aspects. There is first of all the amorphous or sarcodical

[1] Gegenbaur, *loc. cit.*, p. 465.

[2] Leydig, *loc. cit.*, p. 152.

state, that of the amœba, for example; then the cellular form; finally the form in tubes. According to a certain number of histologists and physiologists, there is under all these varied forms an identical substance. Doubtless the form can be dominated by the essence and the same organised substance may, strictly, constitute anatomical elements diversely configured. Nevertheless, almost always, the morphological differences correspond to functions more or less dissimilar. Even the diverse varieties of figurate muscular elements have each, as we shall see by and by, their special mode of contractility. As to the amorphous contractile substance, we can, at the most, only consider it as a first outline of the muscular element. It is semifluid, coagulates at 40 degrees, while the true muscular substance is solid. We can bring it into affinity with the semifluid and vitreous substance contained in the large contractile cells of the hydras, and also with the substance which constitutes first of all the heart of the vertebrated embryon.[1]

It is only in the form called fibro-cellular that the muscular substance acquires a veritable autonomy. It is then represented by fusiform cells, more or less flattened, provided with one or two nuclei and capable of compressing or of swelling themselves out by contraction. There are often fine granulations round nucleus. The length of the fibro-cell can vary from some hundredths of a millimètre to half a millimètre. These fibro-cells are found in great number in man and in the mammifers, in the muscular coats of the intestine, of the veins, and of the arteries, in the excretory conduits, under the skin in diverse regions, especially under the pilose follicles, which they put in movement, under the scrotum, and so on. This type of contractile element is found with variations in all the vertebrates, and in a great number of invertebrates. In both cases it usually

[1] Recently it has been pretended that the sarcode is semifluid, even in the striated muscular fibre, and that this substance is contained in tubes. And it has been said that a dermatoid was seen marching in every direction in the very heart of a muscular fibre as if in a liquid.

co-exists with fibrous contractile elements, which are themselves of varied species.

According to the strict interpretation of the cellular doctrine generally admitted in Germany, every muscular fibre comes originally from cells, which are simply developed in length by end being joined to end. The nuclei, which we observe here and there along the course of the fibres in the adult, are in that case, the last vestiges of that primitive cellular state.

Let this be as it may, the typical muscular fibres are of two histological species, connected however with each other by a series of transitory forms. They are *smooth* or *striated.* The smooth fibres must be considered in man and the superior vertebrates as fibro-cells very much elongated. In many invertebrates, on the contrary, they seem to be very fine threads without nuclear enlargements and without divisions. It is those fine filiform elements, cylindrical and smooth, which constitute exclusively the muscular apparatus of the inferior invertebrates. Among the mollusks, we already in certain organs, for instance in the pharynx of certain gasteropods, meet with those muscular fibres striated crosswise, which constitute the voluntary muscular apparatus of the arthropods (crustaceans, arachnida, insects) and of the vertebrates.

This striated muscular apparatus or apparatus of the animal life, resolves itself, in final analysis under the microscope, into very thin fibrils, about the thousandth part of a millimètre in diameter and offering alternately in the direction of their length transparent parts and dark parts of the same extent. These fibrils juxtapose themselves in larger or smaller numbers to form *striated muscular bundles,* contained in an elastic transparent sheath called sarcolemma or myolemma. In the bundle the fibrils are placed in such fashion, that that there is nearly exact correspondence between their bright parts and their dark parts, whence the striated appearance of the bundle. The volume of these bundles is very variable, according to the animal species, and in the same species according to the regions. It is these bundles which in vertebrates constitute the flesh, properly so called, the muscular mass, obeying the orders of the will, obeying the volitions formulated in the brain. It is, in effect, a constant characteristic of the striated muscular fibre, and likewise of the fibro-cell, that of being in relation, in contact more or less immediate, with central nervous masses. On the contrary, the elementary sarcodic substance contracts of itself, obeying direct exterior impressions. In the same way the vibratile cilia oscillate independently of all nervous incitation, even in the superior animals, for even the death of the individual does not suspend their movement. In certain inferior animals, in which the nervous system is not yet differentiated, there already exist smooth muscular fibres contracting as spontaneously as the sarcodical substance.

In the vertebrates, in which the two systems of smooth fibres and striated fibres are largely represented, there is between them a certain division of labour. The smooth fibres, or fibre-cells, appertain in a general manner to the apparatus and to the organs of nutritive life, while the striated muscles are the servants of the conscious will, the muscles of the life of relation. Thence the classical division into muscular system *of the organic life* and muscular system *of the animal life.*

This division, however, is true only to a certain degree. Here, as everywhere in biology, there is no absolutely determined distinction. We find striated fibres in the intestine of the tench, in the gizzard of birds, in the pharynx of some gasteropods. Finally, the wall of the heart of all vertebrates is formed of striated fibres. But those cardiacal fibres have an arrangement in some sort transitory. They anastomose into each other, as do the smooth fibres in certain invertebrates.

To the anatomical differences of the two orders of fibres, correspond differences of functions, physiological differences. Contractility, that is to say, the shortening and the widening of the fibre, is effected according to different modes in the smooth muscles and in the striated muscles. In the first it is slow and durable: in the second it is prompt and instantaneous, and does not continue longer than the excitation itself which has provoked it. This law is verified in the diverse apparatus of the vertebrates; it is verified still more visibly in invertebrates. Those of them, for instance the mollusks, whose muscular system is almost entirely composed of smooth fibres, move slowly. On the other hand, the arthropods, which have a striated musculature, have movements vivacious, vigorous, and precise.

Contractility is a property inherent in the muscular fibre, and can be brought into play by diverse excitants. It exists in the inferior invertebrates, and in the vertebrated embryon when it has not yet a nervous system. It persists in the superior and adult vertebrate, when the nervous system has been killed by a special poison.

Like all the organic properties, contractility has nutrition for necessary basis. It cannot be accomplished without influencing the double assimilation and disassimilation movement indispensable to the maintenance of the life of the muscular tissue. Every contraction corresponds to a more energetic oxydation, to a more active assimilation, to the formation and the elimination of disassimilated products. The result is manifested immediately, in the vertebrates, by changes in the colour, the composition and the temperature of the blood which comes from the muscle, the muscular venous blood. If a muscle is at rest, the venous blood which comes from it is almost as rutilant as the fresh and oxygenised blood brought by the artery. The phenomenon is more marked still in the case of paralysis, in certain maladies which produce muscular atony, in syncope. The explanation is that the muscle, in its state of rest, is at its minimum of consumption, of life, of contraction; it absorbs nothing more than is strictly necessary to its maintenance.

On the other hand, when the muscle contracts, it wears itself, it expends itself; it absorbs and eliminates more. Hence the venous blood which comes forth from it becomes immediately black.[1]

[1] Cl. Bernard, *Leçons sur les Propriétés des Tissus Vivants*, pp. 220-274.

We know that this change of coloration is the indication of important chemical mutations. Thus, the venous muscular blood which, in the state of repose, contains only 6,75 per cent of carbonic acid more than the arterial blood, contains 10·79 per cent. after the contraction.

An experiment of Matteucci has demonstrated in another manner that muscular contraction corresponds to an oxydation of the fibre. In two glass vessels of the same size the same number of frogs, skinned and prepared, was suspended. In one of these vessels the frogs were suspended to metallic hooks, through which was made to pass an electric current, which determined contractions. At the end of five minutes the frogs were taken out of the vessels, into each of which the same quantity of limewater was poured. Thus resulted in the two vessels precipitates of carbonate of lime. But the precipitate was very slight in the vessel where the frogs had remained immobile; on the contrary, it was very abundant in the other.

All oxydation, in some degree energetic, has, for effect, a certain development of heat. Thus it can be demonstrated, that during contraction the muscular venous blood is hotter than the arterial blood.

But the production of carbonic acid in the living tissues results from and is accompanied by many other chemical mutations; also we see that the contractions alter the composition of the muscular juice bathing the fibre. In the state of repose, when the muscle is not fatigued, this muscular juice is abundant and its reaction is neutral or alkaline. The existence of this liquid and alkaline medium is necessary to contractility. In effect, a muscle into which a liquid even slightly acid is injected, loses the faculty of contraction.

The alkaline muscular juice of the muscle in repose contains oxygen, creatine, creatinine and other analogous substances, which are products of oxydation of the albuminoids.[1] We find therein also sugar, lactic acid and potash. It is this last substance

[1] Cl. Bernard, *loc. cit.*, pp. 226, 227, 170.

which gives to the muscular juice its alkaline reaction. In proportion as a muscle is fatigued its reaction becomes less and less alkaline, and finally passes into acidity. At the same time the muscle furnishes a quantity greater and greater of soluble substances. According to Helmholtz, there is in a fatigued muscle 0·73 per cent. of parts soluble in water: there is only 0·65 per cent. in a muscle which has rested.

Some hours after death the muscle becomes rigid. It is then spoken of as in a state of cadaveric tenseness, which seems due to the spontaneous coagulation of the contractile or syntonine substance; for we can provoke the rigid state by plunging a muscle into a liquid at 45 degrees.[1]

At the outset we can also make the rigidity disappear by letting a sanguineous current pass into the vessels.

As the muscle in cadaveric rigidity generally offers an acid reaction due to the lactic acid, it has been believed that the coagulation of the syntonine was due to the presence of the acid; but Cl. Bernard saw in crabs the muscles of the tail in a state of cadaveric rigidity, though they still presented an alkaline reaction. Finally, he observed that the acid reaction is often lacking in the bodies of animals that have died after a long abstinence.

We have already remarked that muscular irritability is a property inherent in the muscular element, wholly distinct from the excitant which brings it into play, without excepting the most important of those excitants, the nervous system. The fact is demonstrated either by killing the excito-motory nervous fibres, as we do, for instance, by means of the curaré poison, or inversely by abolishing the contractility of the fibre, with the help of certain substances which do not act on the nerves. The muscular poisons the most in use are the sulpho-cyanuret of potassium, the sulphate of copper, the sulphate of mercury, and also certain organic poisons which act first of all on the

[1] Cl. Bernard, *loc. cit.*, p. 230.

cardiacal muscular tissue; for example, the upas tieuté, digitaline, and so on.

From what precedes it results that there is no essential difference in the contractions, whatever the excitant may be which has provoked them. Little it matters whether this excitant is mechanical, chemical, physical or physiological. But there are individual differences in the muscles; all are not equally sensitive to the same excitants. The contact of an irritant chemical substance, a slight puncture made with the point of a scalpel, and so on, determines the contraction of most of the muscles, when laid bare, of the animal life. The smooth fibres of the stomach, of the intestine, &c., appear on the other hand more sensitive to the variations of the temperature. The muscular fibres of the scrotal dartos, the testicular cords, contract, under the influence of a sudden variation of the temperature, more or less. The tissues with smooth fibres of the stomach, of the intestine, seem in a certain measure to revive under the influence of an elevation of the temperature. In effect, if we plunge into an atmosphere of hot vapour an animal which has just died, we see when the body is heated to 20 degrees, the stomach and the intestines execute for half-an-hour and even an hour, peristaltic movements.[1]

Light, which seems to have little or no action on any of the muscles, impresses notwithstanding the contractile fibres of the iris, direct and without the succour of a reflex action. Thus two, and even three, days after death the iris of eels still contracts under the influence of luminous rays, provided care is taken to humect the eye, and prevent its dessication.

But the most important physical excitant, the excitant which acts most surely on the smooth fibres and the striated fibres, is electricity. When a muscle is maintained in suitable conditions of humidity, of temperature, of medium, it contracts under the influence of the continuous electrical currents and of the induced currents with frequent interruptions. It was with the help of

[1] Cl. Bernard, *loc. cit.*, pp. 189, 190.

a feeble continuous current that Helmholtz formerly determined the duration of muscular contractions by means of very simple experiments which had at the time a certain notoriety. He made to pass along a frog's muscle detached from the animal a current, which the fact itself of the contraction interrupted. A galvanometer made it possible to appreciate to a certain degree the duration of the different phases of contraction. The total duration of the contraction was in these conditions 0″,305, and decomposed itself into three parts. In a first phase, which Helmholtz calls *the pose*, and which continued 0″,20, there was no appreciable effect. The action of electricity on the muscular fibre is not therefore instantaneous. In this period of incubation contraction succeeded; it continued 0″,180, gradually augmenting in intensity; finally came a period of slackening, which amounted to 0″,105. If the contractions are thus reiterated a certain number of times, the muscle is fatigued, is exhausted: it contracts with a force gradually decreasing, and the duration of the period of slackening grows longer and longer.

The electrical current seems to act on the muscular fibre by determining in it a change of molecular state. In effect, if the current passes during a certain time without interruptions, we see the muscle contract, especially at the commencement of the passage of the current, and at the moment of its interruption. On the other hand, the induced currents provoke a permanent contraction. With an induced current of 32 interruptions a second, Helmholtz obtained the tetanic contraction of the masseter muscle.

Not only the muscular fibre contracts under the influence of electricity, but it is itself a sort of electric pile, as M. Dubois-Reymond discovered. If, in effect, by the help of a conductor and a galvanometer, we establish a communication between a point of the surface of a muscular fibre and its section, we see that there is an electrical current going from the exterior surface to the section, and that this current ceases during the contraction of the fibre.

This fact is observed alike in the striated fibres and the smooth fibres. But according to the observations of M. Dubois-Reymond the intensity varies in the different muscles. It seems therefore certain that every muscular fibre is electrised positively at its surface, and negatively at its centre.

At the moment when the muscle enters into the state of cadaveric rigidity, the current is interverted; it ceases finally when the fibre commences to decompose.

In like fashion the heart of a frog is electrised, positively at its point, and negatively at its base. Similar phenomena are observed, moreover, in the nervous fibres, and analogous ones have been observed in vegetal fibres. We must content ourselves with mentioning in passing these curious facts, of which no satisfactory scientific explanation has yet been given.[1]

Of all the excitants of muscular contractility, which we have just passed under review, there is none which equals in power the special physiological excitant, the nervous fibre. The excitants, not themselves physiological, act on the muscle with a hundred-fold energy when they borrow the aid of the nervous threads which are distributed in the muscle. It results, in effect, from the observations and the measurements of Matteucci, that during its contraction a muscle accomplishes a mechanical labour at least 27,000 times greater than the chemical labour, whence has come the excitation of the nerve.[2] Now in the complex animals all the muscles receive nervous fibres, connected with the central nervous system, and the influence of these nervous centres on the muscular movements is the greater the more perfect the animal is. But in order to form an idea of the relations of the muscular system and the nervous system, it is indispensable to study the last and the noblest of the general properties of organised matter, that is to say, innervation.

[1] Cl. Bernard, *loc. cit.*, pp. 205, 207.

[2] Matteucci, *Théorie Dynamique de la Chaleur* (*Revue Scientifique*, 1866, n°. 51.

BOOK VI.

OF INNERVATION.

CHAPTER I.

THE NERVOUS SYSTEM IN THE ZOOLOGICAL SERIES.

WE have now arrived at the most elevated point of elementary differentiation of organised matter, and at the most aristocratic properties of that matter. Every living substance is nourished. Certain histological elements are nourished, move and contract themselves. Certain others, the elements of the nervous tissue, possess, besides the fundamental property of nutrition, a whole group of special properties. Thus, the nervous cells can excite the contraction of the muscular elements; they have *motricity.* Moreover they have the consciousness of the action exercised on them by the ambient mediums; they feel pain and pleasure. They can also, by means of the organs of the senses, classify, select the excitations bearing on the extremities of the nervous fibres, with which they are anatomically related. They have *sensibility.* Finally, they can treasure up, combine in a thousand manners the sensations perceived; they can *think* and *will.* It is the totality of these primordial properties which constitutes *innervation.*

As we see, innervation is a very complex whole. Also the anatomical tissue which is endowed therewith, and the systems which this tissue forms, present a number of varieties in structure and in form. In certain beings innervation is rudimentary; it reduces itself to the single faculty of exciting movements and connecting them with each other; that is to say, to *motricity*. In others *sensibility* is joined to *motricity*. Finally, the most perfect animals possess alike *motricity*, *sensibility*, *thought*, in, however, very numerous degrees of perfection and imperfection.

In passing by for the moment the varieties of the nervous elements, we can consider every nervous tissue or nervous system as reducible to two histological types. Invariably, every nervous apparatus is composed of fibres and cells. Always the fibres continue direct with the cells, and always they connect those cells with a motory or sensitive organ. We shall soon have to describe the form and the function of the diverse nervous fibres and cells. At present it is enough for us to mention that the fibres are especially the conductors of impressions and of incitations, while the cells act on the impression which is transmitted to them, either to send it back, or to reflect it simply from one fibre to another, or to transform it into phenomena of consciousness. Finally, the cells can, in their turn, become centres of excitation, and then, without any exterior impression, they provoke movements or combine thoughts.

Like all other organic systems, the nervous system complicates, perfectionates, diversifies itself gradually in the animal series.

No trace of nervous system in the rhizopods, the sponges, in the monerions, the infusoria, unless we admit, with some romantic naturalists, a diffused nervous tissue, disaggregated, or rather not yet aggregated, invisible nervous molecules, infused and latent in a living gangue not yet differentiated. But all this is pure fancy. Neither is there any nervous system in the hydroids, the lucernaries, the anthozoaries. In the medusæ we meet with a nervous ring formed of a cord, following the edge of the disk, and offering, from distance to distance, cellular enlargements called *ganglionaries*. This is already the schema of all the nervous systems, more complex and more perfect, which we are about to pass under review, for there are, even in this rudimentary network, conducting cords composed of fibres, and receptive and excitation centres composed of cells. The existence of a nervous system is problematical in certain radiata. A nervous system seems, however, to exist in the holothuria, namely, an œsophagian, fibrous collar, emitting radii at points of which there are ganglionary enlargements.

M. D. Quatrefages found, in the planaries, a nervous ganglion, situated on a median line, and formed of two small masses joined together.

The rotifera have also a central ganglionary mass, situated on the pharynx. This mass, sometimes divided into two portions, emits nervous filaments.

In the colonies of bryozoaries, every colonist is provided with a cerebroidal ganglion, and besides, as Fritz Müller has shown, there is a system of cords connecting all the individual ganglions, that is to say, a colonial nervous system.

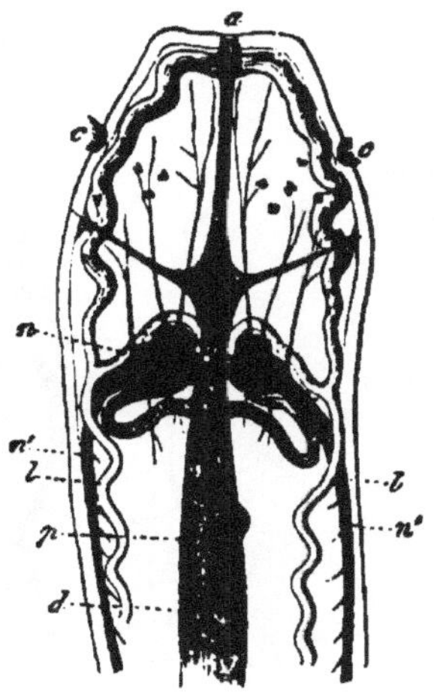

FIG. 55.

Anterior extremity of the body of a næmertian (*Borlas iacamilla*): *o*, opening of the trompe; *p*, trompe; *c*, vibratile fossettes; *n*, cervical ganglion; *n'*, lateral nervous system; *l*, lateral sanguineous trunks, which bend forward in order to join; before this junction they send round the brain a branch which unites with that opposite to form the dorsal vessel (*a*).

The plathelminths have, in this anterior part of the body, two nervous masses, relatively large, and united by a nervous commissure. The position of these ganglions has made them be called *cerebral*. The trompe passes as if into a ring, between the two cords of the commissure. From these ganglions go forth two longitudinal fibrous trunks which follow the lateral edges of the body, and are provided with small ganglionary enlargements emitting nervous filaments (Fig. 55).

This nervous system, so simple, is the rough model of that of most of the articulated invertebrates (worms, annelates, arthropods), especially when the lateral cords approach each other on the ventral face, as is the case with some species.

According to M. Blanchard, the nematoids (ascarides, strongyli filarians, and so on) have, at the origin of the œsophagus, four

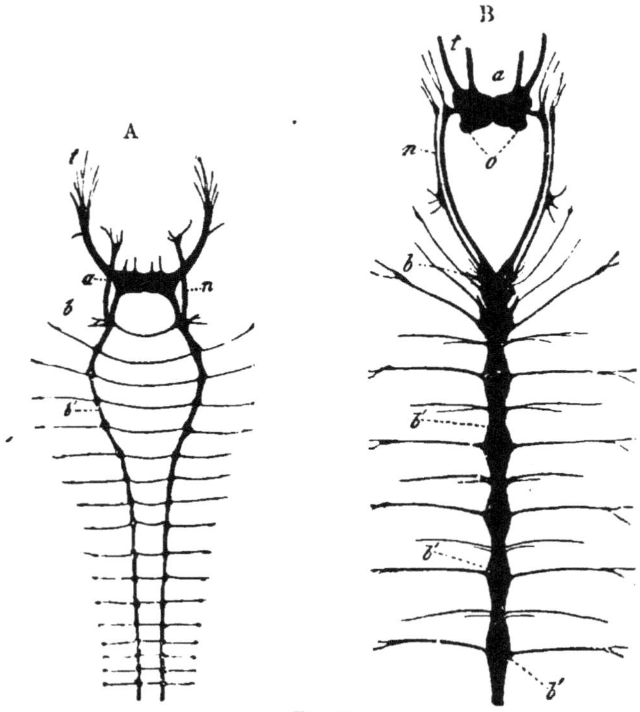

Fig. 56.

A, nervous system of *Serpula contortuplicata* : *a*, pharyngian ganglions superior ; *b*, inferior ; *b′*, ventral trunk ; *n*, nerves of the mouth ; *t*, nerves of the antennæ.
B, nervous system of *Nereis regia* : *o*, eyes reposing on the œsophagian ganglion superior. The other designations as in the preceding figure.

small ganglions, situated two at the right, two at the left, and connected by commissures forming an œsophagian collar.

This œsophagian collar is found in all the annelates and arthropods, but it is usually formed of a super-œsophagian ganglionary mass, sometimes connected by fibrous commissures, sometimes joined together direct. From the inferior ganglion go forth two abdominal cords directed from the front to the back, and bearing each a series of ganglions, whence radiate nervous threads. The ganglions of each cord usually face each other, and thus form, by their juxtaposition, pairs of nervous centres of which each corresponds to a segment of the body (Fig. 57). If many segments

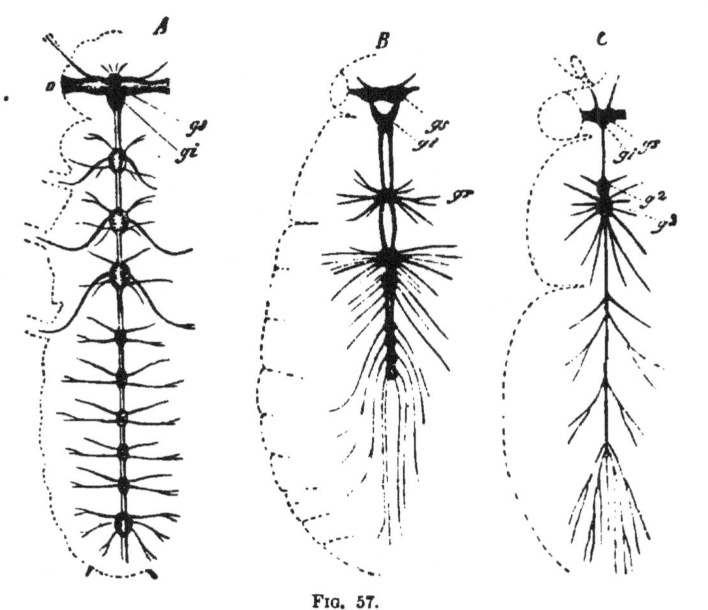

FIG. 57.

A, B, C, NERVOUS SYSTEMS OF INSECTS.
A, *termites*; *B*, coleopter (*dytiscus*); *C*, fly; *gs*, œsophagian ganglion superior (cerebroidal ganglion); *gi*, œsophagian ganglion inferior; *gr*, g^2, g^3, united ganglions of the ventral chain; *o*, eyes.

blend together in the course of the development, the corresponding nervous ganglions do the same, and thus constitute a more important nervous mass.[1]

The more voluminous of the ganglions is the more anterior of the super-œsophagian ganglion. It is called *cerebral* or *cerebroidal*

[1] Gegenbaur. *loc. cit.*

ganglion, and it usually furnishes nerves to certain organs of the senses and regularly to the eyes.

As a rule, the ganglions are constituted exclusively of cells, and the cords exclusively of nervous fibres.

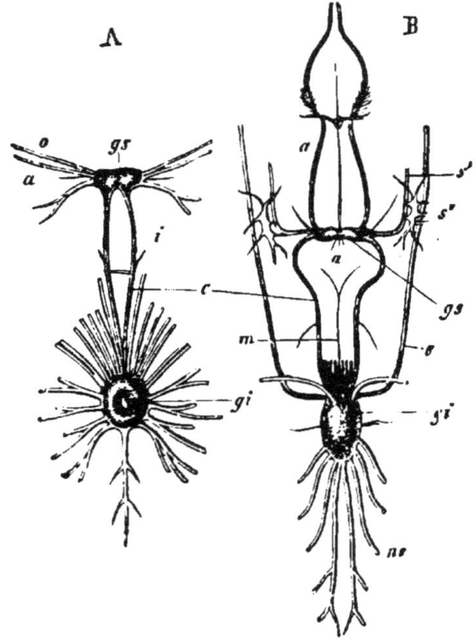

Fig. 58.

A, nervous system of a crab (*Carcinus mœnas*); *gs*, cerebral ganglion; *o*, ocular nerve; *a*, nerve of the antennæ; *c*, œsophagian commissure; transversal connection of this commissure; *gi*, fusionated ventral chain.
B, nervous system of a cirrhipod (*Coronula diadema*) seen in the ventral face: *gs*, *c*, *gi*, as in A; *a*, antennary nerves which distribute themselves over the mantle and the shell. Between them is situated the ocular ganglion, in connection with the brain; *m*, nerve of the stomach; *s*, visceral nerve which unites itself in a plexus *s''* with a second visceral nerve *s'*, coming from the anterior part of the œsophagian ring. The abdominal ganglion emits forward the first cirrus, and rearward (*nc*) the other cirri.

The body of the myriapods, and that of the grub or larva of the insects, being formed of very numerous segments, the nervous system of those animals is also very richly provided with ganglions. On the other hand a concentration, a coalescence, of nervous enlargements, takes place in the arachnida, the insects (Fig. 57), the crustaceans. The cerebral ganglion and the central chain of these last sometimes fuse into a single mass, whence radiate the nervous cords (Fig. 58). This mass is merely divided into two parts by the foramen, the œsophagian ring.[1] There is here a degree of concentration of the ganglionary cells, greater, in some respects, than in the vertebrates themselves. It is one of those paradoxical facts which show how far from absolute our pretended biological laws are. Nervous concentration, which we have accustomed ourselves to consider one of the signs, of the means, and of the results of organic perfectionment, is better realised in the crustaceans than in man.

There is, however, another law which does not seem to suffer exception; it is that of the predominance, greater and greater of the cerebral ganglion, in proportion as the intelligence is developed. One of the reasons of this development is assuredly the greater perfection and the more important office of the organs of the special senses, forasmuch as certain crustaceans, for instance the large-eyed amphipods, have a very voluminous cerebral ganglion, lobed, and emitting optical nerves. In like fashion, and for the same reason, the large-eyed libellulas, many dipterans, hymenopters, the lepidoptera (Fig. 59 *g s*) have a powerful cerebroidal ganglion.[2] Nevertheless the degree of development is closely related to the volume of the cerebral ganglion. The brain of the spinning spiders, of ants, of bees, is remarkable for its volume, and even for its conformation. Though the bee is much smaller than the cockchafer, it possesses a brain much more developed, and relatively three times larger, if we take into consideration the difference of size. The brain of the ant is proportionally larger still. Besides, the surface of these cerebroidal ganglions so much developed is mammilated; we find there bourrelets, something analogous to what are *circumvolutions* in the brains of the vertebrates. In the bee, according to Dujardin, the brain has a very singular form. We perceive

[1] Gegenbaur, *loc. cit.*, p. 348. [2] *Ibid.*

a disk with stellated striæ surmounting like a hood the superior ganglion.[1] According to some experiments of M. Faivre, the cerebral ganglion has, like the cerebral hemispheres of the vertebrates, the property of being insensible to punctures and lacerations.[2]

Already we can discover in many of the arthropods the existence of a special nervous network, destined for the digestive system. In general, this visceral nervous or sympathetic system springs from the central ganglion by a single trunk. Then it ramifies, and its branches are furnished along their course with small ganglionary enlargements. (Fig. 59, *r*, *r'*). Dugès saw this visceral nervous system in the spiders: Andouin and Edwards and others found it in the crustaceans: Lyonnet, Cuvier, Brandt demonstrated its presence in the insects.

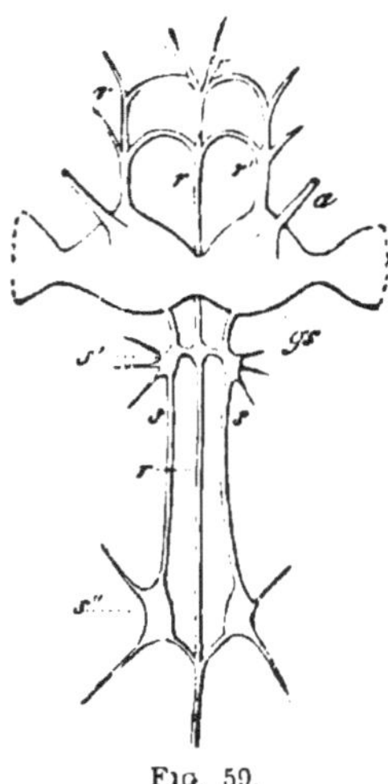

Fig. 59.

Œsophagian ganglion superior with visceral nervous system of a lepidopter (*Bombyx mori*); *gs*, cephalic ganglion superior (cerebroid); *a*, antennary nerve: *o*, optical nerve; *r, impair trunk of the visceral nervous system*; *r', the root springing from the œsophagian ganglion superior*; *s*, nerves in pairs with their ganglionary enlargements *s'*, *s''*.

Essentially, and spite of the apparent irregularity of its general arrangements, the nervous system of the *mollusks* is merely a kind of copy of that of the arthropods. Here we still find the œsophagian ring, emitting from its central portion a ganglionary peripheric nervous system, distributing itself to the diverse organs, but without regularity, without symmetry, as moreover the general conformation of the body demands. The super-œsophagian or cerebral ganglion is naturally very small in the lamellibranchians which have not a head provided with organs of the senses: it is, on the other hand, very large for the contrary reason, in the cephalophores.

[1] *Annales des Sciences Naturelles*, 1850.

[2] Faivre, *Annales des Sciences Naturelles*, 4e série, t. VIII. et IX.

The visceral nervous system of the mollusks, is like their general nervous system, a copy of that of the arthropods.

On the other hand the cerebral ganglion of certain cephalopods has affinity in some respects with the brain of the vertebrates. There exists in these animals a cephalic cartilage, forming a sort of cranian cavity, and hollowed with a fossette destined to be occupied by the cerebroidal ganglion. This cartilage is the rudiment of an orbit and lodges the organs of hearing.

The super-œsophagian ganglion of the mollusks seems also to have special functions. If we remove this ganglion in the snail the animal survives the operation four or five weeks, but remains completely without movement. On the other hand, the extirpation of the sub-œsophagian ganglion kills the animal in twenty-four hours.[1]

The excitation of the cerebroidal ganglion of the mollusks produces little or no effect. It is the same with its galvanisation. But the case is altogether different with the sub-œsophagian ganglion. Its irritation provokes a vigorous muscular agitation: its galvanisation by the continuous currents has often for effect to stop the heart in its state of dilatation, or of diastole, exactly as happens with the galvanisation of the pneumogastric nerves in the vertebrates.

These facts tend to confirm the opinion of the German evolutionists, who wish to connect genealogically the vertebrates with the mollusks. However, if we take into account nothing but the general conformation and distribution of the nervous system, the analogy, remote though it may be, exists rather between the arthropods and the vertebrates. In effect the nervous system of the acranian vertebrates can be strictly considered as a very coalescent ganglionary nervous system. The most imperfect of the acranian vertebrates, the amphioxus, has as central nervous system merely a nodose spinal marrow, that is to say, offering a series of enlargements, each of which corresponds with the origin of a

[1] Vulpian, *Leçons sur la Physiologie Générale et Comparée du Système Nerveux*, pp. 757-761.

pair of nerves. An enlargement which may be compared with the central ganglion of the arthropods terminates, forwards, their spinal marrow. It does not perceptibly differ from the others, but emits five pairs of nerves, among which are the optic nerves and the auditive nerves.

The great difference consists in the complete absence of œsophagian ring. In the vertebrate it is no longer merely the cerebral enlargement, it is the whole central nervous system which is above the digestive system. In saying that the arthropod may be compared with a vertebrate reversed, marching on the back, it has been thought that the difficulty was got rid of: but it merely changed its place. In effect, in supposing that the arthropod is a vertebrate reversed, we certainly put the sub-œsophagian ganglionary chain above the digestive system, but on condition of making to descend below it the sub-œsophagian ganglion, that is to say the analogue of the brain. Better it is, spite of the Hæckelian theories, which are still a little rash, to confess that the genealogy of the vertebrates is far from being elucidated.

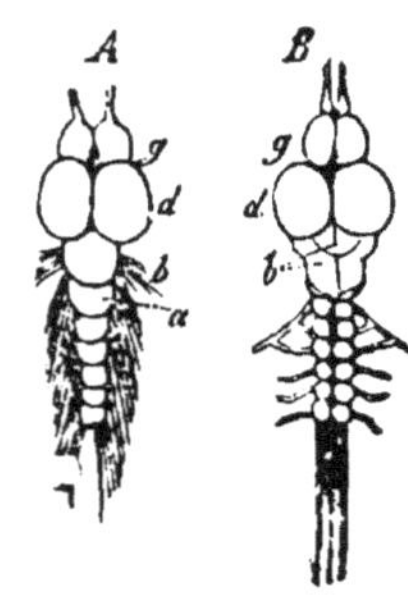

Fig. 60.
A, brain and spinal marrow of *Orthagoriscus mola*. *B*, brain and commencement of the marrow of *Trigla adriatica*.

Let this be as it may, we cannot deny the analogy of the spinal marrow of the craniote or acranian vertebrates with the ganglionary chain of the arthropods. Rudimentary expansions continue to be found even in man, and histologically and physiologically, the ganglionary chain of the arthropods as well as the spinal marrow of the vertebrates are cellular centres endowed with a certain sum of independence.

In the craniote vertebrates the spinal marrow ends, forwards, in an enlarged portion, in a brain properly so called. But here still, the transition is graduated, and we pass with no remarkable suddenness from the type acranian to the type cranian, even if we compare merely the adult animals (Figs. 60 and 61).

If we follow step by step the embryological evolution of any superior vertebrate the transition is more graduated still: for we see the nervous centres begin with the acranian form, then assume by degrees the complete cerebral type, passing through intermediate forms very analogous to those which persist in the fishes and in the reptiles.

In these two last classes the encephalic nervous centres are represented by a series of vesiculiform expansions (Figs. 60 and 61). In like fashion the encephalon of the superior vertebrates is composed first of all of five vesicles, which have been called, going successively from the front to the back, *anterior brain, intermediary brain, median brain, posterior brain,* and *terminal brain* (Figs. 62, 63, 64, 65). With the inferior face of the anterior brain are connected the olfactory *bulbs,* which furnish the nerves of smell.[1]

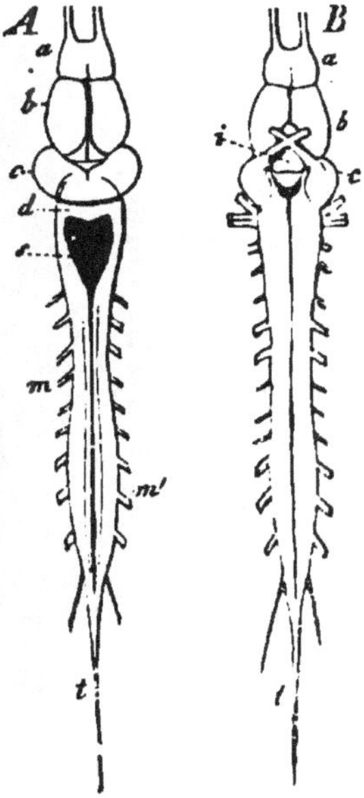

FIG. 61.

Brain and marrow of the frog: *A*, upper aspect; *B*, nether aspect; *a*, olfactory bulbs; *b*, anterior brain; *c*, median brain; *d*, posterior brain; *e*, elongated marrow; *i*, infundibulum; *s*, rhomboidal fosse; *m*, spinal marrow; *t*, its terminal thread.

The anterior brain is divided by an antero-posterior fissure into two hemispheres, which, continuing to increase in size, become the principal cerebral mass, the cerebral hemispheres, and cover the other encephalic expansions, the more the vertebrate is intelligent.

The intermediary brain divides also into two masses, which, as we shall see ere long, play a very important part in the cerebral functionment. These are the optical expansions or *optical layers.*

The median brain reduces itself more and more, and is no longer represented in the human brain by anything but four small tubercles, called *quadrigeminous tubercles.*

[1] Gegenbaur, *loc. cit.* pp. 681-690.

The posterior brain becomes the cerebellum.

The surface of the cerebral hemispheres, at first completely smooth in the embryon of the superior vertebrates and in the inferior vertebrates (marsupials, edentals, &c.), goes into folds in proportion to the evolutive or hierarchical progress. In most of

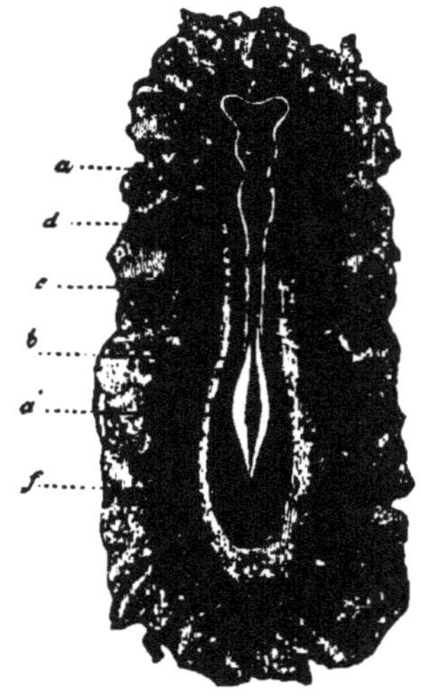

Fig. 62.

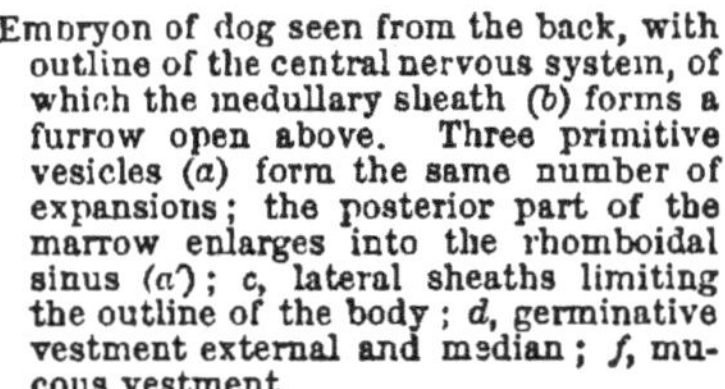
Embryon of dog seen from the back, with outline of the central nervous system, of which the medullary sheath (b) forms a furrow open above. Three primitive vesicles (a) form the same number of expansions; the posterior part of the marrow enlarges into the rhomboidal sinus (a'); c, lateral sheaths limiting the outline of the body; d, germinative vestment external and median; f, mucous vestment.

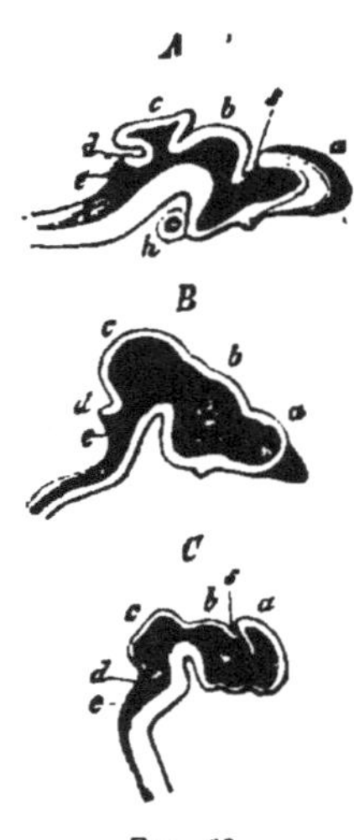

Fig. 63.

Vertical sections through the brains of some vertebrates: *A*, young *selacian* (heptanchus); *B*, embryon of adder; *C*, embryon of she-goat; *a*, anterior brain; *b*, intermediary brain; *c*, median brain; *d*, posterior brain; *e*, terminal brain; *s*, primitive cleft; *h*, hypophysis.

the mammifers it is moulded into cords sinuous, hemicylindroidal, juxtaposed, called cerebral circumvolutions.

The circumvolutions, gathered always into regular groups, are usually so much the more flexuous, so much the more complicated, the more the animal is intelligent. They are, for instance, very richly developed in the elephant, the arthropoid apes, and in man. But here there are exceptions, and the cerebral hemispheres of the beaver, like those of all the rodentia, for instance, are almost smooth, while in the sheep exists a somewhat complex system of circumvolutions.

We have seen that in the arthropods the cerebroidal ganglion has the privilege of furnishing always the optical nerves, and

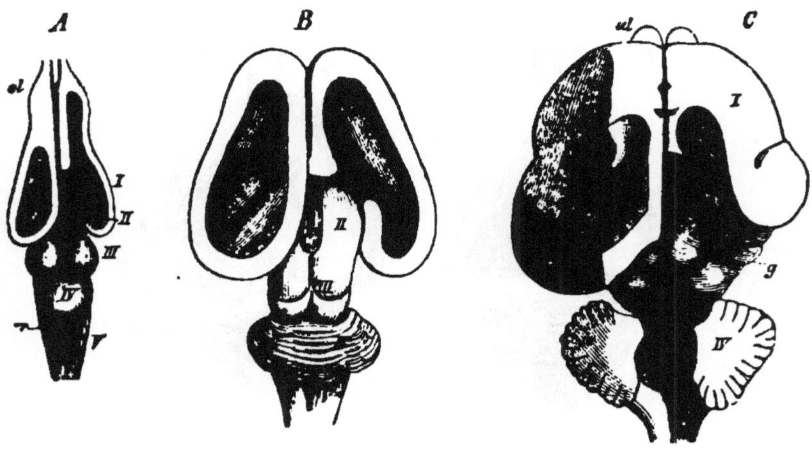

FIG. 64.

Differentiation of the anterior brain; *A*, brain of *tortoise*; *B*, brain of a fœtus of *calf*; *C*, brain of a *cat*. In *A* and in *B* has been removed on the left the roof of the cavity of the anterior brain; and on the right the four-pillared fornix. In *C* has been removed on the right side all the lateral and posterior portion of the anterior brain, and on the left enough to permit the curvation toward the base of the Cornu Ammonis to be seen. In all the figures *I* represents the anterior brain; *II*, the intermediary; *III*, the median; *IV*, the cerebellum; *V*, the spinal cord; *ol*, the olfactory bulb (its communication with the cerebral cavity figured in *A*); *st*, striated body; *f*, four-pillared fornix; *h*, large foot of the hippocampus; *sr*, rhomboidal sinus; *g*, geniculated protuberance.

sometimes other nerves of the special senses. In like fashion, in the vertebrates, the olfactory, optical, auditive, gustatory nerves, in short all the nerves of the special senses, have their original centre in the encephalon (optical layers), that is to say in the grand conscious nervous centre. Most of the other nervous branches have, at least apparently, their origin in the spinal marrow.

Like the arthropods and the mollusks the vertebrates have a special nervous network for the apparatus of the vegetative life. This network has been called in its ensemble *sympathetic nervous system*, and we are obliged, to render what follows intelligible, to give a succinct description thereof.

Here still there is no essential difference between the superior invertebrates and the vertebrates. The dissemblances bear on

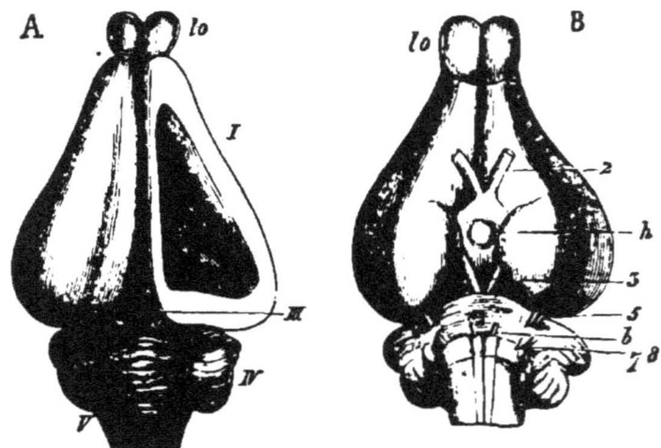

Fig. 65.

Brain of rabbit; A, upper aspect; B, nether aspect; *lo*, olfactory lobe; *I*, anterior brain; *III*, median brain; *IV*, posterior brain (cerebellum); *V*, elongated marrow; *h*, hypophysis; 2, optical nerve; 3, oculo-motor nerve; 5, trigeminus nerve; *b*, abductor nerve; 7, 8, facial and acoustic nerves. In A has been removed the roof of the right hemisphere to show the interior of the lateral ventricle, the striated bodies which are found therein before and behind the four-pillared fornix with the commencement of the *Hippocampus major*.

the exterior arrangement, on the morphology. In sum, in the invertebrates and in the vertebrates there is a special nervous network destined for the organs, systems, and apparatus of the vegetative life: namely, for the digestive tube, for the respiratory organs, for the circulatory system, for the genito-urinary organs. In both the invertebrates and vertebrates this system has its origin, or at least its roots, in the great nervous centres. Following the distribution of the sympathetic branches to their furthest extremities, we see that their fibres are distributed wherever there are vegetative contractile elements, smooth fibro-cells. In ultimate analysis the great sympathetic can be denominated nervous system of the vegetative muscles. Such at least is its physiological characteristic. We shall see that histologically it is pre-eminently composed of special nervous fibres, like the office which is assigned to them. With these special fibres are blended a small number of fibres similar to those of the nervous system of the animal life. These last are probably charged with the sensitive function. They gather the impressions on the surface of the mucous membranes, and so forth.

The sympathetic network is remarkable, morphologically, for the very great number of ganglionary expansions which exist on its plexus. Always in the vertebrates, it is connected with the nervous centres, the spinal marrow, the brain, by numerous roots. These roots throw themselves into the ganglions which are connected with each other by cords, and whence set forth other ganglionary cords, which go to form networks, complicated plexus in the viscera.

The sympathetic system is not much developed in fishes. It seems even to be wholly absent in the more inferior fishes, and to have its place supplied by simple intestinal ramifications.

We shall have to return to the very interesting functions of the great sympathetic; but before going further it is needful to expound with some detail the histological structure of the diverse parts of the nervous system, and of the diverse types of nervous systems, of which we have just given a general morphological description.

CHAPTER II

OF CELLS AND OF NERVOUS FIBRES.

All nervous system resolves itself under the microscope into cells and fibres. All the nervous cords are almost exclusively composed of fibres. On the other hand, the cells exist in very great numbers in all the distended parts, in all the central or ganglionary masses of the diverse types of nervous systems.

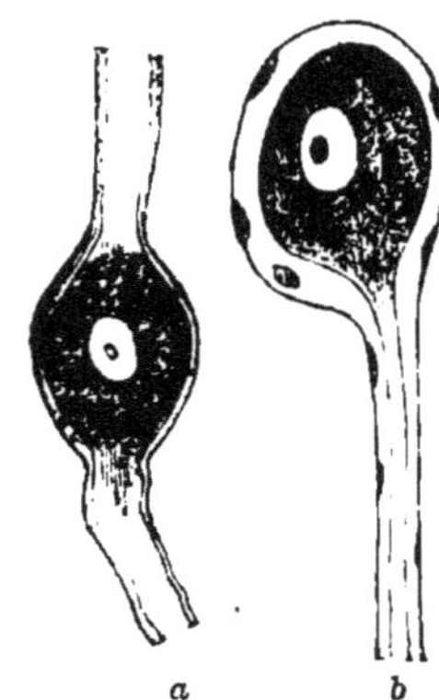

Fig. 66.

a, Bipolar nervous cell of the ganglion of the trigeminous nerve of the trout, with a thick sheath, a granulous content, a vesicular nucleus and a nucleole; *b*, unipolar ganglionary cell of man, with a thick sheath made of connective tissue and containing nuclei.

Though there are diverse varieties of cells and of fibres, yet as the differences concern only the details, it is possible to give of them all a general description.

The nervous cells are corpuscles of a somewhat irregular form, and are more or less spheroidal (Fig. 66). They have a wall, a content, a nucleus, and a nucleole. This last, of a brilliant and yellowish colour, is included in a large transparent and spherical nucleus, which itself is surrounded on all sides by a granulous and solid substance. Usually a somewhat thick cellular membrane covers the whole. According to M. Ch. Robin this membrane is lacking in the cells of the nervous centres of the superior vertebrates. The diameter of the nervous cells varies considerably. It is on an average from 0^{mm},020, to 0^{mm},050.

The nervous cells continue always with one or more fibres. They are thus called, according to the number of these fibrous prolongations, *unipolar*, *bipolar*, *multipolar*. Cells called *apolar*, without nervous prolongations, are long admitted: but their

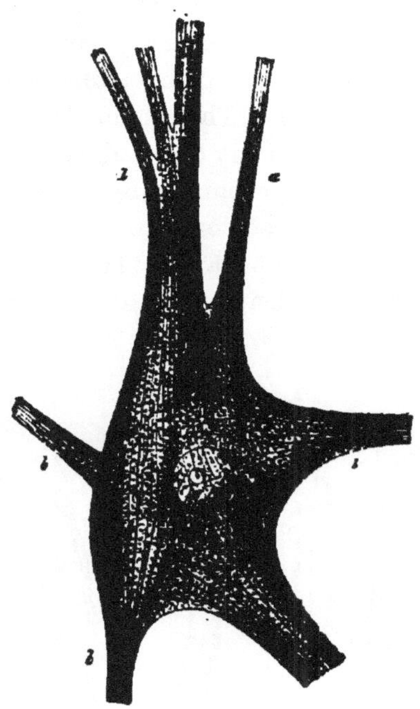

FIG. 67.

Ganglionary multipolar cell of the spinal marrow of the ox, with a rounded nucleus and a nucleole: *a*, cylinder axis; *b*, prolongations of the cell finely striated and fibrillary.—Very greatly magnified.

existence is very problematical. The nervous cells having a greyish tint, which they owe to their contents, the regions of the nervous centres when they are gathered together in great number have the same colour. Thus is constituted the grey *nervous substance*, which we find on the surface of the brain and in the central part of the spinal marrow in the mammifers, and so on.

The nervous fibres continue direct with the cells. They are in some sort tentacles which these project to connect themselves with the other organs.

The fundamental part of every nervous fibre is a very thin filament solid and flexible (Fig. 68, *h*). This filament, called *cylinder axis*, *cylindraxis*, is chemically constituted by a quaternary azotised substance. It continues direct with the granulous content of the nervous cell. This filament is the truly essential part of the nervous fibre: it is the soul thereof. It is the conducting portion, the telegraphic wire which puts the nervous cell in relation with the other organs.

In the complete fibre the axile filament is enveloped by a protecting manica. It occupies then the centre of a tube, of

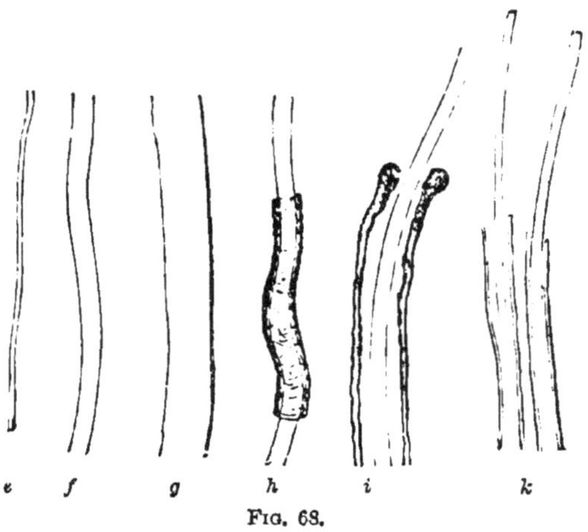

FIG. 68.

Nervous fibres magnified 350 times; *e*, fine fibre; *f*, fibre of middle size; *g*, broad fibre with dark edge, and in the fresh state, taken from the nerve of a rabbit; *h*, fibre of the spinal marrow of man; visible therein are the bright cylinder axis and the contracted sheath; *i*, analogous fibre of the brain of man; *k*, passage from the fixed cerebral fibre to the sheathed fibre; fibre taken from the brain of the torpedo.

an elastic sheath, filled with a sort of viscous oil. This substance, which Kölliker has compared with turpentine, but which belongs rather to the family of fat bodies, is transparent. Like all transparent bodies, rich in carbon, it strongly refracts the light. Also, under the microscope, the nervous fibres with oily envelopments seem limited by two dark pàrallel lines. For this reason the name of *fibres with double contour* has been given to them (Fig. 68).

It is these fibres which, united in bundles with a general fibrous envelopment, constitute all the nervous cords, the nerves properly so called, at least those of the animal life, those which in man and the superior vertebrates distribute themselves to the skin, to the organs of the special senses, to the muscles. Seen with the naked eye, all the regions or portions of the nervous system in which they dominate are white. This white tint is due to the oily substance of the fibres with double contour.

Physiology has demonstrated that certain of these fibres seem to transmit to the muscles motory excitations arising in the nervous centres : that certain others, on the contrary, carry from the periphery to the nervous centres the sensitive excitations. On this account, the first have been called *motory fibres*, the second *sensitive fibres*. There is however no decided anatomical difference between these two orders of fibres. But there is seen on the course of the nervous fibres, shortly before they reach the nervous centres, a ganglionary expansion, a nervous cell (Fig. 69 *p*).

In most of the nervous cords, motory fibres and sensitive fibres are mingled and confounded under the general neurolemma. We have however to except the nervous cords, specially transitive, which proceed to the organs of the senses.

We have seen that usually the special sensitive nerves proceed direct to the cerebroidal ganglion of the invertebrates and to the brain of the vertebrates. The others, the mixed nerves, take their course, in the arthropods, to the ganglions of the abdominal chain, and in the vertebrates to the spinal marrow (Fig. 69, *q*, and Fig. 71). But in this last case the two orders of fibres separate a little, before arriving at the marrow, into an anterior, cylindrical root and into a posterior ganglionary root, that is to say, expanded at a point where are found united the cells of all the sensitive fibres of the nervous cord (Fig. 69).

It is only by the presence of this cellular expansion at a point of their course that the sensitive fibres are distinguished from the motory fibres. In a general manner they seem also to be less voluminous; but the diameter of the nervous fibres of every kind is a very variable thing.

A third variety of fibres exists especially in the cords of the great sympathetic nerve. These are the *grey*, *gelatiniform*, or

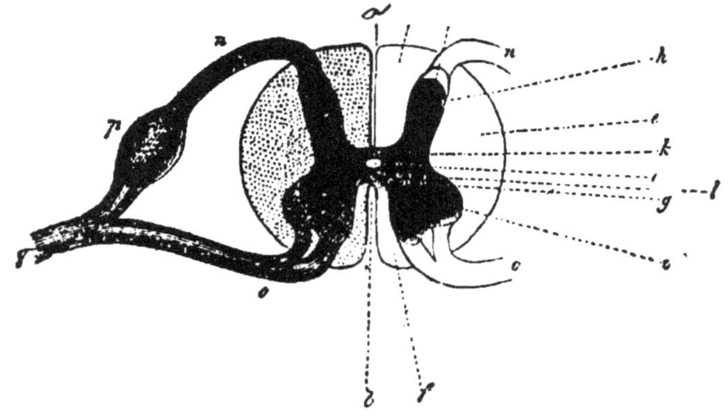

Fig. 69.

Section magnified and schematic of the spinal marrow of man at the commencement of the lumbar region; *a*, posterior fissure; *b*, anterior fissure; *c*, central canal; *d*, posterior cords; *e*, lateral cords; *f*, anterior cords; *g*, anterior commissure of the white substance; *h*, posterior horns; *i*, anterior horns; *k*, posterior commissure; *l*, anterior commissure of the grey substance; *m*, gelatinous substance; *n*, posterior roots; and *o*, anterior roots of the nerves; *p*, ganglion of the posterior root; *q*, mixed nerve from the union of the two roots.

Remak's fibres. They constitute almost the totality of all the cords or grey threads of the great sympathetic, and are on the other hand rare or absent in the white branches of the same system. These fibres are thin (0^{mm},003), pale, greyish, bestrewn with fine granulations of the same colour, and even with elliptical nuclei (Fig. 70). They dominate in the threads of the great sympathetic, and are lacking in the others. As M. Ch. Robin remarks, they seem to be fœtal nervous fibres. In effect all the white fibres of the animal life assume in the first phase of their histological evolution the form of gelatiniform fibre : and it is the same with the fibres which regenerate themselves in the adult after a lesion. According to Remak, who discovered them, the gelatiniform fibres differ especially from the others by the absence of the oily envelopment, the medullary substance. Moreover they are constituted also by an axile filament and a delicate envelopment. The lack of oily envelopment is the cause of their grey colour.

These diverse degrees of perfection in the structure of the nervous fibre in the vertebrate, suggest naturally the idea of

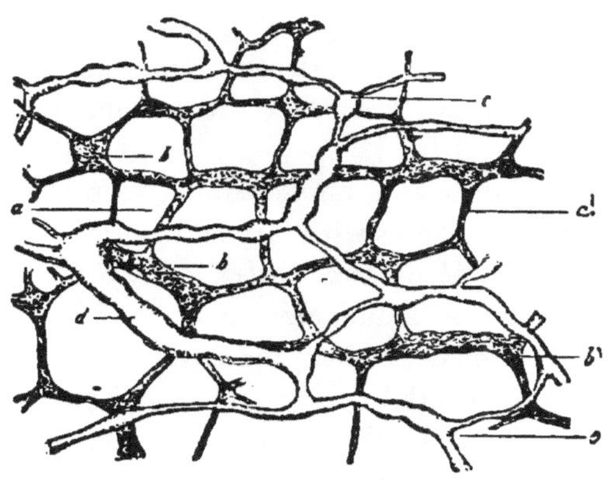

FIG. 70.

Ganglionary network of the muscular membrane of the small intestine of the guinea pig; *a*, nervous network ; *b*, ganglion ; *c* and *d*, lymphatic vessels.

an evolution. No doubt some imagination is needed to find, as Hæckel finds, all the genealogical links from man to the *gastrula*, that is to say to the animal reduced to be nothing more than a simple digestive pouch. But it is certain, nevertheless, that if we classify hierarchically all the types of the animal kingdom, we see, from the foot to the top of the scale, the vegetative life adjoin to itself little by little the animal life, which gradually expands in its turn. Now in the superior vertebrate all these phases have left their impress behind, without speaking of the embryological evolution which reproduces them all in epitome.

The superior mammifer is a sort of summary of the entire kingdom. In him are combined all the tissues, all the apparatus scattered through the entire series: he has a special nervous system, but he possesses nevertheless a portion of the ganglionary system of the invertebrates, and in him as in them this ganglionary system is constituted especially by gelatiniform fibres.

In effect the nervous fibre with double contour is not common among the invertebrates. It is absolutely lacking in the inferior invertebrates: it appears, but it is still rare, in the arthropods. Even the nervous system of the inferior vertebrates, the cyclostomes, is constituted especially by granulous fibres, enveloped with a delicate sheath and destitute of oily covering. In this case, as in many others, is verified the current saying: *Natura non facit saltus.*

There is no sudden bound, no *hiatus*, between the grey nervous fibres and the nervous fibres with large diameter. The connecting link is formed by the *thin nervous tubes.* These are nervous fibres with double contour, of a diameter much smaller than that of the other fibres. Some of them moreover seem to appertain to motricity, others to sensibility: for certain of them are provided with a cellular expansion, which is also of small size. The thin nervous tubes are met with to some extent everywhere in the nervous system: but they abound especially in the great sympathetic. This authorises us to consider them as the first stadium of perfectionment of the gelatiniform fibres.

The complete fibres despoil themselves habitually of their accessory parts, in their terminal portions, and reduced then to their axile threads, they have a considerable affinity with the gelatiniform fibres.

According to recent researches of Dr. Luys, the nervous cells and even their nuclei appear, when much magnified under the microscope, to turn themselves into intricate fibrils (Fig. 67). But the experiments of Dr. Luys having been generally made on specimens hardened by chromic acid, it is possible that this fibrillary severance is merely the result of the chemical agent, and till there is more convincing proof it is wise to consider the cell and the nervous fibre as the ultimate elements of the nervous tissues.

On the whole, every nervous system, invertebrated or vertebrated, resolves itself into a number more or less great of cells, and into a number more or less great of fibres, which connect the cells or end therein.

The regions of the system where the cells accumulate in great number are the nervous centres. The parts almost wholly composed of fibres form the nervous cords, and if we embrace at a glance the ensemble of the kingdom, we see that the cellular centres are the more voluminous and the less numerous, the more exalted the animal is in the hierarchy.

It is in the superior vertebrates and especially in man that the cellular concentration attains its maximum (Fig. 65). It is here also that unfold themselves in all their plenitude the special properties of the nervous system, sensibility, motricity, and thought.

We have not to describe here in their infinite details the nervous centres of the superior mammifers. It suffices to figure them to ourselves schematically. The spinal marrow (Figs. 69 and 71) is essentially a column of grey substance, that is to say of multipolar cells, emitting or receiving three orders of fibres: sensitive fibres radiated in all the organs, and especially the skin; motory fibres, setting forth from the cells, to arrive at the muscles; finally, intermediary fibres, serving solely to connect the cells with each other, to solidarise them, to make of all the grey substance of the marrow a harmonious whole, able, in a certain degree, to vibrate in unison.

As the mixed nerves, before arriving at the marrow, break into posterior or sensitive bundles, and anterior or motory bundles (Fig. 69), men of science have considered as sensitive the posterior cells of the spinal marrow, which receive direct the sensitive fibres. On the other hand they have regarded as motory the anterior cells, whence go forth the motory fibres. It has been remarked at the same time that the cells, probably sensitive, are of smaller dimensions.

But the cells of the spinal marrow are not merely solidarised among themselves: they are also solidarised with the central cells. In what concerns the anatomical relations of the spinal marrow and of the brain, and also the repartition in the encephalon, of the cells, and of the fibres, we are indebted to the brilliant labours of Dr. Luys for a picture of the whole subject as simple as it is fascinating.[1]

According to this anatomist all the sensitive fibres of the spinal marrow, whether they have or have not encountered on their passage the cells of this first centre, lead into, first of all, two masses of cells, situated at the inferior part of the brain, the *optical layers* [*Thalami nervorum opticorum*]. Then having passed through this secondary centre, they radiate toward the surface of the brain, the grey covering of the circumvolutions. This covering is formed of numerous layers of triangular cells, superposed by veins, like geological strata, having all their ridges at high point and being all connected by fibrous points of union.

The sensitive fibres, radiated from the optical layers, traverse from below to above all these cortical layers, to reach the more superficial strata, formed of cells analogous by their volume to the sensitive cells of the spinal marrow.

Beneath these layers of small cells, we find beds, superposed, of cells more or less voluminous in proportion to their distance from the periphery of the brain. The last strata, the deepest, are constituted by voluminous cells, analogous to the cells, called motory, of the marrow.

From these last cells set forth descending fibres, which all converge towards two cellular masses, situated also toward the base of the brain, in the vicinity of the optical layers. These masses of grey substance have been called in anatomy *striated bodies.*

[1] J. Luys, *Rech. sur le Syst. Nerv. Cérébro-spinal*, Paris, 1865.

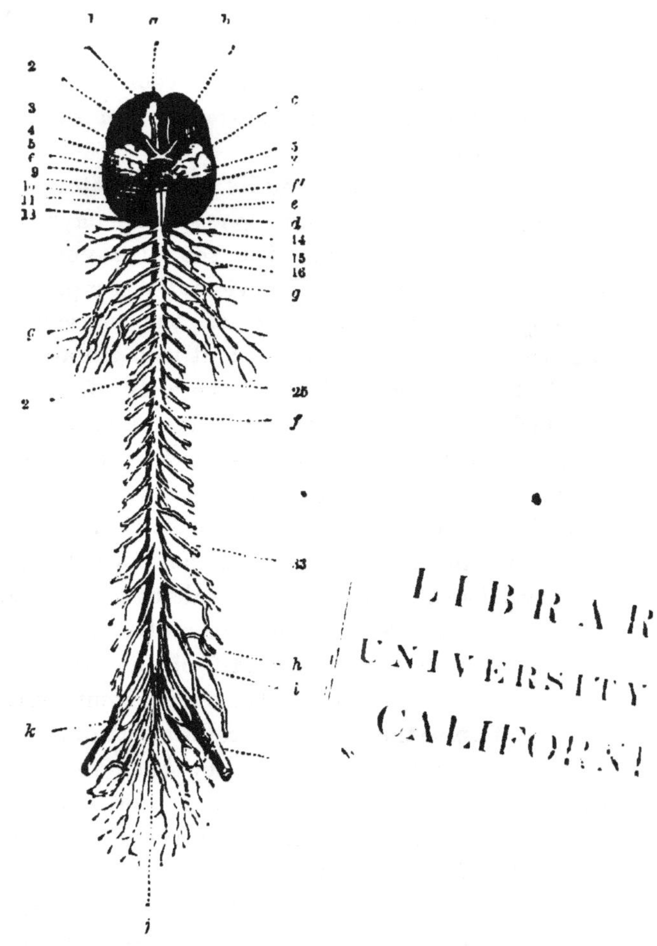

FIG. 71.

Central nervous system of man seen from the ventral face: *a*, brain; *b*, anterior lobe of the brain; *c*, median lobe; *d*, posterior lobe nearly covered by the cerebellum; *e*, cerebellum; *f'*, elongated marrow; *f*. spinal marrow; 1, olfactory nerve; 2, optic nerve; 3. oculo-motory nerve; 4, pathetic nerve; 5, trigeminous nerve; 6, abductor nerve of the eye; 7, facial nerve and auditive nerve; 9, glosso-pharyngial nerve; 10, pneumogastric or vagus nerve; 11, accessory nerve and hypoglossal nerve; 13 to 16, the first four cervical nerves; *g*. nerves of the neck forming the brachial plexus; 25, dorsal nerves; 33, lumbar nerves; *h*, lumbar and sciatic nerves uniting to form the lumbar plexus; *i*, the last nerves forming what is called the horse's tail (*Cauda equina*); *j*, impair terminal nerve of the spinal marrow; *k*, sciatic nerve.

From the striated bodies travel descending fibres, which encounter on their path the motory cells of the spinal marrow, then radiate thence into the nervous cords and threads, and end their journey at the contractile elements, the elements of muscular tissue.

We can thus follow along its whole anatomical career and in all its physiological metamorphoses the impression received by the terminal extremity of a sensitive nervous fibre. If, for example, a hard body strikes violently any point of the cutaneous envelopment, the molecules of the nervous fibres, harshly touched, enter from point to point into vibration; the shock communicates itself first of all to the cells of the marrow, then to those of the optical layers, then to the cells of the cerebral circumvolutions. No doubt this shock is modified in a fashion not known in its passage through these nervous centres. Let this be as it may, on reaching the superficial cells of the circumvolutions, the shock, the vibration, the molecular movement, whatever may be the form thereof, awakens in these cells an altogether special phenomenon, a phenomenon of consciousness, or *sensation*. But we are only yet at the half of the circuit. The sensitive nerves which have incurred the shock communicate in their turn the agitation to the subjacent cellular strata. These last cells, a little larger than the superficial or sensitive cells, a little smaller than the deep, or motory cells, are probably the thinking cells. In these the molecular shock is transformed into ideas. The ensemble of these thinking cells constitutes the soul of the organism. They take account of the causes of pain, combine the means of preventing the return thereof, and their decision, communicated to the deepest cortical layers, is there metamorphosed into *volitions*. In effect the deep cells of the cortical layers are motory, or rather *volitive*. They ordain the muscular movements necessary to prevent the return of the painful shock, and to ward off danger. The command is transmitted along the convergent central fibres, then through the cells of the striated bodies and of the spinal marrow. Finally, by the motory cells of the peripheric nervous cords, this command arrives at the muscles charged to execute it. The cycle is then complete, and the mechanical stimulation of the extremities of some sensitive nervous fibres has, like a succession of gun-discharges, determined a sensation, a ratiocination, a volition, movements.

Such is the complete series, but it is not always so distinct. The one or the other, even the one and the other of the two first periods, may be lacking. That depends on the organisms and the organs. Assuredly each of the three stages of this nervous circulation desires a small special study. We have therefore now to occupy ourselves with motricity, with sensibility, and with thought.

CHAPTER III.

OF MOTRICITY.

In the case we have just mentioned at the end of the preceding chapter, the series of phenomena may, spite of its complexity, be summed up in a brief formula. A peripheric excitation is transmitted by nervous fibres to multipolar cells, which transform it and send it back under the form of motory incitations along other nervous threads. We may in a sort of rough way compare the agitated nervous cell with a mirror which reflects an incident ray. But when the shock is transmitted through the brain, its reflection is not accomplished so simply, and along the path are awakened phenomena altogether special, conscious psychical phenomena. It is wholly otherwise when the conscious nervous centres do not come into play. Then everything comes to pass silently; the molecular shock comes from the periphery, and is sent back thereto, but its passage is not seen. There is thus truly what is called in physiology a *reflex action.* As a very great number of nervous phenomena of every order may be regarded as simple reflex actions, it is indispensable to speak with some detail of these primordial acts of the physiology of the nervous system.

Three principal cases can present themselves: the centre, which is the seat of the motory reflection, is either a simple multipolar cell, or a ganglionary nervous centre, or the spinal marrow of a vertebrate.

The first case is observed in certain rudimentary protozoaries, in which the nervous system is reduced to some scattered cells, forming with a fine fibre a network scarcely perceptible. The second is found normally and frequently in a number of invertebrates with a ganglionary nervous system. In the vertebrates it is the ordinary mode of functionment of the sympathetic nervous network. In effect it is by virtue of unconscious reflex acts, going on in the sympathetic ganglions (Fig. 72), that the motory incitations are transmitted to the smooth fibres of the intestine, of the stomach, of the bladder, of the sanguineous vessels; and, in sum, almost all the movements of the apparatus of the vegetative life in the vertebrates are achieved by virtue of reflex ganglionary actions.

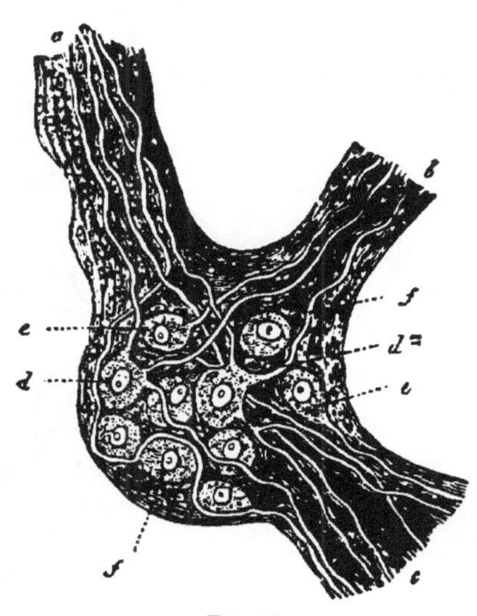

FIG. 72.

Peripheric ganglion of a mammifer; schematic outline: *a*, *b*, *c*, three nerves proceeding from the ganglion; *d*, multipolar ganglionary cells; *e*, unipolar cells; *f*, apolar cells.

The cells of the spinal marrow (Fig. 73) do not seem, spite of what has been pretended, endowed with conscious sensibility; nevertheless they are very active, and the spinal marrow must be considered one of the most important reflex centres, especially for the muscles of the life of relation. It holds in effect under its sway a very great number of muscles, which owe to it that permanent half contraction which is called *tonicity*, and which is the cause of the constant contraction of the sphincters, of the antagonist muscles; and so on. This tonus depends so much on the spinal marrow, that it is instantaneously abolished by the destruction of the marrow, whence the nerves of the muscles indicated derive their motricity.

The same effect moreover is produced by the simple section of the corresponding sensitive nerves. The muscular tonus results consequently from a veritable reflex action, in which the brain does not at all participate, forasmuch as, by its separation from the encephalon, the spinal marrow, so far from losing its reflex power, possesses it on the contrary in a very high degree.

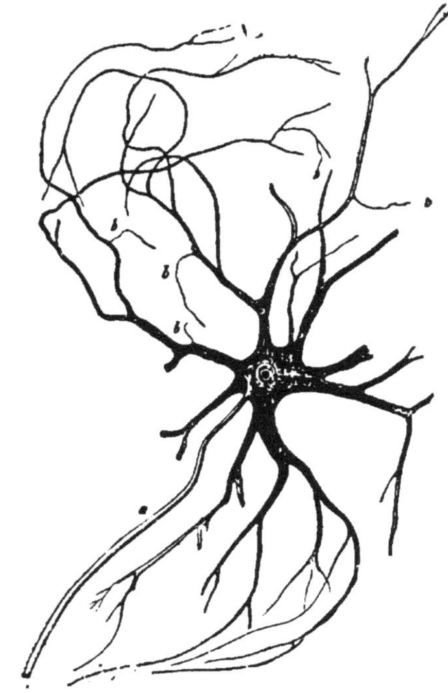

Fig. 73.

Multipolar ganglionary cell of the spinal marrow of the ox: *a*, cylinder-axis; *b*, prolongations of the substance terminating in fibres excessively fine.

If, for instance, we pinch the foot of a decapitated frog, the animal immediately withdraws it with much more energy than he did in his state of integrity. If we pinch with more force the reflex action is irradiated on the other members. In a frog poisoned by strychnine and decapitated, the smallest contact at any point of the skin determines convulsive movements almost general, while in the perfect animal, we can sometimes touch the central end of a cut nerve without exciting the smallest reflex movement.

In many vertebrates decapitation by no means hinders reflex movements, complex and associated, having all the appearance of voluntary movements. All the world has heard the celebrated experiments of Flourens spoken of, so often repeated and varied since. Deprived of its cerebral hemispheres, a pigeon still flies when we throw it into the air, swallows grain when we put it into its mouth; and so on. The experiment is always the more striking, the more the vertebrate is inferior, and the less brain it consequently possesses. For instance, a fish without brain swims as well as a normal fish. If we pinch the skin near the anal orifice of a decapitated frog, we see the posterior feet of the animal directed towards the irritated point, then extended suddenly, as if to repel the aggressor. If we pinch laterally the skin of the posterior part of a triton, consisting mainly of the trunk and of the two posterior members, this fragment, when pinched, curves laterally as the intact animal would do to withdraw the irritated point from the irritant body.[1]

Analogous facts have been observed by M. Ch. Robin on the body of a decapitated man. He says:—"The right arm of the executed man, being extended obliquely on the side of the trunk, the hand being 25 centimètres apart from the hip, I scratched the skin of the breast with a scalpel, on a level with the aureola of the mamma, to an extent of from 10 to 11 centimètres, without interfering with the subjacent muscles. We saw immediately the great pectoral muscle, the biceps, the anterior brachial muscle, and the muscles covering the epitrochlea contract successively and rapidly. The result was a movement of approach of the whole arm toward the trunk, with rotation of the arm on the inside, and demiflexion of the forearm

[1] Vulpian, *Leçons sur la Physiologie Générale et Comparée du Système Nerveux*, p. 417.

on the arm, a true movement of defence, which projects the hand from the side of the chest to the pit of the stomach."[1]

It is assuredly the grey, that is to say cellular substance, which serves as centre to these reflex acts: for it is indispensable and it is sufficient that these substances exist, for these reflex actions to be produced. After making almost complete hemi-sections of the marrow Van Deen still obtained reflex actions of the four members. We have seen Professor Schiff interrupt by many alternate hemisections the continuity of the white substance of the marrow in a cat, which did not hinder him from still obtaining contractions of the iris, by pinching the animal's tail.

As the celebrated experiments of Legallois have demonstrated, the diverse regions of the marrow correspond each to a special region of the body. They have a certain independence, and after a section can live a long time, by retaining their reflex excitability, provided their capillary circulation is not fettered. This sort of federation of the diverse regions of the spinal marrow confirms the opinion which regards this nervous centre of the vertebrates as a fusionated ganglionary chain.

We know moreover that apart from all vivisection, there are in the marrow normal functional centres. Such is the *nodus vitalis* of the superior vertebrates, that joint some millimètres in thickness, situated at the origin of the marrow, and whose destruction has for result instantaneous death, due to the abolition of the respiratory muscular movements. Such is also the cilio-spinal centre, situated as high as the two first nervous roots of the dorsal marrow, and holding under its empire almost all the capillary circulation of the head. Such finally is the genito-spinal centre, situated in the lumbar region in man.

Like all the tissues, the nervous tissue only works on condition of being fed: also its vitality is intimately dependent on the integrity of the sanguineous circulation among its elements. But more than all other tissues, it is intermittent in its mode of

[1] Ch. Robin, *Journal de Physiologie*, Paris, 1869.

functionment. A large expenditure of motricity exhausts for a time the nerves and the marrow. For them as for the muscles it has been demonstrated that their chemical reaction, neutral in the state of repose, becomes acid after labour.

To sum up, the spinal marrow is an unconscious nervous centre, a grand source of reflex acts. It probably holds under its dominion a quantity of associated movements, harmonious in appearance, but which in reality are for the most part effected without the intervention of the conscious centres, of the cerebral centres. This is why the young of many vertebrates run about as soon as they are born, when the cerebral hemispheres act little or not at all, in any case cannot consciously command complicated movements, which the young animal has not yet learned to execute.

The brain, on the other hand, is charged to accomplish reflex acts of another order, which often trouble in their execution the unconscious acts of the marrow. This is why this last nervous centre acts with more energy and is more excitable when by vivisection it has been in some sort delivered from the intervention of its mobile neighbour.

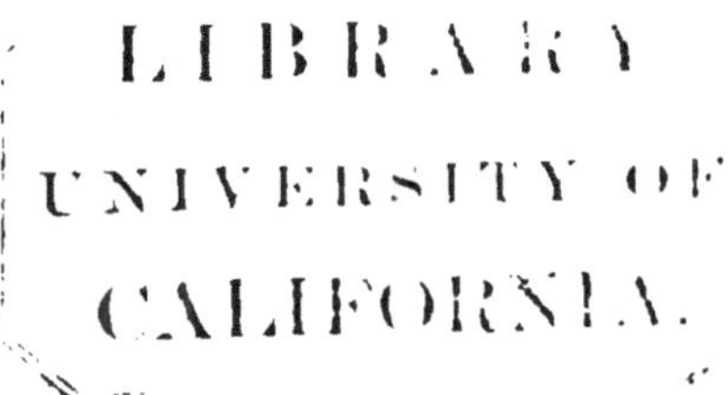

CHAPTER IV.

PROPERTIES OF NERVOUS FIBRES.

THAT the cellular element is, in every nervous tissue, the important agent, that which plays the principal part, is beyond doubt; nevertheless it would give an incomplete idea of the nervous fibre to see in it only a conducting thread, which may be motory or sensitive indifferently, and solely by reason of its terminal insertions. For example, it is not because the optical nerve leads to the eye that it transmits or awakens only visual sensations, since, after the section of this nerve, all excitation conveyed to its central point arouses in the nervous centres exclusively special sensations. We must therefore attribute to the various orders of nervous filaments peculiar properties, inherent in their very structure, or rather in their molecular constitution. This diversity is disclosed even by the manner in which the nervous fibres die. Each order of fibres has its peculiar kind of death. If, for example, we kill the nerves by stopping circulation, we see the sensitive element die first, and losing its properties from the periphery to the centre. Then the motory element succumbs, but from the centre to the periphery. The functional and nutritive centre of the motory fibre is, then, the centre in which it ends. The fact is still more clearly demonstrated by the section of a motory nerve. We see then, as in the case of the stoppage of the circulation, this nerve die from the centre to the periphery. To obtain movements by exciting the cut nerve, we must in fact bring excitation to bear upon a point nearer and nearer the terminal extremities, till we reach the muscles themselves. This gradual abolition of motricity corresponds, in this case, to an appreciable alteration in the constitution of the nerve, that is to say, a progressive coagulation of the medullary substance, the myeline of the nervous fibre. But, while the peripheric tronçon of the nerve thus dies little by little, the central end, which is always in continuity with the central cell, lives intact, preserving all its properties.

How is this curious influence of the central cell upon the fibre exercised? What is, essentially, the character of the molecular shocks transmitted by the central cell to the motory fibre? These are questions which still wait for an answer. Though electricity is produced in the nervous tissue as in every living tissue, it is not likely that the molecular shock which runs along the motory fibres is of an electric nature. In effect, a simple ligature, placed upon its course, is sufficient to stop it completely, by disorganising the nervous fibre at a single point, whilst it opposes no obstacle to the passage of an electric current. Finally, the rate of propagation of the motory excitation along a nerve is extremely slow. In fact it is proved by the experiments of Helmholtz, Dubois-Reymond, &c., that in the motory nerve of a warm-blooded animal this rate is only from 50 to 60 metres in a second. In a nerve cooled to zero it diminishes by nine-tenths, and we can even arrest all conductibility in the sciatic nerve of a frog by applying ice to one point of this nervous trunk.[1] After that, it is very natural to see the speed of nervous transmission retarded in cold-blooded animals, whose temperature is at once lower and more variable than that of warm-blooded animals. In effect, in the frog, for example, this speed does not exceed from 15 to 20 mètres in a second.

[1] E. Onimus, *De la Théorie Dynamique de la Chaleur dans les Sciences Biologiques.* Paris, 1866, p. 70.

If we are still ignorant as to what the molecular vibration, formerly called *nervous influx*, is, in its essence, we are not better instructed as to the manner in which this vibration is transmitted to the contractile muscular elements. By what mechanism do the dissociated nervous fibres, divested of their protecting envelopment, when they are insinuated between the muscular fibres of the animal life, between the fibro-cells of the sanguineous vessels and the apparatus of nutritive life, succeed in modifying the state of these elements, determining their contraction, and that with much greater energy than would be exercised by an excitation bearing upon the contractile element itself? One more unsolved problem to be added to so many others.

In most questions relating to the method in which the nervous tissues live and act, we must therefore content ourselves with recording plain facts, of which we are unable to give any satisfactory explanation. If physiologists are unable to tell us why the nervous cell excites the motory fibre, how the latter transmits this excitation to the muscular fibre, neither can they explain how curaré specially kills the motory nervous fibre, sparing the sensitive fibre. Certainly we have here a fresh proof of radical diversity between these two orders of nervous conductors.

The wourara, or curaré, of which the South American Indians make use to poison their arrows, has, in fact, the very curious property of selecting, to some extent, the motory nervous fibres. Once introduced into the circulatory system of any vertebrated animal, in a very small quantity, it strikes with paralysis the whole of the motory network, leaving perfectly unimpaired the whole of the sensitive network. Its action is first disclosed by slight convulsive movements, followed by the progressive extinction of all the movements of the life of relation. The effect of curaré, whilst more manifest and prompt in vertebrated animals, nevertheless is not peculiar to them. In fact, curaré also acts upon the aquatic larvæ of insects, upon naïds, and mollusks, but slowly. It produces no apparent effect upon planaries, asterias, and fresh water polypi.[1] In short it is specially upon the white fibres of animal life that its action is prompt and sure. The gelatiniform fibres resist it better. In man even, the sympathetic nerve is difficult to touch, and those organs which, like the heart, receive simultaneously white and grey fibres, are the last to be smitten. As the chief difference between the white fibres and those named after Remak seems to consist in the presence of a manica of myeline in the first, which is absent in the others, we should be tempted to believe that the curaré acts specially upon this medullary substance; but in this case, as in so many biological phenomena, the conditions are too complex to lend themselves readily to such simple explanations. In fact, it is from the periphery to the centre that curaré effects the toxication of the nerve; the peripheric extremities of the nerve are first and perhaps solely struck, since a motory nerve, separated from the marrow by section, is still poisoned by the curarised blood which bathes its extremities.[2] Now exactly by penetrating into the muscles, the nervous motory fibres divest themselves of their oily envelopment, and assume in some degree the aspect of the fibres of Remak. It has been supposed that curaré did not provoke profound lesions in the nervous fibre. In fact, the living nervous fibre, like the muscular fibre and certain vegetal fibres, produces an electric current, going from the surface to its centre of section; now the nervous fibre, physiologically destroyed by curaré, is still the seat of the ordinary electric phenomena; consequently, life has never quitted it, and the nutritive exchange is always taking place. Here then we must admit either delicate and invisible molecular perturbations, or perhaps a simple depression of the nutritive energy, sufficient to abolish the special function, but not yet reaching the fundamental basis of life.

[1] Vulpian, *loc. cit.*, p. 201.

[2] Cl. Bernard, *Rapport sur les Progrès de la Physiologie*, p. 19.

We cannot pass over in absolute silence the electric properties of the nerves; but we shall speak of them briefly. It seems to us that, in physiology, far too much importance has been attached to this secondary question. People have allowed themselves to be drawn too far by the idea and desire of establishing a connection between the electric agent and the nervous agent. But all identification between these two forces, or rather these two modes of molecular vibration, is manifestly erroneous. It is true that the electric currents bring nervous motricity into play with great energy, but many other mechanical, physical, and nervous excitants do as much.

As to the electric current which, as M. Dubois-Reymond has shown, is established between two electrodes placed, one on the external surface of a nerve, the other on its sectionised surface, it is not peculiar to the nervous tissue, and always establishes itself from the external surface to the section; it is simply the result of nutritive chemical actions. Also, it appears indifferently and in the same manner, whether the nervous fibre is motory or sensitive. A nerve which has just been crushed with blows of a hammer, and has lost its motricity, nevertheless engenders an electric current; for, in spite of the violence of the lesion, which has destroyed the form of the nervous fibres, their substance still lives some time, or at least certain parts of this substance; for example, the nervous sheaths and the neurolemma.

Nevertheless a nervous current is established throughout the length of the nerve, and is perceptible in the galvanometer, when we have excited by a continuous current a portion only of this nerve, not comprehended in the circuit of the galvanometer. This is what M. Dubois-Reymond has too pompously named the *electro-tonic force* of the nerves. Is this faculty of electric irradiation peculiar to the nervous fibres, as M. Dubois-Reymond would assert? It does not seem that this so-called special property has been sufficiently sought for in the other fibrous tissues, and Matteucci has demonstrated it, though with more difficulty, in a piece of cotton soaked in a conductor liquid.

As to the action of electricity upon the motory nerves, it also presents some interesting peculiarities.

As a general rule, every interrupted current, or every strong continuous current, applied to a motory nerve still in relation with the muscles, provokes a muscular contraction, more or less tetanic.

On the contrary, the application of a feeble continuous current causes one contraction at the moment of closing, and another at the moment of opening. In the interval there is no appreciable effect.[1]

According to M. Cl. Bernard the current acts only by abruptly changing the electric condition of the nerve. In fact, in certain cases, electricity causes the muscular contraction to disappear. The galvanisation of the pneumogastric stops the heart in the act of dilatation, of diastole. The tetanic contraction of a frog's foot, following the application of sea salt to the nerve, ceases when the two points of galvanic pincers are placed upon the lumbar nerve. Finally, all medical men, however little acquainted with electro-therapeutics, know that in man the passage of a continuous current often causes the contraction of certain muscles to disappear instantaneously. For our own part, we have several times had occasion to verify this interesting fact.

Long and very frequently reiterated excitations abolish, at least for a time, the excitability of the nerve; but if, for example, we have employed a continuous current, it is then sufficient to reverse its direction to awâken this exhausted contractility, and the experiment may be repeated many times. These *voltaic alternatives*, as they have been called, show plainly that there is something altogether special in nervous motricity, and that there exists in the inmost constitution of the axillary cord a mobility, a molecular instability, which no other tissue possesses.

[1] Vulpian, *loc. cit.*, p. 71-73.

The persistent vitality of the fibrillary nervous tissue is not less remarkable.

A nerve exposed to the free air dries up, and soon loses its properties, but, within a certain time, it may recover them by simple imbibition, in the same manner that certain desiccated infusoria revive.

If we cut the anterior rachidian or motory roots of a mixed nerve, the peripheric end will not be long in losing its motricity, and it gradually wastes even as far as its finest terminal ramifications. If we cut the posterior or sensitive root, above the ganglion, it is the central end which wastes.

Under such conditions, the nervous fibres lose their excitability, in the mammifers, at the end of about four days, whilst the irritability of the muscles lasts nearly three months. The structure of the nervous fibre is therefore altered, but not fundamentally. In effect, the nerve has only lost its oily envelopment; but its axile cord lasts and persists, apparently intact, covered only by the Schwann's sheath, shrivelled and folded round it. Professor Schiff has found the axile filament, still intact, five months after the section of a nerve; M. Vulpian has seen the same thing after six months. But things do not remain in this state. At the end of an indefinite time, shorter if the animal is young, longer if the animal is old, or of variable temperature, or cold-blooded, the nerve is restored, the anatomical and physiological continuity is re-established between the tronçons of the cut nerve, if they have been kept in contact; it is sometimes re-established even when there has been a resection of a nervous segment of from 1 to 2 centimètres. In this last case, we see a bundle of grey fibrils, budding, shooting at the central end, and joining at the peripheric end. At the same time, myeline is reproduced in the Schwann's sheaths, and, finally, the entire nerve is restored; for a long time, however, its fibres have a very small diameter.[1]

This property of restoration is constant, inherent in the

[1] Schiff, Vulpian, Phélippeaux.

essence of the nervous fibre ; since even nervous fragments transplanted and grafted under the skin of dogs waste and are regenerated, as they would be after a simple section.

How many things are still unexplained in this singular nervous restoration! How can we comprehend the influence of the motory and sensitive cells upon the fibres? Why, some days after the section of the motory nerve, does the motricity of the nerve disappear, since the axile filament, which is the essential part of the fibre, apparently preserves its integrity? It is generally admitted that the genesis of nervous fibres is only possible in the embryon; it would necessitate, it is said, very complex blastemas, which the simple process of nutrition is powerless to produce. Nevertheless, in cases of nervous restoration, after resection, there is generation of nervous fibres. Finally, the nervous medullary substance, or myeline, is a ternary, hydrocarbonised body; now, in the case of nervous regeneration, it is necessarily secreted, either by the Schwann's sheath, or by the axile cord. Here then is a fresh case of synthesis of a carbonised ternary matter in an animal tissue.

CHAPTER V.

OF SENSIBILITY IN GENERAL.

However accustomed we may be to see the adage *Natura non facit saltus* ceaselessly verified, and from whatever point we view the world, to meet with graduated series of facts, we are compelled to recognise in conscious sensibility a quality altogether peculiar, manifesting itself suddenly, and without any link to connect it with the other properties of inorganic or organic matter. Without doubt, there are degrees in sensibility. In following the animal hierarchy, we see this property, at first confused, gaining little by little in intensity, in clearness, then subdividing into various departments; but there is, nevertheless, a point where sensibility abruptly starts up, for between unconsciousness and consciousness there is no bridge. Here then we are in the presence of a new fact, quite as peculiar and unknown in its essence as gravity is. But just as the want of any clear idea on the subject of the inner nature of gravity has not hindered natural philosophers from studying and formulating the laws of this great property of matter, so, though ignorant as to what nervous sensibility is in itself, we can indicate its modes and conditions.

Conscious sensibility is a property inherent in certain nervous fibres and cells. This is a general fact, which can be verified from the highest to the lowest degree of the animal scale. Nevertheless, we must here point out the apparent exception in the case of the infusoria. Many of them, in fact, according to the evidence of our microscopes, have no nervous system, yet seem to experience sensations. We see them move, to all appearance voluntarily, avoid obstacles, &c., &c. Must we, to explain these paradoxical facts, admit, with some physiologists, the existence of an amorphous nervous substance, impregnating as it were the entire bodies of these animals? There is nothing to authorise such hypotheses. In the whole animal kingdom we see the precision of the sensitive perceptions correspond to clearly defined nervous apparatus, aided by special organs called *organs of the senses*, having the office of choosing amongst the shocks, the movements, the vibrations, &c., in short, the physical impressions of the exterior *medium*, all that can awaken in the nervous centres of the conscious being the tactile, olfactory, gustatory, auditive, and visual sensations. How admit that special sensations, which, to be produced in the superior animals, require diverse and complex apparatus, may, nevertheless, be experienced in a state of organic confusion and indivision? It is much more simple here to accuse the imperfection of our means of investigation, and to reserve till we have more ample information those exceptional cases, destined probably to be included one day in the general law.

With the aid of this restriction we can now sketch the picture of nervous sensibility. This sensibility has its seat in the nervous cells and fibres, of which we have already given a brief description. That there may be nervous cells specially sensitive having consciousness of impressions exercised upon the terminal extremities of their fibres, no one has ever attempted to deny.

It has not been the same with the nervous fibres, properly so-called, and upon the authority of some experiments, either erroneous or insufficient, the indifference of the nervous fibres was for some time believed in. According to this theory, the nervous fibres are all identical. It would be necessary to regard them as simple conducting threads, indifferent to the kind of excitations which they transmit. Sensitive, when they bring an organ of sense into communication with the sensitive central cells, the nervous fibres become motory by simple displacement, by the single fact of connecting a muscle with the motory cells. No trustworthy experiment comes to the support of this theory, which is brilliantly refuted by the evident speciality of the optic, acoustic, &c., nerves. In fact, every irritation which touches either the course of these nerves or their central portion, after section, only awakens in the nervous centres special sensations, glimmerings, sounds, &c. In the same way in the vertebrates the posterior roots of the rachidian nerves are sensitive; furthermore, every excitation bearing upon any point whatever of their course between the teguments and the nervous centres determines sensitive perceptions, touch, pain, &c.

There are then systems of fibres and cells physiologically quite different from the motory cells and fibres, in spite of great analogy in form and structure. Besides, this sensitive nervous system is habitually armed, at the extremity of its fibrous irradiations, with special apparatus destined to collect, to concentrate upon the terminal nervous fibres excitations from without. These apparatus are the organs of the senses, which we must now briefly describe throughout the animal series.

We shall merely mention in passing the protozoa. They have no organs of the senses, any more than the rhizopods, the sponges, the infusoria, since these rudimentary organisms are absolutely destitute of nervous system, and the organs of the senses are simply excitatory apparatus, means of reinforcing exterior impressions, instruments by the aid of which the sensitive nervous system palps the ambient medium. Can it be said that all the special senses may be concentrated in one, that of touch? We cannot deny that there is some truth in this generalisation. Under the apparent diversity of special sensations, there is always the same cause, namely, the shock, the agitation of the sensitive nervous extremities by molecules belonging to the exterior medium; but as each sense selects from these multiplied excitations those which correspond to its anatomical construction, we must admit special senses.

If certain physiologists have wished, with some appearance of reason, to reduce all the senses to one, others, on the contrary, have capriciously tried to multiply their number. The ancients admitted five senses; touch, taste, smell, hearing, and sight.

Buffon wished to add the genesic sense. Ch. Bell has imagined a muscular sense, another sense adapted to estimate weight, consistence. According to Carus, supported in this by several contemporary physiologists, there is a special sense for temperature. Many physiologists admit doloriferous nerves. But the sensitive enumeration of the ancients is surely the most simple and accurate; there is only a special sense where there is a special apparatus to gather certain impressions to the exclusion of others. The appreciations of weight, pressure, temperature, consistence, pain, &c., evidently belong to the domain of the sense of touch; otherwise it would be necessary, by analogy, to subdivide the other senses, those of sight, hearing, smell, taste, into a crowd of distinct senses, corresponding to all the varieties of colours, sounds, odours, and savours. If in certain maladies the perception of pain is abolished, while touch seems intact, it is simply because the nerves of touch have become insensible to certain agitations, as the eye sometimes becomes incapable of perceiving such and such a colour.

But if the five classic senses suffice to represent the great departments of sensibility in man and the vertebrated animals, it is not impossible that it may be otherwise with the invertebrated animals. The aërian waves are only sonorous to the human ear within certain limits of number, length, and rapidity. In the same way the chemical rays of the solar spectrum awaken in us no sensations. But sensitive apparatus differently adapted from ours may probably perceive those vibrations which are insensible to us. The inverse is more probable still. Many of the invertebrated animals seem, from the point of view of special senses, much less favoured than man and the higher mammifers, and consequently may be destitute of one or more of our five senses. This question of comparative physiology has been little studied yet. Nevertheless from some experiments made by M. P. Bert we may infer that as regards visual perceptions there is no radical difference between man and insects, in spite of the dissemblance of the perceptive organs.

CHAPTER VI.

OF TOUCH.

TOUCH is the most simple of all the senses, the least differentiated. The tactile sensations result from a shock, a pressure, an agitation bearing almost directly upon the extremities of the sensitive fibres. Where these fibres are very numerous, tactile sensibility is strongly developed. It is habitually at certain points of the teguments, skin, or mucous membrane that these agglomerations of sensitive nervous terminations are found. Ordinarily then these terminal extremities are clothed with an envelopment, a kind of hood; sometimes they are terminated by a nervous cell. There are then what are called *corpuscles of touch.*

The sense of touch is in some measure distributed everywhere among the higher mammifers; but it exists, more or less developed, in the whole of the animal series. As it is the least perfect of the senses, it is also the one which offers the least variety in the different animal classes, compared amongst themselves.

In the hydral polypi and the anthozoaries, the tentacula which surround the mouth are considered as tactile organs. The annelates, the hirudinates, have as organs of touch tegumentary cells, which have the form of bristles, baculi, in connection with sensitive fibres. According to Leydig, these tactile cutaneous organs are sometimes, in the hirudinates, grouped in large numbers at the bottom of cupuliform depressions.

In the *echinoderms*, the tentacula situated in the neighbourhood of the buccal orifice, and receiving the nerves, are considered as tactile organs.

The mollusks have as instruments of touch cutaneous cells with setiform prolongations, disseminated where the body is not covered with hard pieces.[1]

When we can trace the cutaneous nerves of the mollusks, in the transparent species, for example, we see that these nerves are clear, pale, offering here and there ganglionary expansions. It seems then that here the sense of touch is exercised by grey fibres.

The articulated appendages or *palpi* situated near the mouth in the crustaceans, the arachnida, and the insects, are tactile organs. In the crustaceans there are, on the antennæ and other appendages, filiform prolongations, "tactile baculi," which also exist in the myriapods, the insects.[2] Besides these special organs, we are forced to admit, in the crustaceans, the arachnida, and the insects, sensitive organs disseminated in the skin. In effect, in spite of the envelopment, the varnish of chitine, which covers them, these animals feel strongly the contact of the exterior bodies at any point whatever of their own bodies. The soft membrane which lines this bed of chitine is therefore sensitive, as is also the subungual dermis of man.

The general structure of the skin is perceptibly the same in all vertebrated animals, but nevertheless the manner of termination of the nervous fibres varies. Often the fibres end in expansions, some of which are called *corpuscles of Paccini;* others, *corpuscles of touch.* The first are observed in man, birds, &c. In the corpuscle of touch, the sensitive fibre conducts to an ovoid nucleus. The corpuscules of touch, in man, have a special form, which is nevertheless met with also in the hand of the monkey, and in the lingual papillæ of the elephant. Every one knows that these corpuscles are extremely numerous

[1] Gegenbaur, *loc. cit.*, pp. 122, 302, 479. [2] *Ibid*, p. 361.

on the palmary and plantary faces of the hand and foot of man, where, by their juxtaposition, they trace curved lines, crowded and regular. On the palmary face of the ungual joint of the forefinger, Meissner counted eight hundred corpuscles of touch in the space of a square line.

These corpuscles are of ovoid form, and each of them is constituted by a dermic bed, covering a small mass of conjunctive tissue. A sensitive nervous cord penetrates the corpuscle at its base. It is certain that these tactile corpuscles are apparatus which reinforce sensation, but are not indispensable to it. The different tactile sensations are still experienced by the skin in the regions where the corpuscles are wanting, but much less clearly. The internal surfaces of the terminal joints of the fingers, for example, feel the two points of a compass at a distance of seven-tenths of a line, while on the level of the dorsal spine a distance of twenty-four lines is required to keep the two sensations from being confounded.[1] At the tip of the tongue, on the contrary, the two sensations are still received at a distance of half a line.

[1] E. J. Weber, *De Subtilitate Tactus*, in the work entitled: *De Pulsu, Resorptione, Auditu, et Tactu. Annotationes Anat. et Physiol.*, Lipsiæ, 1834.

CHAPTER VII.

OF THE SENSES OF TASTE AND OF SMELL.

THE fact of the extraordinary tactile sensibility of the point of the tongue proves conclusively that the cutaneous corpuscules tactile are not special sensitive apparatus. In effect these corpuscles do not appear to exist in the lingual mucous membrane.

The sense of touch has not therefore in man, in whom it has especially been studied, any localisation well defined, any specialisation very distinct. We see it also transforming itself in the tongue, passing gradually into the sense of taste. For the papillary projections of the point of the tongue which are so tactile are at the same time gustatory. To convince ourselves thereof we have merely to excite them with a feeble electrical current. The effect however is the same when we merely excite the parts of the point of the tongue which are destitute of papillæ.

So far we know nothing positive with regard to the seat or even the existence of taste among the invertebrates and even among many of the vertebrates. The buccal mucous membrane of fishes is covered with small teeth, pointed and hooked, and seems very ill organised for gustation. The tongue, which in the vertebrates seems to be the special, or the most special, seat of taste is rudimentary and dry in many reptiles. Nevertheless the chelonians and the lizards, which chew their food, have a tongue, soft, rich in papillæ.

So far as taste is concerned birds seem to be very badly provided. They probably swallow without tasting: for their tongue, usually destitute of muscular tissue, is dry and cartilaginous.

It is among the vertebrates that the sense of taste is developed, but very unequally, according to species. The rodents, which feed on fruits or animal substances, have a soft tongue, without appendices. Those, on the other hand, which gnaw roots and barks, have on the tongue a kind of integument, garnished sometimes even with denticulated scales. Certain carnivorous animals, especially of the genera *Felis* and *Hyæna*, have also a tongue bestrewn with papillæ conical, voluminous, and covered with a horny sheath.

In dogs, monkeys, men, the tongue is voluminous, musculous, flexible. On the superior surface and on the edges the mucous membrane which crosses it is made rough with dermic projections, rich in nervous threads. These papillæ, denominated, according to their varying aspect, *caliciform*, *fongiform*, *corolliform*, *hemispherical*, seem to have for result and for function to multiply the surfaces of contact with the sapid bodies. These last act only in the state of solution. Their molecules severed and in sufficient number must put themselves into intimate contact with the gustatory papillæ. Sapidity has very different degrees of energy, according to the substances. Water containing in solution a hundredth of cane sugar is insipid, while a solution of sea salt to the extent of an eight-hundredth has still savour, and it is enough for a solution to contain a thousandth of sulphurohydric acid or of sulphate of quinine to be still sapid. We can always, indeed we must, view in their relations the sense of taste and the sense of touch: but taste is a special tactility, less gross, more varied also, for the number of savours perceptible by man is extremely numerous. The division of the sensitive labour is for taste more evident than for touch. It seems as if there were special nervous threads for certain savours. Many sapid bodies, especially salts, produce different sensations, according as they are applied to the anterior part or the posterior part of the tongue.[1]

The essential structure of the gustatory papillæ is still imperfectly known. Certain anatomists think that the nervous fibres terminate freely in the papillæ. Others are of opinion that they anastomose with fine filamentous prolongations emanating from epithelial cells. It is supposed that this arrangement has been observed on the tongue of the frog (Billroth), an animal however in which taste is probably very little developed.

In touch, sensibility is brought into play by the simple shock or contact of bodies. The sense of taste, on the other hand, is only impressionable by solutions, molecules severed in a liquid. But smell demands a matter more attenuated still. In effect, liquids impregnated with odorous substances awaken no olfactory sensation, at least in man, when they bathe the organ of smell. Substances alone in the state of gases, of vapours, or of fine particles suspended in the atmosphere, can provoke olfactory sensations, at least in the superior mammifers.

In the animal series the anatomical sense of smell is rather better known than that of taste. Leydig found, in the hirudinates, at the bottom of cupuliform depressions, compact bundles of rigid prolongations, analogous to the tactile baculi. He met with analogous organs in the cephalopodal mollusks. Thinking that he recognised therein olfactory organs, he called these prolongations *olfactory baculi*. He gave the same name and attributed the same function to analogous bundles existing on the pair of anterior antennæ of certain crustaceans.[2]

In all vertebrates, in which the sense of smell is better and more certainly known, it is also constituted by depressions of varying form and volume situated in the head; these cavities are usually clothed with vibratile epithelium. In the amphioxus

[1] J. Guyot et Admyrault, *Archives Générales de Médecine*, 2e sér., t. XIII., p. 51.

[2] Gegenbaur, *Anatomie Comparée*, p. 190.—Leydig, *Histologie Comparée*, pp. 250, 361.

the organ of smell seems to exist spite of the absence of brain. It is represented by one or two depressions with vibratile epithelium.[1]

In every class of vertebrates, man not excepted, the nervous olfactory threads are composed of pale fibres, finely granulous, without envelopment of myeline: in short of grey fibres, which all, in the craniote vertebrates, emanate from two special and intracranian nervous expansions called *olfactory bulbs.* These masses of nervous substance, very little developed in man, are much more so in the other vertebrates. In many mammifers and vertebrates, moreover, the sense of smell is much more developed than in man. It controls the sense of sight, often comes to its aid, and sometimes takes its place. The olfactory bulbs of many mammifers are very voluminous, situated in advance of the brain: often they are hollow, and communicate with the lateral ventricles. The mole, which is almost blind,[2] has enormous olfactory bulbs. In many birds and reptiles also the olfactory bulbs form an important cerebral expansion. The olfactory bulb of fishes is often as large as the cerebral lobe, properly so called. Sometimes even it is much larger, and this conformation is no doubt in relation with the difficulty of olfaction in an aquatic medium.

The olfactory nerves are nervous threads, proceeding in great number from those bulbs, and ramifying into the superior portion of the nasal cavities. The mucous membrane of those cavities, habitually clothed with vibratile cilia, is destitute thereof precisely in the portion, somewhat limited, where it receives the olfactory threads. There it is furnished with a cylindrical epithelium, whose cells terminate in fine filaments, which go to lose themselves in the dermis of the mucous membrane. Beneath these epithelial cells are other special cells, surmounted each

[1] Huxley, *Anatomia Comparata dei Vertebrati.* Italian translation of E. Giglioli. Gegenbaur, *Anatomie Comparée*, p. 709.

[2] On the contrary, the mole is said to have very sharp sight, though its eyes are peculiarly placed.—*Translator.*

with a thin bacillum (or bacillarium), with a transparent and crystalline extremity (Fig. 74). At their inferior part, these cells emit nodose filaments, which continue with the fine ramifications of the olfactory nerves.[1]

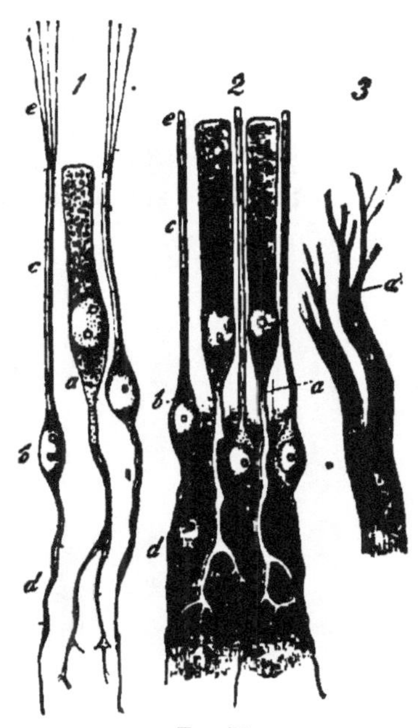

FIG. 74.

Microscopical elements of the olfactory mucous membrane: 1, of the frog; *a*, cylindrical nucleated epithelial cell, terminating at its base with a ramified filament; *b*, nucleated cell of the olfactory bacillum, *c*, which terminates at its extremity with a bundle of long vibratile cilia, *e*, while the cell bears at its base a fine nodose filament, which continues with the fibres of the olfactory nerve.—2, of man: *a*, the epithelial caudal cells, between which are the cells, *b*, with their olfactory bacilla, *c*, their terminations gemmate, *e*, and their internal filaments, *d*, communicating with the fibres of the olfactory nerve.—3, fibres of the olfactory nerve of the dog, dividing themselves into very fine fibrils.

The sense of olfaction, even in man, is of very great delicacy. An air charged with a ten-thousandth of its volume of vapour of essence of roses has nevertheless for us an extremely appreciable odour. The keenness of the sense of smell is far greater still in many animals, capable of catching numerous scents which escape us.

In reality, spite of the volatility of scented substances, the manner in which these substances enter into relation with the olfactory elements does not differ essentially from what takes place for the sense of taste. The olfactory mucous membrane being constantly lubricated by a secretion, it is always in the state of solution that the scented particles come into contact with it. Also the groups of olfactory and gustatory sensations are very near neighbours, and it is not without difficulty that physiologists are able to distinguish them. In sum, smell is akin to taste, and taste is akin to touch. Strictly, we

[1] C. Vogt, *Lettres Physiologiques*, p. 419.

may consider these three varieties of sensibility as three modes, three degrees of the same sense.

From the point of view of dignity, of psychological nobleness, touch, taste, and smell are coarse, inferior senses. In man they have no special memory. The sensations and impressions which they procure leave no durable traces, and the imagination by itself is powerless to revive them.

With the two special senses of which we have now to speak it is far otherwise.

CHAPTER VIII.

OF THE SENSE OF HEARING.

BETWEEN the three preceding senses, which may be called *tactile senses*, and the senses of hearing and of sight, there is an important difference in the mode of generation of the sensations. For the tactile senses, the direct contact of the bodies capable of exciting the sensation is indispensable. On the other hand, for hearing and sight, the sensitive organ is agitated indirectly by simple vibrations transmitted to fluid or solid media. As I have not here to write a treatise on acoustics, it is enough for me to recall that any sonorous body is only a body of which the molecules vibrate more or less regularly, and communicate to the ambient media movements which from point to point go to agitate the auditory organs.

The physiology of the invertebrates is still so imperfectly known that we are unable to say whether in a number of them the sense of hearing exists. Guided by anatomical facts and analogies, men of science have thought that they recognised auditory organs in all the groups of the invertebrates. In worms the auditory organ seems to be a vesiculiform capsule, full of liquid, and containing solid concretions, analogous to those which we meet with in the internal ear of the vertebrates. These *otoliths* are put into vibratory movement by cilia lining the wall of the auditory sac.[1]

Certain echinoderms seem to be provided with auditive

[1] Gegenbaur, *loc. cit.*, p. 197.

vesicles, in which float powerfully refracting homogeneous granulations. These vesicles receive nerves, and sometimes even rest on the central ganglions of the nervous system.[1]

There is the same structure of the auditory organs in the mollusks. Those organs are always small bags full of liquid and containing otoliths formed of an organic basis, impregnated with calcareous substance. These otoliths sometimes amount to many hundreds (Fig. 75).

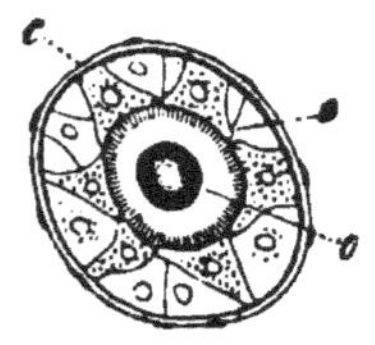

Fig. 75.

Auditory organ of *Cyclas*: *c*, auditory capsule; *e*, epithelial cells provided with cilia; *o*, otolith.

The auditory organs of the arthropods are more varied, and are known only in some divisions of the crustaceans and of the insects. In the crustaceans, the auditory vesicles are sometimes open. The closed vesicles contain otolithical concretions fixed by fine hairs regularly arranged. According to Hensen, auditory hairs, free, sometimes exist outside of the vesicles. These hairs even enter into vibration, isolately, when musical sounds are produced. Each of these vibrates in the water according to a special musical sound.

The organs of hearing proceed in the vertebrates from an invagination which is produced from the two sides of the head at the commencement of the embryonary life. This vesicle, at first in extensive communication with the exterior, closes little by little. The most inferior of the vertebrates, the amphioxus, seems destitute of auditory organs.[2] According to Schultze, the auditory organ of the cyclostomes appears to be first of all a vesicle containing a rounded otolith.[3]

Most fishes have as organs of hearing ampullæ full of liquid and imperfectly divided by an incomplete partition, on which the nervous terminations are spread. The internal face of the wall

[1] Leydig, *loc. cit.*, p. 316.

[2] Huxley, *Anatomia Comparata dei Vertebrati*, p. 71. (Italian translation of En. Giglioli.)

[3] *Entweilung von Petromyzon Planeri*, Harlem, 1856.

is lined with cylindrical epithelial cells. Beneath are found other cells sending bacilla between the epithelial cells. At this base these cells emit fine filaments which seem to be in relation with the nerves (Fig. 76). It is a structure very analogous to that of the olfactory membrane.

In man and the superior vertebrates the auditory organ is composed, as is well known, of three parts, namely:—the external ear, comprehending the pavilion of the ear and the external auditory conduit, shut by the membrane of the tympanum: the median ear, composed of the cavity of the tympanum, communicating with the throat through the Eustachian tube and traversed by the chain of the ossicles; finally, the internal ear, constituted schematically by an ampulla full of liquid, and containing otoliths. It is also known that this internal ear is subdivided into *vestibule*, *semi-circular canals*, and into a part rolled spirally, the *cochlea*. The fundamental portion of all this complicated apparatus is evidently the internal ear, in which are formed the terminal threads of the auditory nerve. The other parts are accessory, and are more or less lacking in many vertebrates. The internal ear is simplified more and more in proportion as we descend in the series. The aquatic mammifers and birds have no external ears. The median ear gradually disappears in reptiles and the amphibia. Already in the inferior vertebrates the cochlea loses its usual shape. Instead of the spiral cochlea birds

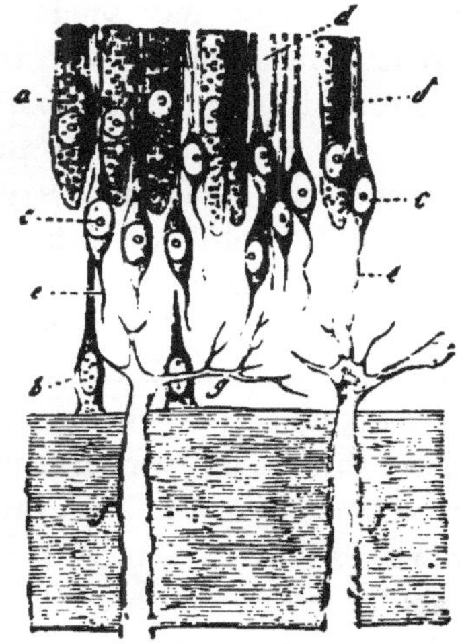

FIG. 76.
Microscopical preparation taken from the partition of the auditory ampulla of the clavated ray (*Raja clavata*): *a*, cylindrical nucleated cell, forming the internal epithelium: *b*, nucleated cells terminating in fine filaments and resting on the cartilage traversed by the nervous fibres, *f*, which terminate in very fine ramifications, *g*; the ramifications continue very probably in the fine filaments, *e*, with which commence the nucleated cells, *c*, of which the terminations in auditory bacilla, *d*, are placed between the cells, *a*, of the epithelium.

E E

have a small pyriform sac. Deserving of remark is this last fact, completely in contradiction with the modern theories of the office of the cochlea in man, in whom it is supposed to have the appreciation of musical sound for function. Such an organ ought to be extremely developed in singing birds. There is no cochlea in the internal ear of reptiles and of fishes. The most inferior fishes have not even any semicircular canal, and consequently their ear has much affinity with that of the invertebrates.

From this rapid anatomical summary it manifestly results, that the truly essential part of the organ of hearing is the internal ear, that is to say a vesicle full of liquid holding in suspension otoliths, and into which lead the terminal threads of the auditory nerve. The shock of the aërian sonorous waves puts in movement the liquid of the auditory ampulla, and the vibrations of the molecules of this liquid agitate and impress the threads of the auditory nerve. In the superior mammifers and in man these threads are very numerous; also the auditory ampulla has grown complicated; it is furnished with appendices, with circular canals, and especially with the labyrinth, that is to say, with a tubular cylindrical prolongation, rolled on itself spirally, and the windings whereof go on gradually decreasing (Fig. 77). The nervous auditory threads penetrate into the stem of this cochlea, and come forth therefrom successively, spreading themselves in the partition which separates from each other the spiral windings. These nervous threads of the cochlea are composed of nervous fibres, which, at first white, and with double contour, have continuance with the ganglionary cells, become afterwards fine and pale, and end their course in small cells emitting each an extremely fine filament, a sort of auditory bacillum.

As the spiral windings go on decreasing, the length of the terminal nervous threads diminishes also gradually, and their regular arrangement recalls that of the strings of a harp or of a piano, with which they have often been compared. An accessory organ, the organ of Corti, lodged in the spiral partition itself, has been compared to the fingerboard of this living piano. But the nervous fibres do not vibrate after the manner of strings, and the portion of each of these fibres, included in the partition of the cochlea, represents only a very small part of the total fibre. The analogy is therefore forced. There is nothing to prove that in this order of decrease each of these fibres has the duty of seizing a special tone higher and higher, and the construction of the cochlea may be nothing more than an organical device having the

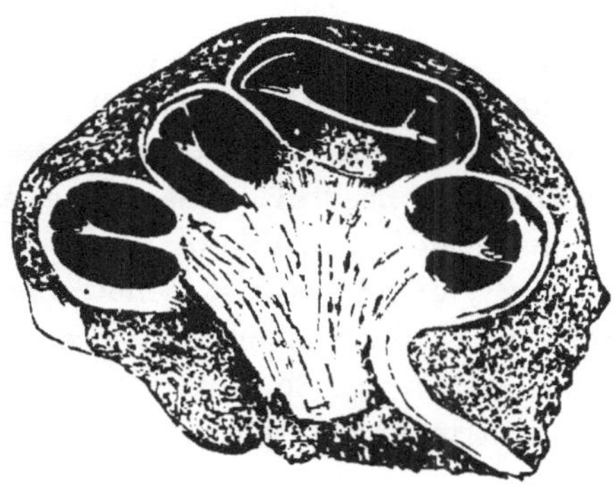

FIG. 77.

Vertical section of the cochlea of a fœtus of calf. The central stem or columella and the spiral lamina are not yet ossified. We see distinctly in each spiral winding the three cavities as well as the thickening due to the organ of Corti.

advantage of furnishing to the extension of the nervous threads a surface relatively large under a small volume. It is not the less certain that the ear of man discerns musical tones and semitones with great facility; but observations of pathological anatomy prove that this faculty of appreciating musical tones can persist after the destruction of the cochlea. Finally, as we have had occasion to remark in passing, singing birds have no spiral cochlea.

Let this be as it may, the ear can seize an infinite variety of tones of every kind, apart from the musical sounds, the number of which is somewhat limited. As regards the wealth of the sensitive field, hearing is very superior to the three tactile senses, and is inferior only to the most delicate, the most intellectual of the senses, the sense of sight.

CHAPTER IX.

OF THE SENSE OF SIGHT.

Already we have remarked how in going from touch to taste, from taste to smell, from smell to hearing, the excitants of the special sensations, the modes according to which the exterior medium impresses the organs of the senses, always go on attenuating themselves. At the outset it is the shock, the coarse contact: then it is the sapid particle; afterwards comes the odoriferant effluvium; finally, the simple vibration, the sonorous nerve. From hearing to sight the series continues. In effect, to excite the optical nerve, an undulation of gaseous liquid or solid molecules is no longer needful. The normal excitant of the sense of sight is the most impalpable of all: it is the vibration of the ether, a body, doubtless, not penetrable, as was long believed, for everything which is material is extended and impenetrable, but at least a gas so rarefied and light, that for our rough clumsy instruments ot physics it has no appreciable weight.

The eye, whatever may be the type of its structure, may be considered, when it is furnished with its essential parts, as a transparent and refractive apparatus, suitable for concentrating the luminous rays on the expansions of the optical nerve. But the eye is far from being always complete, and it is very curious to see it perfecting itself little by little as we ascend the animal series. How many pages overflowing with a bombastic admiration have been written to vaunt the structure pretendedly marvellous of the eye in man and the superior vertebrates! It has been praised as a perfect instrument, necessarily the work of an intelligent artificer anxious to adapt means to ends, and so forth. We now know that, considered as an optical apparatus, the eye is a tolerably good instrument, but by no means perfect. Moreover, comparative anatomy and embryology prove to us that, spite of its complicated construction, the eye is like all the other organs, merely the result of a slow labour of perfectionment and accommodation.

In effect, the rudimentary eye is only a simple drop of black pigment, reposing on nervous elements. Such it is in the inferior medusæ, in which we find at the base of the tentacles pigmentary spots. These spots are the first rudiment of the eyes: for sometimes we meet therein with crystalline rods. In other cases the construction of the optical apparatus takes an additional step, and we encounter in the masses of pigment a transparent and refractive body (Fig. 78). In certain species, nervous bundles penetrate manifestly into the capsule.[1]

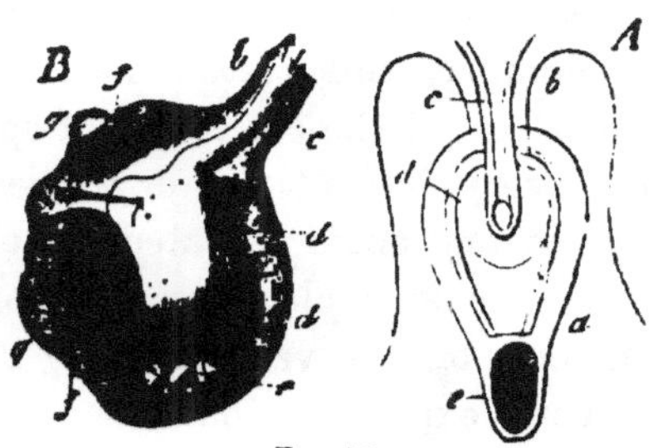

FIG. 78.
Marginal bodies of acraspodal medusæ: *A*, *Pelagia noctiluca*; *B*, *Charybdea marsupialis*; *a*, free part of the edge of the animal placed between the scolloped fringes of the disk; *b*, peduncles; *c*, canal thereof; *d*, ampulla; *e*, sac with crystals; *f*, pigment; *g*, body in form of lens.

Many planaries have first of all in the embryonary state pigmentary spots in the place where are to be at a later period eyes with *rods* or *crystalline cones*. The absence or the presence of the organs of vision is often subordinated to the kind of life. Eyes, for example, are sometimes lacking in the annelids which live in the darkness. These eyes have not, in this case, been the object of selection or differentiation, for they are useless.

[1] For all questions of evolutive anatomy, consult especially the *Manuel d'Anatomie Comparée* of Gegenbaur, one of the few treatises composed in harmony with the results of transformism.

The eyes of the asteriæ are situated at the extremity of the radii. They are spherical bodies reposing on a nervous mass. They are surrounded by an envelopment of pigment, and covered with an epithelial cuticle.

Visual organs are met with in the mollusks at all degrees of development. They are absent in the fixed mollusks. Certain of these last, which in the state of mobile larvæ had eyes, lose them by organic degradation when they have become immobile. Certain species of lamellibranchians have as organs of vision sometimes pigmentary spots, sometimes brilliant organs, disseminated on the edge of the mantle and receiving nerves.

The eyes of the *cephalopods* and *cephalophores* are always two in number, and placed usually at the base of the tentacles, sometimes at the extremity of special tentacles. They are moreover very irregularly developed. Sometimes they are simple pigmentary spots situated on the œsophagian ganglion; sometimes, for example, they are complex organs, comparable with the eyes of the vertebrates. They then rest on a ganglionary nervous expansion, have a retina, a layer of optical bacilla, separated from the deep layer of the retina by a pigmentary stratum. On this retina rests a sort of vitreous body, in front of which is a lens, covered by a cornea. A thin integument covers all the apparatus. The eye of the sepia is constructed on this plan, but with great perfection (Fig. 79).

In sum, the complete eye is composed first of all of optical bacillaria, that is to say of small special organs, destined to receive, or rather to select the luminous waves, and to transmit the molecular agitation to the optical nervous threads. That is the essential part of the organ of vision. In advance of these bacillaria, refractive media concentrate the luminous rays, and thus intensify the impression. Finally, a pigmentary envelopment isolates, more or less perfectly, the sensitive elements.

How remarkable soever the compound eyes of many arthropods may at first sight appear, these organs do not differ essentially from those of the other zoological groups. The eye of the arthropods is, schematically, composed of a crystalline bacillum, wrapped in a pigmentary mass, and covered exteriorly with a transparent chitinous lamina. This lamina connects with the general tegumentary envelopment, and sometimes bombates within so as to form a plano-convex lens. Sometimes this

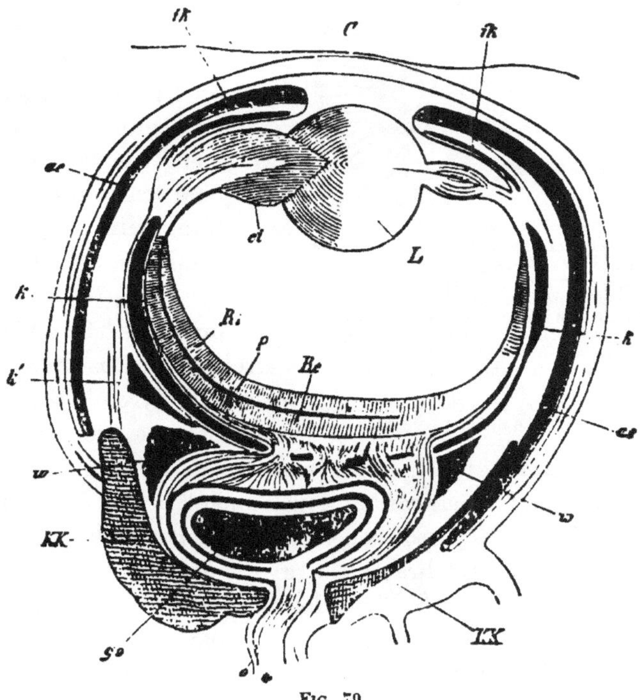

FIG. 79.

Schematic horizontal section of the eye of a *sepia* ; *KK*, cephalic cartilage ; *C*, cornea ; *L*, crystalline humour ; *ci*, ciliary body ; *R*, internal layer of the retina ; *p*, pigmentary layer ; *o*, optical nerve ; *go*, ganglion of the optical nerve ; *k'*, cartilage of the globe of the eye ; *ik*, cartilage of the iris ; *w*, white body ; *ae*, external argentine layer (after Hensen).

chitinous crystalline substance bombates also outwards, and then it is bi-convex. This simple eye, with a simple bacillum, exists in the inferior crustaceans.

In the arachnida many bacilla group themselves, cling together ; they are habitually covered with one and the same cornea. But it is in the superior crustaceans, and especially the insects, that this compound eye attains its maximum of development. Then it is constructed by the aggregation of a great number of simple eyes, sometimes many thousands, radiating round a nervous expansion, pressed against each other, which gives them a hexagonal form. The external chitinous tegument, the cornea, has then the aspect of a hexagonal network, exactly resembling tulle (Fig. 80 *B*). This has been called the *facetted* eye (Fig. 80).

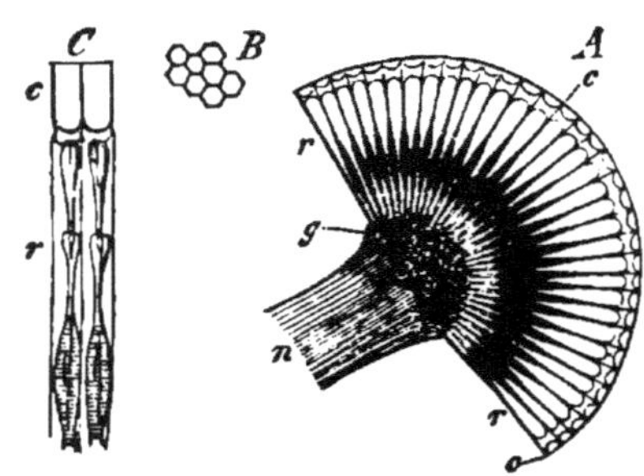

FIG. 80.

A, schematic section through the compound eye of an *arthropod*: *n*, optical nerve; *g*, its ganglionary expansion; *r*, crystalline bacilla coming forth from the ganglion; *c*, facetted cornea formed by the teguments, each facette, in consequence of its internal convexity, appearing to be a refractive organ (lens).
B, some facettes of the cornea seen on the upper side.
C, crystalline bacilla (*r*), with the corresponding cornean lens (*c*) of the eye of a coleopter.

In principle the eye of the vertebrates does not differ essentially from that of the invertebrates and even of the arthropods. At the outset it is represented in the adult amphioxus, and even in the first stages of development in the cyclostomes, only by a pigmentary spot reposing on the nervous centres.

In all the other vertebrates the eye is first of all constituted, in the embryological state, by a double vesiculiform expansion of the intermediary cerebral cell. These expansions join themselves to the epithelial cells of the epidermis, which at this point multiply, modify themselves, repel the vesicular nervous wall, depressing it into the form of a cup. Then the epidermoidal expansion which penetrates into this depression differentiates itself there and forms the crystalline humour (Fig. 81). At the same time, the expansions of the optical nerve spread themselves over the segment of the ocular globe thus formed, and their terminal threads connect themselves with the retinian bacilla in very large number.[1] Finally, we have the complete eye of the superior

[1] Huxley, *loc. cit.*

vertebrate (Fig. 82), that is to say a vitreous globe, receiving in the rear a voluminous nerve, and containing a retina reposing on a pigmentary layer, then divers refractive media, which go from back to front; the *vitreous humour*, the *crystalline*, enchased in a contractible and vascular apparatus, called ciliary process, and protected in front by a coloured contracted screen, the *iris*. This screen pierced in the centre by a circular opening, the *pupil*,

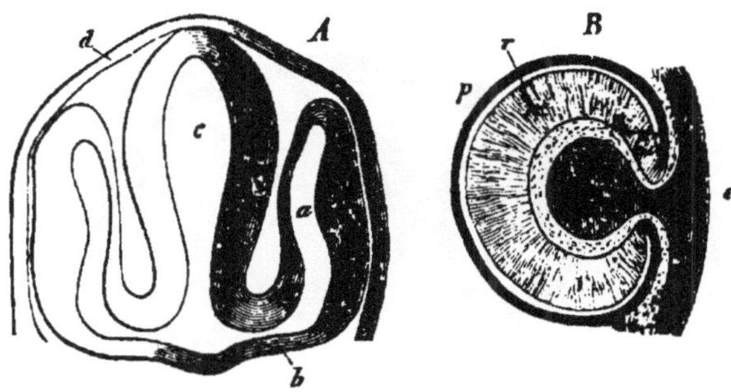

FIG. 81.

A, transversal and vertical section of an embryon of *fish*; *c*, brain; *a*, primitive ocular vesicle; *b*, its stem by which it communicates with the medullary tube; *d*, dermic layer. *B*, ulterior state, formation of the secondary layer; *p*, anterior wall (pigmentary layer); *r*, posterior wall (layer of the retina) of the secondary ocular vesicle; *e*, cornean layer (epidermis) emitting into the secondary vesicle the crystalline humour; *l*, the vitreous body behind.

merges into a liquid medium, the *aqueous humour*, which bathes also the anterior face of the crystalline on the one hand, and the posterior face of the *transparent cornea* on the other. We have not here to describe in all its details the complex structure of the eye of the vertebrates. In sum this retina is formed of bacilla connected behind with cells, into which the optical nervous fibres seem to pass (Fig. 82).

If we are willing to disregard accessory parts, the eye of man does not differ essentially from that of the arthropods. Each of the bacilla of the human retina is comparable with one of the voluminous optical bacilla of the insects, &c., and the eye of the mammifers may be brought into relation with the compound eye, with single cornea, of many arachnida. We still therefore find here an essential uniformity under an apparent variety.

But we can generalise still more. In effect, there is a great analogy of conformation in the terminations of the sensitive nervous fibres. For the optical, auditory, olfactory fibres the morphological resemblance is striking. Each of these fibres ends, at the periphery, in a cell connected with a special organ, or *bacillum*, whose function probably is to select among the

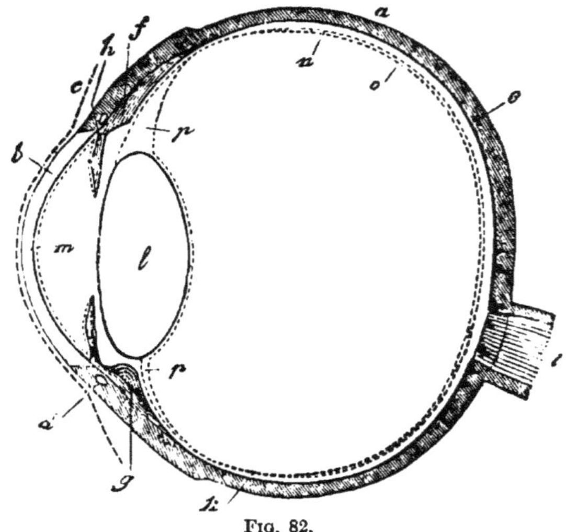

Fig. 82.

Transversal and horizontal section, magnified, of the globe of the eye; *a*, sclerotic; *b*, cornea; *c*, lamella of the conjunctiva passing over the cornea; *d*, circular vein of the iris; *e*, choroid with its pigmentary layer; *f*, ciliary muscle; *g*, ciliary process; *h*, iris in the middle of the pupil; *i*, optical nerve; *k*, anterior edge of the retina (ora serrata retinæ); *l*, crystalline lens surrounded by its capsule; *m*, membrane of Descemet lining the interior chamber of the eye; *n*, internal layer of the retina; *o*, membrane of the vitreous body; *p*, canal of Petit.

molecular vibrations of the ambient world those which are adapted to the special sensibility of the fibre which it terminates. The organs of the senses with complicated structure, such as the eye and the ear, are only accessory apparatus, which render more easy the function of the bacillum, by concentrating on it such or such a kind of excitation. In effect in proportion as we

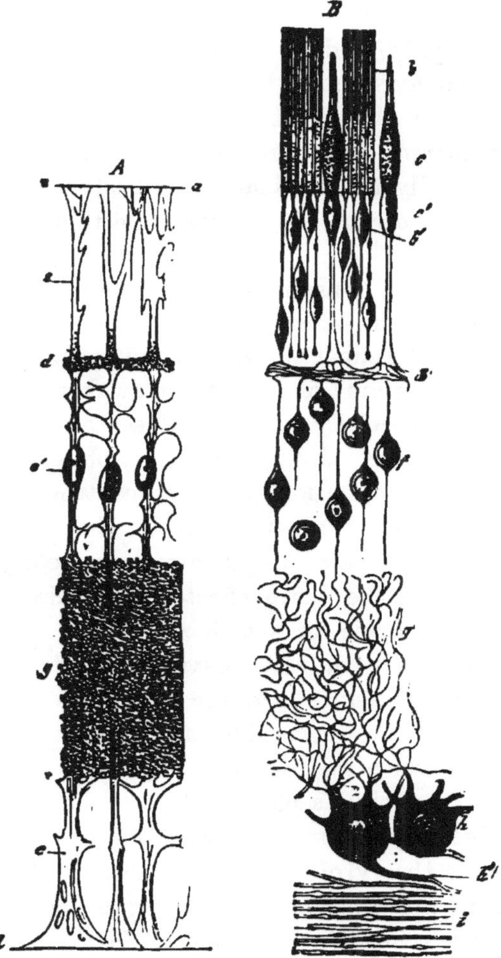

FIG. 83.

A, *Elements of the retina.*—The trellis, isolated, of the supporting fibres: *a*, membrane, extremely fine, which separates the bacillated layer from the other layers; *e*, vertical supporting fibres, with lateral prolongations and intercalated nuclei (*c'*) visible especially in the median layers; *d*, intermediary layer; *g*, molecular layer; *t*, internal limiting membrane, where the supporting fibres have their origin.

B, The nervous elements, isolated, of the retina: *b*, bacilla with external granules (*b'*); *c*. cones with their granules (*c'*); *d*, intermediary layer with fibrils extremely fine; *f*, layer with internal granules; *g*, trellis of nervous fibres, very fine, in the molecular layer; *h*, ganglionary cells; *h'*, nervous fibres which proceed thereto; *i*, fibres of the optic nerve.

descend in the animal series, we see the apparatus, the organs of the senses simplify, and even disappear, leaving desquamated the nervous terminations and their bacilla.

As to the tactile fibres and the gustatory fibres, which are only a variety thereof, their function is much more simple ; also they have not usually any terminal cells. Neither do they lead to true bacilla, but to corpuscles, dermic projections, appendices of varied form.

The nature of the sensations transmitted to the conscious nervous centres by the sensitive fibres is, however, altogether independent of the terminal appendices and apparatus. In effect any sensitive nerve excited on its passage awakens in the nervous centres solely and specially the kind of sensation which it has the function to provoke. Every excitation, physical, mechanical, chemical, and so on, any lesion, puncture, contusion, section of a sensitive nerve, have as repercussion in the nervous centres sensations tactile or painful, gustatory, olfactory, auditory, optical, according as the nerve affected appertains to touch, to taste, to smell, to hearing, or to sight.

It remains for us now, after having passed in review the peripheric organs of sensibility, the apparatus which are the gatherers of sensations, to consider the receptive centres, the nervous parts, in which the mechanical agitation of the terminal sensitive fibres determines a physical fact, a phenomenon of consciousness. Recently some unsuccessful efforts have been made to apply here the theory of the transformation of forces. But there is not here any transformation of the molecular movement in the habitual sense of the expression. When a luminous undulation comes to agitate the nerves of the retina, the central sensation which results therefrom is by no means a direct transformation of the ethereal vibration. This undulation has been simply the occasional cause of a whole series of physiological phenomena, the nutritive oxydation of the tissues being the real cause. The mechanical agitation of the sensitive nervous extremities plays in sensation the part of a trigger in a firearm.

Now no one dreams of seeing in the deflagration of the powder and the propulsion of the ball which succeed the shock from the discharge of a musket, the direct transformation of the movement operated by that discharge. It would be just as little logical to wish to find in the conscious phenomenon of sensation the mechanical agitation which has brought into play the sensitive apparatus. In the firearm and in the nervous system, the exterior shock has simply determined a perturbation of equilibrium ; it has let loose forces which neutralised each other ; it has acted as *a force of disengagement.*

CHAPTER X.

OF THOUGHT.

We are now about to explore a biological department which has for centuries been the exclusive domain of metaphysical phantasy, and which is still its sanctuary, its last refuge. The fact is, that scientific investigation could not undertake with any success the analytical study of conscious biological facts without being armed at all points. It was first of all necessary to acquire, by patient and innumerable researches, a just idea of organisation and of life; it was necessary to prove that the living world and the mineral world are only diverse modes of an identical substance; to watch, step by step, the circular movement of the *material* molecules; to see them *vitalised*, then mineralised, in the nutritive vortex; to distinguish and co-ordinate the great organic properties. At the same time, it was necessary to unravel the anatomical structure and texture of all living beings, to resolve them into systems, apparatus, organs, tissues, and anatomical elements. That done, it was possible to classify the functions, as we have classified the organs, and especially to connect the fundamental properties of the living substance with the tissues. We have seen that all the anatomical elements grow, are nourished and reproduced; but that on this general substance are grafted properties peculiar to each histological species, that the epithelial cell secretes, that the muscular fibre contracts, that the nervous fibre conducts the motory and sensitive incitations; finally, that all the phenomena of consciousness have their seat in the nervous cells. In effect, wherever there are phenomena evidently conscious, there are nervous centres. Thought is assuredly a property of the nervous cell. There are conscious nervous cells, and without them there is no psychical phenomenon humble or sublime. This is the idea which the Greek expressed in an ingenious allegory, by making Minerva spring from the brain of Jupiter. The existence of these nervous cells is a necessary condition to the production of the phenomena of consciousness, but the mode in which they are grouped is only a secondary condition of thought. The latter exists with all its principal modes in the nervous ganglionary chapelet of the ant as well as in the brain of the mammifers. Nevertheless, it is a biological law, that the concentration of nervous cells in great masses is favourable to the development and intensity of the psychical phenomena. The caterpillar has more ganglions than the perfect insect; the more intelligent the insect is, the larger the cerebroidal ganglion. Finally, in the vertebrated animals, the intellectual development is large in proportion as the encephalon is more voluminous.

Must we consider the nervous ganglions of the invertebrated animals as so many little brains? Has each swelling of the central nervous chain, in insects, for example, the faculty of feeling, thinking, wishing, on its own account? Is all this ganglionary concatenation a nervous federation, each member of which enjoys a certain independence, and presides over the conscious life of the segment of the body which corresponds with it? In this case, there would be a kind of regional division of conscious nervous labour, and the cerebroidal or superœsophagian ganglion would scarcely have any advantage over the others but that of presiding over vision, and of being in consequence intrusted with the guidance of the whole cohort drawn up behind it.[1] A celebrated experiment of Dugès seems at first to solve the question affirmatively: "I rapidly cut off," he says, "with scissors, the protothorax or *protodère* of the *Mantis religiosa;* the posterior

[1] Dugès, *Physiologie Comparée*, t. I. p. 343.

tronçon, still supported upon its four feet, resists the impulsions by which we try to overturn it, rises up and resumes its equilibrium if we overcome this resistance, and at the same time, by the trepidation of the wings and elytra, testifies a lively feeling of anger, as it would during the integrity of the animal, when provoked by being touched or threatened. But this posterior tronçon contains a considerable part of the chain of ganglions. We can carry on this experiment in a more striking manner: the long corselet, which has been detached from the other segments, contains a bi-lobed ganglion, which sends nerves into the arms or anterior feet, armed with powerful claws (snatching feet); let the head be detached, and this isolated segment will live nearly an hour with its single ganglion; it will agitate its long arms, and is quite capable of turning them against the fingers of the experimentalist who hold the tronçon, and seizing them so as to inflict pain. This single thoracic, or *deric* ganglion, then, *feels* the fingers which press the segment to which it belongs, *recognises* the point by which it is held, *wishes* to free itself from it, and *directs* towards it the members which it animates."[1]

At first sight, the experiment of Dugès seems decisive; but the co-ordination of the movements, their apparent intention, do not necessarily prove that they are conscious. Many very complicated reflex, but perfectly unconscious movements co-ordinate perfectly. We observe many of these apparently voluntary movements, even in the vertebrated animals, and even in the mammifers, where nevertheless the coalescence of the nervous centres is incomparably greater. A fish, or a frog, deprived of the brain, still executes a series of movements, apparently combined. It was the same with the pigeons whose brain Flourens had removed.[2] Finally, reflex, co-ordinate movements still take place in a decapitated human corpse. The spinal marrow seems

[1] Dugès, *Physiologie Comparée*, p. 337.

[2] Flourens, *Recherches Expérimentales sur les Propriétés et les Fonctions du Système Nerveux*, &c., Paris, 1824.

to be a source of automatic nervous activity, an unconscious nervous centre. It may then be the same with the nervous ganglions of the arthropods, &c., in which, besides, the preponderance of the cerebroidal ganglion is much less evident, and surely much less absolute than in the brain of vertebrated animals.

In the latter, the brain is the principal, and very probably the only conscious organ. In traversing the series, from lower to higher, we see the cerebral hemispheres grow large in proportion as the species is better endowed and more intelligent. There are first simple nervous ampullæ, scarcely to be distinguished from the other nervous intracranian expansions (see Book I., ch. VI). Then gradually these cerebral vesicles grow, and end by dominating and covering, more or less completely, the other encephalic nervous ganglions; their surface, almost exclusively formed of nervous cells, becomes furrowed, is folded in flexuous digitations called *circumvolutions*. The more intelligent the animal, the more numerous generally are these circumvolutions. Their object, like that of the *pecten*, which thrusts, in the eye, from back to front, the retina of certain birds, is to multiply surface under a small volume. At the same time that their number and ramifications augment, the grey substance, the cellular cortex, which covers them, and is the conscious part of the brain, augments in thickness. In man, we must estimate at many thousands the number of nervous cells superposed in strata in a square millimètre of cortical cerebral substance, and the number of these strata is greater at the anterior part of the brain, in the frontal lobes, which seem to be the headquarters of intelligence.[1]

In a preceding chapter, we have summarised the beautiful anatomical systematization of the cells and fibres of the human nervous centres, for which we are indebted to M. Luys. We have seen that thousands of sensitive and peripheric fibres of

[1] J. Luys, *Etudes de Physiologie et de Pathologie Cérébrales*, p. 11. Paris, 1874.

each half of the body, first conduct to a cellular nucleus, *the optic thalamus;* that thence they radiate towards the cortical cells of the circumvolutions; that other fibres start from these cortical cells, and converge towards another cellular nucleus, *the striated body,* whence finally irradiates the whole system of peripheric motory fibres, connected, in addition, with the cells of the cerebellum, which seems to be simply a co-ordinating centre of the movements. As the cortical cells of the two cerebral hemispheres are united by transverse fibres, the result is that the hemispheres, the two optic layers, the two striated bodies, form a complete system, the different parts of which are anatomically and physiologically connected. If the man is a healthy adult, then the instrument is harmonic, and vibrates accurately: the circumvolutions are swollen, and dilated; their summits rise equally to the surface of the hemispheres; a cortical, cellular layer, several millimètres in thickness, covers them. On the contrary, in senility, insanity, and mental maladies, certain circumvolutions give way and break down; they are notes which no longer sound.[1]

On the whole, every nervous system, however little developed, in the invertebrated animals as well as the vertebrated, may be traced to a conscious cellular part, in continuous relation with two fibrous systems, the one afferent, through which sensitive excitation is conveyed, the other efferent, by which motory incitation is transmitted. The schema of such a system would be a conscious cell furnished with a single afferent fibre and a single efferent fibre.

But the mode of operation of a system thus arranged is evidently reflex action, and, in effect, there is not a central nervous act, from the protozoon to man, which cannot be traced to reflex acts; thus it is infinitely interesting to follow through all its principal phases the gradual transformation and complication of the reflex nervous act.

[1] J. Luys, *Le Cerveau* (*Revue Scientifique*, No. 35, 1875).

First of all, the reflex action is absolutely unconscious. There is an agitation of the afferent fibres, excitation of the cells, which react upon the efferent fibres. At a higher degree, the nervous cell sensibilises: it becomes conscious of the vibration of its molecules; it experiences the *sensations* of touch, taste, &c., more or less varied and numerous according as the organ is more or less perfect. At the same time it has *impressions* of pain or of pleasure. At this stage the conscious being is still very inferior; each sensation, each impression, dies as soon as it is born; there is no chain of conscious phenomena, no link, no relation in the psychical life. But everything changes, when the nervous cell preserves the trace of the reflex act of which it has been the centre. It is, so to speak, impregnated with it, as certain phosphorescent substances catch the luminous rays, as a plate of prepared collodion stores up the luminous waves.[1] From that moment, the conscious phenomena are linked together; those which come last find in the nervous centres the echo of those which have preceded them. Speaking in the language of the psychologists, we may say that the faculties are now born. The traces of past sensations and impressions become *recollections;* there is *memory*. Then these recollections are disunited, group themselves capriciously, forming complex pictures, fictitious as a whole, though formed of old sensations and impressions; this is *imagination*. But from persistent impressions of pain and pleasure spring *desires* to escape the former, to feel the latter. Impressionability, sensibility, imagination, gather round these desires, and are more or less vigorously incited by them. This co-ordination of impressions, sensations, images, with the attainment of one object in view, becomes ratiocination, and the faculty of effecting this co-ordination is what psychologists have called *understanding, intelligence, reason;* in the same way that the conscious result of every confrontation, every comparison, amongst themselves, of impressions, sensations, &c., is called idea, thought. Finally, every desire preceded and accompanied

[1] J. Luys, *Recherches sur le Système Nerveux*, etc., p. 270.

by reasoning, by a relative estimate of the motive, becomes a *volition;* whence the *will* of the psychologists.

But behind all this labyrinth of psychical phenomena there are simply reflex acts, transformed sensations and impressions. Moreover, all this mental labour, of which the excessive complication in man has so long defied observation, is simply the result of the special properties of the nervous tissue. In effect, every sensation is accompanied by an elevation of the temperature of the nerve and a disturbance of its electric condition, by a negative oscillation of the nervous current. Besides, in order that it may take place, a considerable time is necessary, corresponding to an elevation of temperature in the cells which are its seat,[1] coinciding with a sur-oxydation, a waste of the substance of those cells which eliminate a much greater quantity of phosphates, &c., &c.[2]

Finally, comparative anatomy, anthropology, pathological anatomy furthermore lend their valuable co-operation to physiology; they show us that the moral and intellectual faculties are completely in subjection to the nervous centres; that they follow with docility their variations, more or less, to better or to worse; that they develop, diminish, or change with them. The phenomena of consciousness are, then, throughout the animal kingdom, not excepting man, functions, acts of the nervous cell. Upon this point doubt is impossible.

But we are not confined to this simple declaration. A young Italian physiologist, Dr. Mosso (of Turin), has just opened up to experimental psychology a track altogether new, by providing it with an apparatus which may aptly be called a *psychometer.* Our readers will surely be grateful to us for giving here a summary of the still unpublished work of Dr. Mosso.

It has long been known that there was a certain connection between the contractility of the vessels, especially of the capillary vessels, and certain agitations of the nervous centres, for example,

[1] Schiff, *Archives de Physiologie,* t. II., 1870, and Lombard, id., t. II., p. 670.

[2] Byasson, *loc. cit.*

that impressions, strong emotions, modify the beatings of the heart, the colouration of the cheeks, &c. (see p. 220). It was this interesting fact which served M. Mosso as a starting-point. He formed the design of finding a practical experimental means of measuring with some precision this influence of the nervous centres upon the vaso-motory nerves. For this purpose he made use of a large glass cylinder, closed at one of its extremities, open at the other, a kind of large cylindrical flask, sufficiently wide to accommodate easily the hand and entire forearm. A large armlet of caoutchouc fixes the tube to the bend of the elbow; at the same time shutting the upper orifice, while exercising only the most moderate compression. The wall of the cylinder is furnished with three orifices. One, which is closed with a stopper, serves to fill the vessel with tepid water. A second opening permits the introduction of the ball of a thermometer, which indicates the temperature of the water. To the third is adjusted a curved tube, very analogous to the hæmodynamometer of Poiseuille. This tube communicates with the cavity of the cylinder, and is so arranged that a small column of water rises therein. Finally, it is furnished at its upper extremity with a movable needle, adapted to note the oscillations of this column upon the paper of a registering tambour. Things being thus arranged, it is easy to see that the slightest sanguineous afflux in the forearm will diminish the volume of the member, and consequently will cause a part of the liquid column of the hæmodynamometer to flow back into the receiver. Every congestion, every dilation of the vessels will produce an inverse effect. In both cases the needle will write upon the registering paper the oscillations of the liquid column.

With this apparatus Dr. Mosso has been able to prove that every cerebral phenomenon has its repercussion upon the peripheric circulation. The following are the principal facts which he has observed with regard to man during wakefulness and slumber.

In the waking state, every sensation, every moral or physical impression, every intellectual labour is accompanied by strong contractions of the peripheric vessels, and the degree of contraction is proportioned to the effort made. Thus, while a young man was translating successfully Latin and Greek, the liquid level sank less during the translation of the former than of the latter, because the translator knew the Latin language well, the Greek imperfectly.

The facts observed during slumber are not less curious. Already, in certain cases of loss of substance by the bones of the skull, Blumenbach first, and afterwards Dr. Pierquin had remarked that dreaming corresponds with a congestion of the cerebral substance : "A woman," says Pierquin, "had in consequence of a syphilitic affection lost a large portion of the hairy scalp, of the bones of the skull, and of the dura-mater. The corresponding portion of the brain was exposed. When her sleep was without dreams, the brain was immobile, and remained in its osseous case. But when agitated by dreams, the turgescent brain projected from the skull. This turgescence was evidently in this case the result of vascular excitation."

The observations made with the apparatus of M. Mosso fully confirm that of Pierquin. At the beginning of sleep, there is a peripheric sanguineous afflux ; the vessels of the members dilate. Any sound whatever always provokes, in the vascular peripheric system, considerable contractions in the sleeping man. Dreams are always accompanied by a peripheric contraction, even when they leave no remembrance. The return of conscious life, at the end of sleep, is always preceded by a vascular contraction on the periphery of the body.

A general fact springs out of these observations, namely, that there is an alternacy, or rather antagonism between the action of the brain and that of the other organs. The active congestion of the first involves the relative anæmia of the others, and this gives us the explanation of some noteworthy facts ; for example, the sedative action of intellectual occupations upon the physical functions and instincts, the debilitating influence of mental labour upon the general constitution, the difficulty there is in being at once a man of action and a man of thought.

It is evident that the psychometrical process which we have just described is one of the most fortunate applications of physiology to the study of the cerebral functions. This process will surely be perfected and largely utilised, and we may predict that it will bear yet more fruit.

As we have already remarked, sensorial excitation at the terminations of the sensitive nerves only provokes a rupture of equilibrium, setting at liberty forces which counterbalance each other. It is the spark which kindles the powder. Doubtless, it is very hazardous to assert, with M. Wundt, that the sensation grows like the logarithm of the excitation which produced it.[1] The facts of biology, with their diversity, their infinite variability, are very ill adapted to the inflexibility of mathematical formulas. But leaving logarithms aside, it is unquestionable that sensation is not contained in the peripheric excitation, and that its intensity increases more rapidly than that of the external cause which has provoked it.

Sensibility is, then, a property inherent in the nervous cell, and the sensations, or to speak more generally, the facts of consciousness, are phenomena which interpose between the afferent and efferent currents of the reflex action, and which, once produced, persist, have a kind of separate existence, can revivify, combine, aggregate themselves to fresh impressions and sensations, are awakened according to the impulsion of the desires, forming finally a mental amalgam which constitutes the psychical individuality. But definitively, in spite of the extreme complication of the mental microcosm, the muscular movements which follow our volitions are only the last point of a reflex series, of which the direct or indirect origin is always an excitation bearing upon the extremities of the sensitive nerves.

Surely this is one of the most important services rendered by

[1] Wundt, *Menschen und Thierseele* (analysed by Th. Ribot, in the *Revue Scientifique*, 1875, No. 31).

modern physiology, to have reduced to such a simple formula the apparently entangled skein of the psychical phenomena; but the labour of analysis has been carried still farther, and after having demonstrated that every conscious movement is the result of a reflex nervous action, the fact has been established, that there is in the brain a localisation of the reflex properties.

The division of labour is already made manifest in the optic layer, where, in man, we can isolate four clusters of nervous cells, four receptive centres.[1] The first of these nuclei, the anterior, is the *olfactory centre*, highly developed in the animal species, which possess large olfactory nerves. Behind is the *optical nucleus*, the most voluminous of all in man, and, on the contrary, very little developed in the species which live habitually in darkness, as the mole. The third nucleus, or *median centre*, from front to back, receives the non-special tactile fibres. Finally, the fourth is the receptive centre of the acoustic fibres, the *acoustic centre*. The special office of these nervous centres is established: 1st, by comparative anatomy, which shows each of them to be more voluminous when the corresponding sense is more developed in such or such an animal species; 2nd, by pathological anatomy, which proves the coincidence of their isolated atrophy with the disappearance of the senses whose sensitive agitations they receive; 3rd, by experimentation, since M. E. Fournier has succeeded, by making irritant injections in the web of the optic layer, in destroying at will this or that category of sensitive impressions.

In the cortical substance of the hemispheres, perceptive departments of sensation and impression answer to these optical receptive centres. On the brain of individuals who had suffered amputation long before, Dr. Luys has proved local atrophies of circumvolutions.[2] In these later times, Dr. Ferrier[3] has observed

[1] Arnold, *Içones Cerebri et Medullæ Spinalis*, Turici, 1858. Luys, *Revue Scientifique*, No. 37, 1875. [2] Luys, *ibid.*

[3] Ferrier (*Progrès Médical*, 1873, No. 28, and 1874, No. 1) *Recherches Expérimentales sur la Physiologie et la Pathologie Cérébrales.*

that in exciting by electricity this or that region of the grey cerebral cortex of animals, a contraction of this or that group of muscles is caused; that we can at will make the eyes, the tongue, &c., move. Mr. Robert Batholon, Professor in the Medical College of Ohio, has obtained the same results by experimenting upon a man whose brain was laid bare by a large loss of osseous substance. Professor Schiff himself has seen that in animals the cerebral substance grew heated in such or such a locality according as it was agitated by such or such a category of sensorial excitations.[1] We may compare this nervous mechanism to a piano, the sensitive peripheric fibres being the key-board, the isolated centres of the optic layers the hammers, and the various cortical regions the cords.

A physiological note responds to each touch: here a secretion, there a palpitation, a contraction or a dilatation of capillary vessels; elsewhere a sensation, such or such a sensation; somewhere else, by reflex action, such or such a movement.

The more voluminous the peripheric nervous cords are, the more developed will be the nuclei of the optic layers, and the stronger the current of the sensations and impressions, and the more vigorously agitated will be the cortical perceptive centres, the nervous elements which have consciousness of sensations, which weigh them, compare them, register them; consequently the more difficult will be their ponderatory labour. If at the same time these cortical layers have little surface and depth, in other terms, if the cerebral hemispheres are little developed, the animal or the man will be peculiarly instinctive; will instantly and blindly obey the actual impression. If, on the contrary, the perceptive centres dominate, then the being will be intelligent, reflective, master of itself. It is by virtue of this law that the inferior vertebrates, in which the nervous intracranian vesicles, olfactory, optical, &c., are as voluminous as the cerebral vesicles (anterior brain; see Figs. 60, 61), have a rudimentary intelligence.

[1] M. Schiff, *Archives de Physiologie*, 1870.

It is easy enough to explain why a particular man, having otherwise little intelligence, is nevertheless endowed with this or that sensitive aptitude. It is enough that the external ear be well shaped, the nucleus of the optic layer which corresponds with it voluminous, the fraction of cortical substance in relation with this nucleus rich in cells, in order that the individual, though otherwise ill endowed, should have musical aptitudes. We thus comprehend singular facts, which have seemed abnormal to many observers; for example, that many idiots have shown a taste and even an aptitude for music.

The importance of these generalisations will escape no one. They shed light upon the most obscure, the most mysterious domain of biology. They snatch psychology from the hands of dreamers to give it a true scientific basis; they undermine a number of deeply-rooted prejudices, of myths illegitimately revered; they are the true and solid foundations upon which the science of the moral man will one day rise.

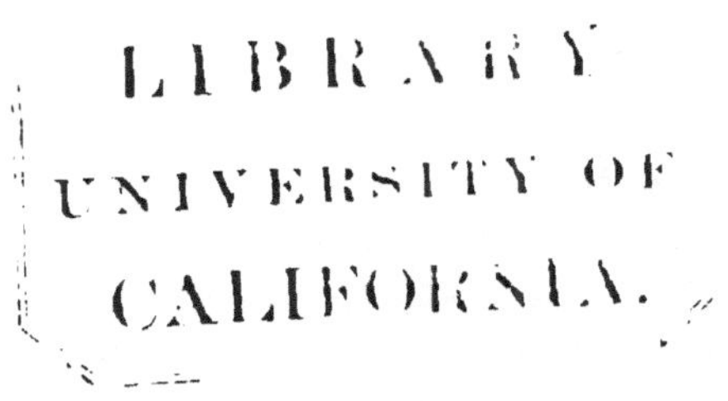

BOOK VII.

OF THE PHYSICAL FORCES IN BIOLOGY.

CHAPTER I.

OF ORGANIC HEAT.

CHEMICAL reactions, which have not for corollary a certain development of heat, are noted as rare exceptions. But life is only maintained by perpetual material exchanges, by combinations and decombinations, incessant molecular mutations; we must, then, expect to see organised bodies present, during their life, calorific combinations altogether peculiar to them.

A general fact already results from the thermometric observations gathered in the two living kingdoms; namely, that the elevation of the temperature is habitually so much the greater in proportion as the organic structure is more complex, more differentiated, more perfect.

Plants nearly follow the thermal variations of the ambient medium. This is the conclusion at which M. Rameaux, of Strasburg, and also M. Becquerel have arrived.[1] The first of these observers made holes in the trunks of trees, into which he introduced thermometers; the second made use of very sensitive thermo-electric apparatus. These observations, however, must only be accepted with caution. We must first of all throw aside hibernal observations. In winter, the life of the vegetal is

[1] *Des Phénomènes Physico-Chimiques.*

reduced to its minimum; it is a hibernant slumber much more profound than that of animals. The nutritive exchanges are then, in most cases, too unimportant to determine an appreciable elevation of the temperature. As to estival observations, they must also be cautiously judged. The plant has not, like even the lowest animal, a perfected circulatory system, which everywhere equalises the temperature. In the vegetal, each region, each tissue, each organ, enjoys a sufficiently large independence; the federative rule is there much more accentuated than in the animal.

The centre of trees, the hardened ligneous part, is half mineralised, and any thermometrical instrument introduced into this half dead region can indicate little or no elevation of temperature. It would be necessary to take in summer, during full vegetative movement, the temperature of the cambium, of the intermediary layer between the aubier and the bark. It is certain that, at all times when, in any part whatever of a vegetal, the phenomena of nutrition, .of development, of transformation, acquire a certain degree of intensity, the local temperature rises much. It is sufficient to signalise, in support of this assertion, the thermometric phenomena which accompany germination and florescence.

However, it is specially in the animal kingdom that the thermic independence of the organised being is clearly indicated. But here again the distinctive temperature of the animal is higher, less subject to exterior thermometric fluctuations, in proportion to the perfection of the species. Professor Valentin, who has noted the numerical proportions of the organic temperature in the principal invertebrated groups has found on an average the excess of the animal temperature over the exterior medium to be :—

In the Polypi	0°, 20
„ Medusæ	0 , 27
„ Echinoderms	0 , 40
„ Mollusks	0 , 46
„ Cephalopods	0 , 57
„ Crustaceans.	0 , 60

Nobili and Melloni, taking observations with a thermo-electric apparatus, have always found in insects a positive temperature of a fraction of a degree, or even of some degrees. Réaumur observed in a beehive a positive temperature of 12°,5, when the thermometer marked 3°,75 outside it.[1]

But if the temperature of invertebrated animals almost always preserves a more or less notable excess over the outer temperature, it varies, absolutely, in very large proportions. Ordinarily, it follows, more or less readily the exterior thermometric oscillations, rising in the day in summer, falling at night in winter. During the latter season, and in cold climates, most invertebrated animals which are not killed by the lowness of the temperature become benumbed, and fall into hibernation, subject as they are tc climateric vicissitudes.

That which we have just said of invertebrated animals might almost be applied to the two lower classes of the vertebrates. Like the invertebrated animals, fishes and reptiles have a temperature which varies with that of the ambient medium, whilst generally higher than the latter; like the invertebrated animals again, most of them are benumbed in winter. Nevertheless, their overplus of temperature is habitually higher than that of the invertebrated animals. Moreover, it is very variable. In fishes it has sometimes been found to be only 0°,20, sometimes 3°,88. In the fishes of the Sea of Marmora, Davy has even observed a thermal excess much more considerable. In taking the temperature of the abdomen and that of the dorsal muscles, he has found the former 6°,11, and the second 7°,22 in excess of that of the sea.

In reptiles the surplus temperature may sometimes be stated as still greater; though sometimes only 0°·04 in the frog, it has been seen to rise to 8°·,12 in the *Lacerta agilis*.

In the vertebrated animals of the two higher classes, in birds

[1] Consult J. Gavarret, *Les Phénomènes Physiologiques de la Vie*, and the article on "Chaleur Animale," in the *Dictionaire Encyclopédique des Sciences Médicales*.

and mammifers, where the respiratory and circulatory systems are better constructed, where the two kinds of blood are not mingled, where respiration is more active, the organic temperature shakes off in some degree the yoke of the outer temperature, and if we except some rare species, which are still subject to hibernation, it is proved that the superior vertebrates enjoy a temperature of their own tolerably high, and only varying to a small extent, in spite of climates and seasons. The temperature of adult and well-fed birds varies from 38 to 45 degrees. This variation of some degrees corresponds with specific, individual, even sexual differences; in fact, Professor Martins has seen, in the duck species, that the temperature of the female was sensibly higher and also more variable than that of the males.

The temperature of the mammifers is lower by some degrees than that of birds, but is invariable, like this latter. Thus, in the Arctic regions, Parry has found the temperature of a fox exceed the ambient medium by 76°,7.

The temperature of the ambient medium oscillates between 36 and 40 degrees. That of man varies between 36°,50 and 37°,50, and differs according to age and sex; it is higher in the child than in the adult, higher in the adult than in the menstruous woman, and lower perceptibly in the old man, during sleep, &c., &c., besides following with sufficient fidelity the oscillations of the respiratory energy. (See Book II., Chap. xiv.)

From the series of facts which we have just enumerated, it results that there are no *cold-blooded* animals in the literal sense of the expression, and neither are there any animals with an *invariable temperature* in the strict meaning of the word. In the whole animal kingdom, the organic temperature, and especially that of the blood, is normally higher than the exterior temperature. Moreover in the whole animal kingdom the organic temperature oscillates and varies. But these thermic variations decrease in proportion as the animal is higher in the series, and in the well-nourished bird and mammifer, the temperature of the interior medium, the blood, only varies within narrow limits;

it depends then much more upon tho biological conditions of the individual than upon the ambient medium. If it resist refrigerant influences, it can also, in a certain degree, resist those of warmth, and maintain itself for a certain time below the exterior temperature when the latter is excessive. The principal means of resistance to heat is evaporation, which proceeds on the cutaneous and respiratory surfaces.

The maintenance of the median temperature of the blood in birds and mammifers is a primordial condition. Thermic deviations, even when very slight, immediately put life in peril. The temperature of the blood cannot with impunity go beyond 43 degrees in the mammifers, and 50 degrees in birds. At 42 degrees the blood may already coagulate in the vessels; at 49 or 50 degrees the muscular substance coagulates; the muscles are then rigid and acid.[1]

As the composition of the albuminoidal substances is much less alterable by cold than by heat, the lowering of the organic temperature is less dangerous than its elevation; at least there is, herein a larger margin. Small mammifers (rabbits, guinea-pigs) make cold rapidly in ice, die when the temperature of their rectum descends to 18 or 20 degrees. But, by proceeding slowly, gradually, it is possible to lower their temperature much more without killing them, since, in marmots, the temperature of the rectum during hibernation is only 4 or 5 degrees.

We know that all animals absorb oxygen in quantities which vary according to determinate laws, with the class, the zoological species, the physiological conditions. We know, on the other hand, that this absorbed oxygen is the principal agent in the chemical mutations and transformations which are at once the effect and the cause of nutrition. Now all combustion is developed from heat. We can, then, taking the foregoing general facts as our sole basis, assign as the cause of the production and maintenance of organic temperatures, the oxydation, the slow.

[1] Cl. Bernard, *Rapport sur les Progrès de la Physiologie*, p. 45, and Hermann, *Eléments de Physiologie*, Paris, 1869.

combustion of living substances, and say with Lavoisier:— "The animal machine is governed by three regulating principles, respiration, which consumes hydrogen and carbon and furnishes caloric; transpiration, which augments or diminishes, according as it is necessary to get rid of more or less caloric; finally, digestion, which restores to the blood what it loses by respiration and transpiration."[1]

But, thanks to the progress of modern physics, chemistry, and physiology, we can follow, step by step, minutely, these important phenomena, indicate what processes organisms employ to produce, according to need, heat and cold, state what organisms are the most active producers of heat; finally, specify the office of the aliments, and apply to the working of living machines the great law of the correlation of physical forces.

[1] Lavoisier, *Mémoires de l'Académie des Sciences*, 1789.

CHAPTER II.

OF THE PROCESSES OF ORGANIC CALORIFICATION.

If we must consider the chemical reactions of nutrition as the principal source of organic heat, if this heat is a function of nutrition, it must augment or decrease according as the exchange, the molecular mutations, are made with more or less activity. Now, every organic functionment has always as its effect, or rather cause, an acceleration of the double nutritive movement of assimilation and disassimilation; consequently the local temperature of each organ must incessantly vary. In effect, this inductive conclusion is confirmed by observation; for it is a general law that every organ grows warm while accomplishing its special function, to grow cold afterwards in the intervals of repose.

This fact can be proved, either by taking directly the temperature of the organs, by the help of thermo-electric needles, or by comparing the temperature of the blood which enters an organ with that of the blood which issues from it. This experiment is practicable, for example, in the arteries and the veins of the glands. It has been made in the kidneys, the salivary glands, and also in the afferent and efferent vessels of the liver. It has thus been proved that the liver is the warmest organ in the economy, that its efferent veins are warmer than its afferent veins; that in glands with intermittent functionment, like the salivary glands, the venous blood issuing from the gland grows always warm when this gland is in a state of activity.[1]

The observations thus made upon the sanguineous glandular vessels have besides brought into prominence one of the most interesting physiological phenomena, namely, that if the oxydation of the living tissues and substances is the principal cause of organic heat, it is not the only one. In effect the glandular venous blood, which is black and oxydised during the repose of the glands, becomes, on the contrary, rutilant when the gland becomes again active, and at the same time its temperature rises. It is then less burned, contains more oxygen, and less carbonic acid. The venous blood of glands with intermittent functionment, like the salivary glands, thus passes alternately from black to red, according as the glands are in a state of repose or in movement. On the contrary, the venous blood of glands perpetually active, like the kidney, is always warm and vermilion. Slow oxydation, organic combustion, is here only a secondary agent of temperature, of which the principal causes are the isomerias, the catalyses, the evolvements taking place in the substance of the glandular organs.[2]

In all the other organs it is oxidation which prevails; the efferent or venous blood is always black, and it is the blacker, the more burned in proportion as the organ has functionated with greater activity. Thus the brain, so rich in capillary vessels, has normally a very high temperature. Professor Schiff has proved directly that the temperature of the cerebral hemispheres rises when they functionate; we know also that these organs are the seat of a kind of vascular erection, of active congestion, while, on the contrary, during dreamless sleep the brain is pale and anæmic. Observations made upon trepanned animals, and also upon man, in cases of loss of substance of the bones of the skull, have put these interesting facts beyond

[1] Cl. Bernard, *Chaleur Animale*, in *Revue Scientifique*, 1872, No. 45.

[2] Cl. Bernard, *loc. cit.*, No. 47.

doubt.[1] It is known besides that intellectual activity in a healthy man and the cerebral sur-excitation in the lunatic are accompanied by a sur-oxydation of the nervous substance, which is betrayed by a greater production and excretion of phosphates. Finally, it has been proved that the venous blood coming from the brain, the blood of the jugular vein, had normally a higher temperature than that of the carotid artery, and that cerebral activity had as corollary a super-elevation of this venous temperature.

Professor Schiff has also, with the aid of thermo-electric needles, seen the temperature rise in excited nervous cords, but rise less in proportion as the animal was near death.

All the anatomical elements, all the tissues, all the organs, co-operate, then, in the production and maintenance of organic heat, without which, on the other hand, they could not continue to live; but in the vertebrates, and probably in the higher invertebrates also, it is certainly the muscular system which furnishes the most powerful calorific tribute. The muscles are, in effect, the seat of very important nutritive exchanges, and besides, in the mammifers, they represent about a third of the weight of the body.

Réaumur, Huber, Newport, &c., have seen the temperature rise in a bocal containing insects, when these animals were put in movement by disturbing them. Dufour has proved that during the repose of the *Sphinx atropos*, the temperature of that insect is only eight degrees above that of the ambient air, while, at twilight, when the animal is in a state of activity and flies, the difference may reach ten degrees. M. Maurice Girard has seen the temperature of the *Acherontia atropos* almost equal that of warm-blooded animals after a prolonged flight. The ambient air being at 23°,4, he has found in the abdomen of the sphinx 25°,5, and in the thorax 32, and even 37 degrees.

[1] Pierquin, quoted by Gratiolet, *Anatomie Comparée du Système Nerveux dans ses Rapports avec l'Intelligence.*—Blumenbach, in *Archives de Médecine*, 1861, t. I.

The muscular temperature in man has especially been deeply studied by M. Becquerel. With thermo-electric needles he has seen the muscular temperature rise 1 degree and some tenths, after violent exercise. At the same time, the venous blood issuing from the muscle, and which, during the period of repose, was moderately black, becomes of an intense black, and its temperature rises. On the contrary, in paralysis by nervous section, when the tonic muscular contraction is itself abolished, the blood issues from the muscle rutilant, and scarcely oxydised.

The production of heat, during muscular contraction, is then manifestly dependent upon oxydation. It is demonstrated also by compressing the nutritive artery of the muscle, which immediately causes a remarkable coldness (Becquerel); by paralysing the muscular system of a dog with curaré, which lowers the temperature of the animal several degrees in an hour (*id.*); by causing the muscles of a frog confined in a bocal to contract by means of electricity, and by proving that these contractions have caused the production of carbonic acid (Matteucci).

This last fact proves, besides, that the muscle can contract, oxydise itself, and engender heat, without being traversed by a sanguineous current. The interfibrillary muscular juice, and perhaps the substance itself of the contractile fibre, are for some time sufficient to produce the indispensable chemical reactions. These reactions, which always necessitate a certain absorption of oxygen, have been variously interpreted. It is certain that the resting muscle is alkaline, and that contraction acidifies the muscular juice. The acid is lactic or sarcolactic acid. The alkaline principle is creatinine, which transforms itself into creatine. According to another explanation, the muscular fibre does not wear itself out through labour (Voit), which is very improbable. There is in the resting muscle an azotised substance, inogen, which, during contraction, evolves into carbonic acid, lactic acid, and an albuminoidal substance, myosine. It is myosine which in solidifying produces rigidity of the muscle. The development of heat, and also the production of mechanical force result from these reactions. There has even been an attempt to prove that the quantity of lactic acid produced was proportional to the muscular labour accomplished, and to the temperature developed (Heidenhain). Oxygen is besides indispensable to contraction, and to the reparation of the muscular forces.[1] In effect, repose alone does not suffice to restore muscular energy; it is necessary, in order to reconstitute the contractile force, that an oxygenised reparatory liquid should circulate through the muscle. In the living body it is the arterial blood which performs this office. We can, however, in a certain degree, substitute for the blood in an isolated muscle the injection of a solution largely composed of oxygen, for example, a solution of permanganate of potash.

If muscular contraction produces heat, inversely heat provokes contractions; it is a peculiar case of the great law of transformation of physical forces, of which we shall shortly speak. The beatings of the heart are accelerated when hot water is injected into a vein, even when the temperature of the blood is very little raised thereby. Inversely, under the slowly graduated influence of cold, the intestines and the heart contract with less and less energy, then finally stop. This is one of the causes of the hibernal sleep.

The chemical phenomena which produce organic calorification are accomplished, on the one hand, in the sanguineous and lymphatic plasmas, since these liquids are living; on the other, in the tissues, in the anatomical elements themselves, the nutrition of which is moreover closely connected with the quantity and quality of the blood which circulates in the capillaries. We must, then, now briefly describe the physiological mechanism which regulates the local sanguineous circulation, the consumption of the sanguineous aliment, and, consequently, the production of organic heat.

The regulating agent of the local capillary circulations is the nervous vaso-motory network, the large sympathetic nerve. After

[1] Hermann, *Eléments de Physiologie*, p. 231.

the section of a nervous sympathetic cord, the venous blood coming from the organs whose circulation was controlled by this nervous branch becomes redder, warmer, and more coagulable. It is less burned, and contains more oxygen, but it has been the seat of active chemical reactions, for it contains less fibrine than ordinary venous blood; besides, it springs from the vein in isochronal jerks, corresponding to the pulsations of the heart.

This is because the capillaries are dilated, and allow the sanguineous current to pass with five or six times more than its ordinary rapidity. Generally, there is in the organ whose vaso-motors are paralysed a sanguineous congestion, and always an elevation of the local temperature. Let the sectionised nerve be electrised, and immediately the capillaries contract, congestion disappears, the local temperature falls, the venous blood becomes again black, and no longer flows except foamingly.[1]

Since by paralysing the vaso-motory branches the elevation of the local temperatures is provoked, we are authorised in concluding therefrom that the nervous network of the large sympathetic performs the function of moderating local combustions, of restraining the nutritive expenditure, of rationing the anatomical elements, of creating coldness.[2]

But we have seen that the sympathetic nervous network is only a dependency of the general nervous system, that it is anatomically connected with the nervous centres, and that there is, in the upper part of the spinal marrow, a general vaso-motory centre. It will, then, be easy to comprehend that excitations, direct or reflex, but having their seat in the nervous centres, can make all, or nearly all, the capillary vessels of the body contract synergetically; and, in fact, by wounding or cutting the spinal marrow at a particular point, we can provoke in the mammifers, for example rabbits and guinea-pigs, a general coldness,[3] which throws these animals into a state resembling hibernal sleep. The

[1] Cl. Bernard, *Revue Scientifique*, 1872, No. 23. [2] *Ibid.*

[3] Cl. Bernard, *Rapport sur les Progrès de la Physiologie*, p. 183.

animal has then become cold-blooded. Moreover, we can obtain the same result by various processes: by directly and gradually cooling the animal,[1] by rendering it immobile for a considerable time, by subjecting it to rocking movements, by covering it with an impenetrable varnish.

The action of physical and psychical sensibility also tells upon the capillary circulation through the medium of the spinal marrow and of the vaso-motory nerves. Experiments made by Professor P. Mantegazza upon man and animals have proved that physical pain provokes a lowering of the temperature. It is often the same with moral pain, sombre passions, prolonged griefs, which sometimes end by impairing the nutrition of the tissues and cause organic degenerations.

Organic heat, then, is the result of nutrition, of life itself. Every organised body, especially every animal, is a fireplace, where diverse substances, mostly ternary or quaternary, burn slowly, developing heat which, in its turn, provokes or forms the exchanges, the chemical metamorphoses necessary to life. A very large proportion of organic heat is engendered by the action of oxygen upon the living solids and liquids. It is this dominant chemical action which gives the impulse to all the others, when it is not directly the only calorific source. Many other chemical reactions take place in the organisms, and also engender heat. "The albuminoidal substances produce decided calorific phenomena, at the time of their hydratation with evolvement, or of their dehydratation with combination. The hydrates of carbon (sugar, starch, &c., &c.) can disengage heat by simple evolvement independently of any oxydation. Finally, the neutral fat bodies can also give heat in evolvement by simple hydratation, as appears to take place under the influence of the pancreatic juice."[2]

[1] Fourcault, *Influence des Enduits Imperméables* (*Comptes Rendus de l'Acad. des Sc.*, t. XVI.).

[2] Berthelot, *De la Chaleur Animale* (*Journal d'Anat. et de Physiol.*, t. II., p. 671).

Heat, then, is produced in every organised body, in a feeble quantity in vegetals and the lower animals, with more energy in the higher organisms. It is produced in all the organs, tissues, and anatomical elements; but very unequally, according to the conditions of time and place. It is the general circulation which levels, in a certain degree, the various local temperatures, which distributes with a certain equality the created heat, which, moreover, thanks to the vaso-motory nerves, regulates the expenditure and temperature of each organ.

The quantity of heat thus produced is variously employed. One part is dissipated and lost by conductibility, radiation, evaporation; another part serves to maintain the organic temperature at a proper degree; finally a last part is directly transformed into mechanical labour, as happens, in steam engines, with the heat developed by the combustion of carbon, for, in spite of their complexity, their mobility, the biological phenomena, like physical phenomena, all are included in the law of the correlation and transformation of forces, as we shall show in the following chapter.

CHAPTER III.

THE TRANSFORMATION OF FORCES IN BIOLOGY.

In its incessant efforts to conquer scientific truth, humanity has proceeded as it did in its mythological conceptions. As the numerous gods of polytheism have little by little given place to monotheism and pantheism, so observation has shown that the physical forces formerly imagined and invoked to explain the universe have undergone incessant transformations, and were only, correctly speaking, diverse manifestations of a single force. We think that scientific language would gain much by finally banishing this word force, which expresses a conception altogether metaphysical of an unknown *something* lurking behind the material elements, and controlling them as a horseman guides his horse. In truth there are in the world only matter and movement, or rather, as movement is only an act of matter, there is only matter in movement. When we say, for example, that heat is transformed into electricity, we are simply understood to say that the atomic vibration called *calorific* changes its mode and becomes the vibration which we call *electric*. When a body, falling freely through space, is abruptly arrested in its fall by an obstacle, and immediately grows warm in proportion to the rapidity of its fall, it is not a force, called *weight*, which is transformed into another force called *heat;* it is the movement of totality of the falling body which is changed into a special vibratory movement of the molecules of this body, &c., &c. All becomes clear if we substitute the word *movement*, which has a clear and definite meaning, for the word *force*, which is nebulous and metaphysical.

Movement is essential to matter, and the atoms of all bodies, as well those of the impalpable ether as those of our hardest metals filling space, are animated with violent and varied movements, which, according to the modes of their rapidity, direction, amplitude, &c., &c., produce upon our senses diverse impressions. But as in their origin these movements are essentially analogous, they are very easily transformed into each other, since for that purpose it is sufficient for them to change their type.

We must here recall the principal points of this great doctrine of the unity of forces, founded by Dr. Mayer, and which has totally revolutionised physics.

The English natural philosopher Joule has demonstrated that every body undergoing mechanical violence, friction, a shock, &c., grows warm in proportion to the gravity of labour expended, that is to say that the movement of totality which has shaken its mass is then entirely transformed into calorific molecular vibrations. We can then determine with exactitude the *mechanical equivalent* of heat, that is to say, find how many kilomètres are necesary to raise by 1 degree the temperature of 1 kilogramme of water. All know that the kilogrammètre represents the labour necessary to raise a weight of 1 kilogramme to a height of 1 mètre. Joule found that 424 kilogrammètres were necessary to obtain this result. The mechanical equivalent of heat in relation to water is, then, according to this calculation, 424. Reciprocally, 1 kilogramme of water, falling freely from a height of 424 mètres, and abruptly arrested in its fall, will give 1 degree Centigrade of temperature.

Furthermore, the amount of heat necessary to raise by 1 degree the temperature of 1 kilogramme of water has been called *calory*.

The calory is, then, about equivalent to 424 kilogrammètres.

But it is easy to show by very numerous experiments that the calorific movement may be transformed into an electric movement, into a luminous movement, into a molecular chemical movement, even into sound, as the beautiful experiment of singing flames proves. Reciprocally, all these movements may be transformed into heat, or into each other. We can, then, affirm theoretically that there are mechanical equivalents for electricity, light, sound, molecular movements which are effected in the chemical reactions, and, as the mechanical equivalent of heat is known, it would suffice to transform these diverse movements into a determinate quantity of heat to have their mechanical equivalent. This labour has not yet been performed with sufficient precision. However, Father Secchi has found approximately that the amount of electricity which decomposes 106 milligrammes of water can raise by 1 degree the temperature of 38 grammes of the same liquid.[1] If then we take as *electric* the amount of electricity capable of raising 1 kilogramme of water from 0 to 1 degree, we shall see, by a simple proportion, that this quantity of electricity is that which can decompose $2^g,789$ of water:

$$0^g106 : 38 :: x : 1000^g : x, \text{ whence } x = \frac{106}{38} = 2^g,789.$$

Consequently, this quantity of electricity is equivalent to 1 calory, or to 424 kilogrammètres.

Similar determinations we can theoretically conceive to be possible with regard to sound and light, but, in spite of several attempts, the mechanical equivalents of sound and light are yet to be determined.

Chemical combinations also obey the great law of the correlation of physical forces. It is clear that every chemical combination or decombination reduces itself essentially to movements of atoms and molecules. In every chemical combination, millions of atoms are precipitated towards each other, clash against each other, till they have reached a state of stable equilibrium. But here the infinite variety of phenomena renders difficult the calculation of

[1] A. Secchi, *L'Unité des Forces Physiques*, p. 328 (2e. édit. 1874).

the mechanical equivalents with relation to water. A step however has been taken in this direction by determining the heat of combustion of a certain number of bodies. The following table indicates the number of calories developed by the combustion, the oxydation of some simple or compound bodies :—

Oxydised Substances.	Kilogrammes of water raised from 0 to 1 degree by combination with the oxygen of 1 kilogramme of each substance.
Hydrogen	34,135
Carbon	7,990
Sulphur	2,263
Phosphorus	5,747
Zinc	1,301
Iron	1,576
Tin	1,233
Olefiant Gas	11,990
Alcohol	7,016

Inversely, when once these bodies are saturated with oxygen, their decomposition by an electric current necessitates the absorption of a force equivalent to the number of calories which develops their combustion.

These observations prove that, in principle, the atomic movements are included in the general law of the unity of physical forces ; but how powerless they are still to permit us to note exactly all the transformations of the molecular movements which take place in the interior of organised bodies !

Without doubt there is, in these last bodies, nothing which does not proceed from the mineral world ; but these elements of mineral production, in ultimate analysis, group themselves in the living organism into compounds infinitely complex, and in a state of incessant metamorphosis. Though these chemical transformations take place in sufficiently regular series, according to a kind of chemical rhythm, nevertheless it is very difficult to follow them through all their phases, and most frequently we are thereby reduced, in order to form an idea of the chemical acts of organisms, to a comparison of the absorbed bodies with the eliminated bodies. But, in the animal machine, the most important chemical acts are oxydations, combustions. The alimentary matters introduced into an animal organism represent molecular systems containing *forces of tension*, that is to say, groups of molecules, in which the atomic attractions mutually balance each other. Once transfused in the nutritive vertex, from contact with the aërial oxygen, these bodies oxydate, their molecular equilibrium is destroyed; they then set at liberty *vival forces*, that is to say, that the molecular attractions no longer neutralise each other; the atomic movements may be communicated to the ambient medium, and consequently be transformed into heat, into electricity, into movement, mechanical, or of totality.[1] In fact these are the three principal methods of transformation of the molecular movements in the essence of the animal tissues. We know that in the nervous and muscular fibres there is an electric current from the surface to the centre, and this current results directly from oxydation; for it is still more clearly produced, as M. Becquerel has proved, in muscular tissue when mangled, which then oxydates, or rather respires, more energetically than in its normal state. We have pointed out, in their place and order, the electric phenomena which take place in the capillary vessels. Analogous electric currents are probably developed in all the living tissues.

It is much easier to prove organic heat than electricity. It is, moreover, the form which the greater part of the vival forces of the free molecular movements takes in the animal machine. We have seen that this heat maintained the bodies of the higher vertebrates at a temperature of from 36 to 40 degrees.

Ingenious experiments have proved that muscular movements as well as organic heat have their source in the chemical combinations of the animal machine. It has been observed that muscular activity corresponded to an augmentation of the con-

[1] In biology, the expression *vival forces* must be taken simply in a metaphorical sense, in the sense of free forces, of *living forces*, as Helmholtz calls them (*Lebendige Kräfte*). Biological mechanik is not yet sufficiently advanced for us to give to the expression *vival force* the value which it has in mechanik, that of the product of the mass by the square of the velocity.

sumption of oxygen proportional to the expenditure of force, and finally that, for a given consumption of oxygen, the developed heat and the muscular labour were in inverse proportion to each other. The force expended in heat is not expended in mechanical labour; and inversely. This is exactly what happens in the experiments of pure physics. For example, when an electro-magnetic motor raises a weight represented by 131 kilogram-métres, it gives, for the same chemical labour in a calorimeter, 308 calories less than in all the experiments.[1] In the same way, every muscular effort corresponds to a greater consumption of oxygen, and an appreciable development of sensible heat; but this quantity of sensible heat is so much less, for the same weight of absorbed oxygen, in proportion as the mechanical labour accomplished has been more considerable. This general proposition is derived from various experiments, amongst which the most important are certainly those of M. Hirn, of Colmar,[2] who has studied, with regard to the production of heat and of power, men ascending upon a revolving wheel, the steps of which moved on under their feet. He noted the quantity of carbonic acid exhaled by the lungs, first in the state of repose, then during labour. He also measured the quantity of air inspired and respired, and simultaneously the total product in kilogrammètres.

The experiments of M. V. Regnault have shown that, in a dog subjected to a mixed alimentation, out of 100 parts of oxygen absorbed, 7 are employed to form water, by combining with the hydrogen of the organic substances. Now the degrees of heat from the combustion of carbon and hydrogen are known; we can, then, the quantity of extracted carbonic acid being given, estimate the amount of heat produced, and compare it with the kilogrammètres of exterior labour. We thus find that man can utilise in exterior labour about a fifth of the total heat developed by internal combustions.

[1] Matteucci, *Revue Scientifique*, 1866, No. 50.

[2] *Bulletin de la Société d'Histoire Naturelle de Colmar*, 4e année, 1863.

The figures obtained by the process we have just described are evidently only approximative. The oxygen exhaled by the lungs under the form of carbonic acid, even when the quantity of oxygen employed in the formation of water is added, according to the observations of M. Regnault, does not represent all the oxygen absorbed. The cutaneous surface also respires and exhales carbonic acid. Finally, a very important part of oxygen serves to form, at the expense of the albuminoidal substances, on the one hand, the immediate azotised, regressive crystallisable principles, such as urea; and, on the other hand, sulphates, phosphates, &c.; all bodies eliminated by the kidneys, perspiration, &c.

To approach the truth more nearly, we must add to this direct study of pulmonary exhalation the comparative study of the aliments and of the excrementitious matters; this is what M. Boussingault has called the *indirect method.*

We can, for example, ascertain the comparative quantities of the *ingesta* and *excreta* of a bird inclosed in a receiver, by placing the receiver in a calorimeter. We thus prove that the heat developed is notably greater than that corresponding to the combustion of the carbon and of the hydrogen. The excess is due, without doubt, to the oxydation of the albuminoids, the true combustible power of which is still imperfectly understood.

Nevertheless, Frankland has given a table of the calorific and mechanical energy developed by the combustion of diverse alimentary substances. According to these statements, fat alimentary matters furnish more disposable heat and force than saccharine and amylaceous matters, and the latter produce more than food composed of beef, veal, pork, or fish.

According to Frankland, 1 kilogramme of dried muscular substance, purified in ether, and transformed into urea by combustion, develops 4368 units of heat.[1]

There are then no respiratory aliments exclusively destined to be burned, thus furnishing heat, and plastic aliments exclusively destined to be assimilated and to furnish useful labour, as Liebig

[1] *Revue de Cours Scientifiques*, 1866-1867, p. 81.

had pretended. All the aliments, all the organic substances become oxydised while furnishing energy. This last proposition has been put beyond doubt by MM. Fick and Wislicenus in their celebrated ascent of the Faulhorn.[1]

These observers, after having previously ascertained the weight of their bodies, and having subjected themselves during thirty-six hours to a non-azotised alimentation, ascended the Faulhorn, taking care to ascertain the current of the urea expulsed during the ascent, and the amount during the six hours following.

Knowing their weight, and the height of the mountain above the Lake of Brienz, they had the sum of the kilogrammètres, of the units of labour. The weight of the excreted urea gave, according to the figures of Frankland, the heat developed by the combustion of the albuminoidal substances, and especially of the muscular tissue. Now, the two experimentalists found that the oxydation of azotised matters could not have supplied, at the maximum, more than the half of the exterior labour accomplished, without taking into account frictions, the contractions of the heart, thoracic movements, &c.

Furthermore, Dr. Verloren has remarked that many insects feed specially upon albuminoidal matters at the time when they accomplish little muscular labour, and, on the contrary, have an alimentation almost exclusively ternary and non-azotised when they labour much. He remarked also that bees and butterflies feed upon substances very slightly azotised.

The preceding facts permit us then to arrive at a proper appreciation of the office and value of the diverse categories of alimentary substances.

[1] *Philosophical Magazine*, t. XXXI.

NOTE.

It is not easy to express in English the distinction between *forces vives* and *forces vivantes*. To lessen, not to vanquish, the difficulty the word *vival* has been employed. Now and then in the translation of this work, French words or expressions have, modified or unmodified, been from necessity used.—Translator.

CHAPTER IV.

OF ALIMENTS.

ALIMENT fills a double office in organised beings: it furnishes substance and energy. The molecular movement is incontestably the very essence of life; every organised and living substance is a vortex incessantly engulfing and expelling new materials; these materials failing, the organism is destroyed, it devours itself. We must then reject the opinion of some modern physiologists who, dazzled by the grandeur of the principle of the unity of physical forces, have ended by seeing only the dynamic side of phenomena, by enthroning spiritualism anew in biology, and pretending, for instance, that the muscular fibre contracts without decaying.

There is no organic functionment without decay of the organs, there is no durable life without the repair of this waste by aliments.

But neither are there clearly defined categories among the alimentary substances. Some furnish more force or movement than substance, others more substance than force; but all furnish at the same time both substance and force, with the exception of water and certain inoxydable salts. Water serves especially as a vehicle, as water of imbibition, of composition. Certain salts, such as the chlorure of sodium, are employed solely in stimulating absorption, in accelerating the nutritive movements. These are mineral aliments.

As to the organic aliments, all are at once respiratory and plastic. The ternary or hydro-carbonised aliments themselves can be fixed, under the form of fat, in the animal tissues, but it is certain that a large proportion of them burn, leaving as the residuum of their complete combustion water and carbonic acid. They evidently represent the great source of the calorific movement. As to the azotised quaternary aliments, they burn for the most part incompletely, it is true, furnishing at once assimilable materials, which lodge for a time in the organism, and also sensible heat and dynamic movement. Moreover, there is no impassable barrier between these two groups of aliments, since Cl. Bernard and Lehmann have proved that the albuminoidal aliments can be partially metamorphosed into sugar, and thus form substances totally oxydable,—*respiratory* aliments. Ordinarily, however, these substances only undergo incomplete oxydations, evolvements, isomeric metamorphoses, catalyses. These varied reactions, which, however, produce heat, leave as excrementitious residua azotised compounds much more oxydised than the quaternary substances whence they are derived. These azotised wastes are composed of leucine, tyrosine, of creatine, of uric acid, lastly of urea, which is the most azotised product of the series. Thus for 1 equivalent of oxygen, albumine contains $3\frac{3}{4}$ equivalents of carbon, creatine 2 equivalents, uric acid $1\frac{1}{2}$, and urea only 1 equivalent.[1]

The value of aliments depends therefore, on the one hand, on the quantity of movement and of force: on the other, on the quantity of assimilable molecules which they can furnish.

A ternary aliment has the more value, the less oxygen, and the more hydrogen and carbon it contains.

A quaternary aliment has the more value, the richer it is in azote, carbon and hydrogen, the poorer it is in oxygen.

Every substance completely oxydised can, from the alimentary point of view, play in the economy only a secondary part (water,

[1] G. Sée, *De l'Alimentation* (*Revue Scientifique*, 1866, No. 35).

sea salt). Likewise urea, which contains still a great deal of azote, has no alimentary value.

Experiments have demonstrated that the animal machine needs alike ternary aliments and quaternary aliments. Diet exclusively non-azotised and diet exclusively azotised equally produce the same disorders as abstinence. The equilibrium between the azote absorbed and the azote expended is maintained on the sole condition that the same equilibrium is also maintained as regards carbon (Voït, Boussingault).[1]

Man requires therefore a mixed diet, compensating at the same time the losses of movement and the losses of substance. The greater the expenditure of muscular force and the lower the temperature, the larger must the proportion of ternary substances be which enter into the alimentation. The inhabitants of hot countries have need of few aliments. On the contrary, the Esquimaux of the Arctic regions eat every day 8 kilogrammes of raw flesh, containing about a third of fat; they drink seal-oil, and swallow pieces of congealed whale-oil. Dr. Hayes states that his European mariners, passing the winter in the polar regions, had to restrict themselves to the diet of the Esquimaux.

But the dynamical value of quaternary substances is nevertheless very considerable. In his *Traité de Zootechnie*,[2] M. A. Sanson has tried to determine the dynamical coefficient of these substances. For that purpose he compared the sum of kilogrammètres furnished in a day by the omnibus horses and the post horses of Paris with the weight of albuminoidal substances contained in their alimentary ration. Deduction made of the portion of albuminoids necessary to the simple maintenance of the animal, conformably to empirical data, he thought that he had nothing more to do than to divide the sum of the kilogrammètres by the number of grammes of albuminoids consumed by labour. By proceeding thus it is found that in the horse 1 kilogramme of albuminoidal substances produces in round

[1] *Annales de Chimie et de Physique*, 3e série, 1846, t. XVIII.

[2] A Sanson, *Traité de Zootechnie*, &c., 2e edit., 1874, t. I.

numbers, 16,000,000 utilisable kilogrammètres. But this figure is evidently far too high, forasmuch as M. Sanson has not taken into account the dynamic value of the ternary substances contained in the ration.

Among all the alimentary substances which man uses, milk alone, which contains in nearly equal proportions proteic matters, fat and sugar, is a complete aliment; and if the animals exclusively carnivorous nevertheless live, it is because they always find in the body of their prey reserves of fatty and saccharine matters.

We have now arrived at the end of our long exposition. The reader will pardon us for having made it especially a long enumeration of facts. Science has in our days abjured the errors of its youth, and nourishes itself with scarcely anything but observations and experiments. Philosophy is on the way to imitate science, of which it is destined one day to be the crown. More and more men are persuaded that without observation and experiment there is no real knowledge.

To terminate our work it remains for us to say a few words of a vast general theory, now almost everywhere admitted. We mean the general law of balancement between the two organic kingdoms. That there is much truth in this view as a whole cannot be denied. It is certain in a general manner that the vegetal kingdom fabricates the aliments of the animal kingdom. But we must not formulate this proposition too strictly. First of all we must regard as withdrawn from the law of alternancy the mushroom family on the one hand, and the fauna of the oceanic depths on the other. Moreover, there is no complete antagonism between the vegetal machine and the animal machine. The comburent function does not devolve solely on the animal. The vegetal tissues oxydise themselves like the animal tissues, like everything that lives. .

Finally, the animal likewise fabricates starches and sugars.

It is still more unwise to fix the attention exclusively on one alone of the nutritive acts of the plant, to take account only of the decomposition of the carbonic acid. No doubt the chlorophillian plant dissevers the elements of the carbonic acid; no doubt, moreover, the chlorophillian function depends strictly on the solar light, and we have a right to regard the labour accomplished by the chlorophillian cell as a dynamic transformation of the radiation of the sun. But the decomposition of the carbonic acid does not represent all the labour effected by the green parts of plants. These organs are assuredly apparatus of organic synthesis; laboratories where are found at the same time both ternary compounds and quaternary compounds. M. G. Ville[1] has shown that seeds sown in calcinated sand and watered with distilled water, and cultivated in a receiver with glass walls yields an excess of azote at the time they have ripened and are gathered. Nevertheless he had added to the sand in which the roots grew nothing but phosphate of lime, some potash, some lime, oxide of iron, sulphate of lime. Moreover, the same experimentalist found normally an overplus of azote in many of our domestic plants when gathered. Finally, he was able to conclude from his observations and experiments, that during germination any vegetal needs to find within reach a certain provision of azote completely prepared. It is in their cotyledons that certain species, for example the leguminosæ, take this quantity of azote, this azote of development. As to the plants less fortunately endowed, they need to find in the soil, in the form of manure, the azote which is indispensable to them, and which they are powerless to procure otherwise. But when the germination is terminated, when the plant is furnished with chlorophillian leaves, it assimilates direct azote taken from the grand atmospheric reservoir. It is then that it forms by synthesis both the starches and the albuminoidal substances which it needs for growth and reproduction. That the green

[1] G. Ville, *Assimilation de l'Azote par les Plantes* (*Revue Scientifique*, 1866, No. 8).

leaves fabricate albuminoidal substances is shown by the single fact that the ascending sap conveys none to the chlorophillian cells, and that nevertheless the descending sap carries some away.[1]

The important chemical elaboration of the chlorophyllian cells is therefore much less simple than at first sight it seems: it is alike a labour of disseverment and of composition, of analysis and of synthesis. No doubt, if it is not completely certain that solar radiation furnishes to the leaf all the mechanical energy which the leaf displays, seeing that every atom and every molecule possess affinities of their own, nevertheless it is the vibrations emanating from the sun which give the impulse to all this mechanical labour, and which are the principal agents thereof. We are therefore justified in a very considerable degree in saying that the plant treasures up solar radiation, that is to say movement, which the animal afterwards utilises while transforming it. The result of the combustions which are accomplished in our tissues and wherever we make fire figure is truly the setting at liberty of energies immobilised by the plant. All the acts, all the movements, all the facts of consciousness of animals and of man are, at least in a considerable degree, transformations of solar energy.

But in this vast conception which contains so much truth, it is needful that force should not cause matter to be forgotten. In effect there is in the universe something besides dynamism. No doubt, organised beings receive from the sun by the help of interplanetary media impulsions, movements, which they utilise and transform. But the substance of these beings has also its own existence, its own sphere of activity.

In reality there is in the universe neither force distinct from matter, nor matter destitute of force. If the dynamik of the

[1] The ascending sap of the vine (the tears of the vine) does not, through heat, yield albuminous precipitate; through distillation a notable quantity of gum is formed therein; appropriate reactions reveal the presence of some saline substances.

universe shows us everywhere a single force, a single movement changing always in mode and in rhythm, without ever being annihilated, there is also everywhere a single extended substance, veritable Proteus, flowing into an infinity of combinations, but eternally, and without being able to grow immobile or to disappear. Definitively, thus, as is superabundantly proved by all the natural sciences and by the long analytical exposition contained in this volume, there is unity of matter as there is unity of force, or rather, movement is inherent in matter, the universe is constituted by a substance in movement.

INDEX.

A.

G.

H.

S.

T.

THE END.

LONDON: R. CLAY, SONS, AND TAYLOR, PRINTERS.

www.ingramcontent.com/pod-product-compliance
Lightning Source LLC
Chambersburg PA
CBHW031933220326
41598CB00062BA/1717
* 9 7 8 3 3 3 7 2 1 4 8 5 2 *